Annual Reports on Fermentation Processes

VOLUME 4

Academic Press Rapid Manuscript Reproduction

Annual Reports on Fermentation Processes

VOLUME 4

EDITED BY

GEORGE T. TSAO

Laboratory of Renewable Resources Engineering
Purdue University
West Lafayette, Indiana

ASSOCIATE EDITORS

MICHAEL C. FLICKINGER

Fermentation Production and Development
Frederick Cancer Research Center
Frederick, Maryland

ROBERT K. FINN

School of Chemical Engineering
Cornell University
Ithaca, New York

1980

ACADEMIC PRESS

A Subsidiary of Harcourt Brace Jovanovich, Publishers

New York London Toronto Sydney San Francisco

ACADEMIC PRESS, INC.
111 Fifth Avenue, New York, New York 10003

United Kingdom Edition published by
ACADEMIC PRESS, INC. (LONDON) LTD.
24/28 Oval Road, London NW1 7DX

Library of Congress Catalog Card Number: 0-275-3790

ISBN 0-12-040304-8

PRINTED IN THE UNITED STATES OF AMERICA

81 82 83 84 9 8 7 6 5 4 3 2 1

Contents

Contributors

Numbers in parentheses indicate the pages on which authors' contributions begin.

M. W. W. ADAMS (267), Department of Biological Sciences, Purdue University, West Lafayette, Indiana

G. I. AMER (67), Department of Chemical Engineering, Virginia Polytechnic Institute and State University, Blacksburg, Virginia

G. C. AVGERINOS (165), Department of Nutrition and Food Science, Massachusetts Institute of Technology, Cambridge, Massachusetts

H. R. BUNGAY (1), Department of Chemical and Environmental Engineering, Rensselaer Polytechnic Institute, Troy, New York

M. L. BUNGAY (1), Department of Chemical and Environmental Engineering, Rensselaer Polytechnic Institute, Troy, New York

J. A. CHAN (209), Chemotherapy Fermentation Program, Frederick Cancer Research Center, Frederick, Maryland

M. A. COUSIN (31), Food Sciences Institute, Purdue University, West Lafayette, Indiana

A. L. DEMAIN (193), Laboratory of Fermentation Microbiology, Department of Nutrition and Food Sciences, Massachusetts Institute of Technology, Cambridge, Massachusetts

S. W. DREW (67), Department of Chemical Engineering, Virginia Polytechnic Institute and State University, Blacksburg, Virginia

C. N. HAAS (1), Department of Chemical and Environmental Engineering, Rensselaer Polytechnic Institute, Troy, New York

N. W. Y. HO (235), Laboratory of Renewable Resources Engineering, Purdue University, West Lafayette, Indiana

F. KLEIN (209), Chemotherapy Fermentation Program, Frederick Cancer Research Center, Frederick, Maryland

R. A. MESSING (105), Research and Development Laboratories, Sullivan Science Park, Corning Glass Works, Corning, New York

L. E. MORTENSON (267), Department of Biological Sciences, Purdue University, West Lafayette, Indiana

R. M. STROSHANE (209), Chemotherapy Fermentation Program, Frederick Cancer Research Center, Frederick, Maryland

R. G. UPCHURCH (267), Department of Biological Sciences, Purdue University, West Lafayette, Indiana

L. C. VINING (123), Biology Department, Dalhousie University, Halifax, Nova Scotia

D. I. C. WANG (165), Department of Nutrition and Food Science, Massachusetts Institute of Technology, Cambridge, Massachusetts

R. J. WHITE (209), Chemotherapy Fermentation Program, Frederick Cancer Research Center, Frederick, Maryland

Foreword

This issue, the fourth volume, represents the beginning of a new era in the Annual Reports on Fermentation Processes series. With the passing of Professor David Perlman, the founding editor, on January 29, 1980, the position of editor has been filled by Professor George T. Tsao. Supporting him now will be two new associate editors: Dr. Michael C. Flickinger and Professor Robert K. Finn. Dave's unique contributions, insights, and devotion to the fermentation field, the Annual Reports, and the Microbial and Biotechnical Technology Division are irreplaceable. The new leadership of Annual Reports plan to build on Dave's initial efforts.

An annual lectureship in honor of David Perlman, to be held at the Annual Meeting of the MBT Division of the ACS, began this year at the Las Vegas Meeting. The address by Professor Arnold L. Demain, from the first lecture, is reprinted in this book.

Robert D. Tanner, Chairman
Division of Microbial and Biochemical Technology
American Chemical Society
% Department of Chemical Engineering
Vanderbilt University
Nashville, Tennessee

Preface

We are all saddened by the death of Professor David Perlman. We must carry on for the sake of Dave, a dear friend, who created so much; it will be difficult to follow in his footsteps.

David Perlman

1920–1980

David Perlman, editor of *Annual Reports on Fermentation Processes,* died January 29, 1980 following nearly three years of a valiant and courageous struggle with cancer.

Born in Madison, Wisconsin, David Perlman grew up and was educated in the academic environment of the University of Wisconsin, where his father, Selig Perlman, the distinguished labor historian, taught for 45 years. After completion of his graduate studies in microbial biochemistry under the tutelage of Professor Marvin J. Johnson and the late Professor William H. Peterson, Dr. Perlman worked for short periods at Hoffmann–La Roche, Inc., and Merck and Co. before joining the Squibb Institute for Medical Research, where he remained until his return to the University of Wisconsin School of Pharmacy as professor of pharmaceutical biochemistry in 1967. From 1968 to 1975 Dr. Perlman served as Dean of the School of Pharmacy while teaching undergraduate and graduate students and coordinating an expanded research program. During his administration, the pharmacy student enrollment doubled, the teaching staff increased 150%, and research space was enlarged 75%. Dr. Perlman subsequently held the Kremers Professorship of Biochemical Pharmacology at Wisconsin.

Over a professional career of some 34 years, David Perlman developed scientific interests of remarkable versatility and productivity, as researcher, educator, scholar, book editor, essayist, historian of applied microbiology, leader of scientific societies, and organizer of more than 30 symposia, conferences, short courses, workshops, and annual programs.

Much of his research focused on the development and/or improvement of fermentation processes for citric acid, 2,3-butylene glycol, penicillin, streptomycin, neomycin, vitamin B_{12}, tetracycline, ascorbic acid, riboflavin, and some aspects of mammalian cell culture. He discovered the enzyme mannosidostreptomycinase, which converts mannosidostreptomycin to the clinically more useful streptomycin.

With his co-workers, Dr. Perlman pioneered microbial hydroxylation of steroids, a technique that led to the development of biotransformation systems for sterols, antibiotics, alkaloids, and organic acids. Twenty-eight patents and more than 350 papers document his manifold research and scholarly works.

Dr. Perlman edited or coedited 28 books, the most recent being *Microbial Technology* (Second Edition), *Advances in Applied Microbiology* (since 1967), and *Annual Reports on Fermentation Processes* (since 1977).

His leadership and achievements in microbial biochemistry and fermentation technology, both in the United States and overseas, have been acknowledged with many honors: a fellowship from the John Simon Guggenheim Memorial Foundation; fellowships in The New York Academy of Sciences, The American Academy of Microbiology, and The Academy of Pharmaceutical Sciences; The James M. Van Lanen Distinguished Service Award and the first Marvin J. Johnson Research Award, both presented by the Division of Microbial and Biochemical Technology, American Chemical Society; the Fisher Scientific Co. Award for Applied and Environmental Microbiology, given by the American Society for Microbiology; the Charles Thom Award for Research in Applied Microbiology, awarded by the Society of Industrial Microbiology; and the Pasteur Award for Applied Microbiology, from the Illinois Section, American Society for Microbiology.

Dr. Káto (Lenard) Perlman, Dave's wife, will continue the joint research study they shared.

The David Perlman Lectureship has been established by the University of Wisconsin Foundation* with memorial contributions. The lecture series is sponsored by the School of Pharmacy, Department of Biochemistry, Department of Bacteriology, and Mrs. David Perlman. In commemoration of David Perlman's distinguished leadership and professional career, the Microbial and Biochemical Technology Division of the American Chemical Society has established The David Perlman Memorial Lecture to be given at the Division's annual meeting. The first lecture was given by Arnold L. Demain August 26, 1980 at the national meeting of the American Chemical Society held in Las Vegas, Nevada. The text of Dr. Demain's presentation is published in this volume of *Annual Reports on Fermentation Processes*.

H. J. Peppler

* University of Wisconsin Foundation
The David Perlman Lectureship
c/o Robert B. Rennebohm
702 Langdon Street
Madison, Wisconsin 53706

CHAPTER 1

PRACTICAL MIXED CULTURE PROCESSES

C. N. Haas
H. R. Bungay

Department of Chemical and Environmental Engineering
Rensselaer Polytechnic Institute
Troy, New York

M. L. Bungay

I. INTRODUCTION

The field of mixed cultures has become too broad to be covered completely in a single chapter. Soil, marine, and freshwater microbiology have begun to focus on mixed culture phenomena because little further progress can be expected from pure culture techniques. Microbial associations are receiving greater attention in studies of the rumen, intestine, oral cavity and skin. In addition, numerous interactions take place between microbial communities and plants or animals. A wide range of mixed culture topics is covered in a book by Brock (13). Discussions of mixed cultures have appeared at intervals (5,16,36,50,63,97). Fredrickson (33) updated previous reviews and provided excellent perspective of mixed cultures from a theoretical and mathematical point of view. This chapter will, in contrast, emphasize the experimental and applied aspects of mixed culture studies. It is our premise that models and stability analyses are important aids in planning experiments and interpreting results, but they have yet to provide close agreement with data for the dynamic behavior of single cultures, much less with infinitely more complicated mixed-culture systems. For this reason we will stress applications and mention only that theory which is supportive.

ISBN 0-12-040304-8

Harrison (43) has reviewed mixed cultures in industrial fermentations with emphasis on single-cell protein. The advantages and disadvantages of such processes are nicely presented; the great drawback of mixed cultures is potential for poor quality control.

A survey of waste treatment is also beyond the scope of this chapter. However, the paucity of information bearing strictly on microbial interactions is good evidence that more research is needed on the microbiology of waste treatment processes.

II. TECHNIQUES

Fredrickson (33) makes the curious claim that the chemostat is the best tool available for studying population interactions under controlled conditions and implies that studies with feedback control systems often have spurious results. A continuous culture undergoing oscillations does not fit the definition of chemostat (95). It is well-known that shocks can induce transient oscillations in microbial cultures, and Sinclair *et al.* (78) have shown how the dynamics of control systems can mask the dynamics of the cultures. A common problem in control systems is delay in detection or distance-velocity lag in metering lines. Bungay (14) pointed out how this can be critical for continuous culture systems with feedback control. As there are means for almost instantaneous detection of several variables such as pH, temperature, and dissolved oxygen, there need be no introduction of artifacts due to the control system so long as delays in control action or abrupt additions of reagents are avoided. Carried to the extreme, Fredrickson's case against control action would rule out thermostats for continuous culture vessels. Instead, it seems logical to employ as much control as possible whenever a control system can be designed for negligible distortion of the dynamic behavior of the biological constituents. On-off control means stops and starts of control action that are likely to induce transient upsets. More sophisticated proportional control, proportional-integral control, or a complicated control algorithm can have gradual action that will not shock the culture. Dilution by the solutions used for control is a possible source of error. Usually these reagents can be concentrated enough to minimize dilution. When the reagents must be dilute to avoid shocking the culture, it may be possible to have the feed stream and the control reagent of identical composition except for the ingredient being controlled. Substitution of the control reagent for the feed permits control without dilution. The message is that control should not be condemned as inherently bad, but good design is essential. As with any research, eliminating unnecessary variables can lead to more definitive conclusions. A simple chemostat may be excellent for some types of experiments, but other arrangements can be superior depending upon the goals of the research.

Resolution and enumeration of population components can be difficult. Many laboratory studies have been contrived using organisms of different sizes so that electronic particle counting was applicable. Sometimes morphological differences are sufficient to permit resolution while counting by microscopic examination of a calibrated chamber. Tedious, less precise differential counting can be performed by plating on solid media. When colonial differences are too slight for discriminating between species, selective plating media must be used. Indirect methods that can be used are analysis of pigments assumed to be associated with particular organisms, or assaying other biochemicals indicative of a given species.

Groups led by Bailey (3) and by Hatch (45) are using flow microfluorometry to analyze microbial cultures. Individual cells pass through a laser beam, and estimation of concentrations of internal biochemicals can be made from the cell's fluorescence. This technique can also be used to resolve and count mixed populations and to follow biochemical changes in the systems. Various methods for studying mixed cultures are reviewed by Lovelock and Davies (57).

For practical processes there are several possible indices of performance such as product concentration, rate of product formation, respiration coefficient and rate of substrate utilization. Usually there are special analytical problems for each process. The indices used for biological waste treatment are rather crude because it is troublesome to identify the many compounds present.

III. NOMENCLATURE

Most of the terminology for biological interactions has fairly general acceptance, and deficiencies when applying these terms to complicated, compound interactions are obvious. However, it has proven relatively easy to overcome any limitations of the nomenclature by devoting more words to the description of a given process. Thus, precise one or two word definitions have not been established for the various possible interactions, but this has in no way been a barrier to progress. Bungay and Bungay (16) proposed a symbiotic index to designate and to quantify interactions, which is clumsy and

TABLE 1. Common Terms for Microbial Interactions

Neutralism	*Lack of interaction*
Commensalism	*One member benefits while the other is unaffected*
Mutualism	*Each member benefits from the other*
Competition	*A race for nutrients and space*
Amensalism	*One adversely changes the environment for the other*
Parasitism	*One organism steals from another*
Predation	*One organism ingests another*
Synergism	*Cooperative metabolism to produce a substance not produced by either alone*

not very useful. Fredrickson (33) develops nomenclature, but falls back upon compounding of the accepted terms for simple interactions with the addition of + or - to designate whether a strain is helped or hurt by the interaction. Fredrickson's nomenclature is logical, in accord with earlier proposals, and better than many other conceivable approaches to the problem. It does not appear timely to criticize nomenclature or to propose alternatives.

Some of the common terms for biological associations are given in Table 1. It is virtually impossible to have a truly single-interaction system because organisms occupy space and affect the chemical and physical aspects of their environment in ways that may stimulate or inhibit other types of organisms present. Nevertheless, an association may closely fit one of the definitions in Table 1 while there is an implicit understanding that elements of other definitions may be occurring. When more than one definition fits, they may be joined for clarification. For example, a competitive commensalism would refer to a system in which the biological constituents compete for space and/or nutrients but one also excretes a substance beneficial to the other.

IV. GROWTH LIMITATIONS

The textbook by Bailey and Ollis (9a) provides an excellent overview of the theoretical foundation and experimental basis for continuous culture of microorganisms. However, there has not been a clear distinction between the stoichiometric limitation of growth and limitation by rates of biochemical processes. Stoichiometric limitation is based on the concept that cell mass cannot be in greater proportion than the components from which it is made. In other words, cell mass consists of carbon, hydrogen, nitrogen, sulfur, phosphorus, and other elements supplied to the culture to become cell mass or be consumed by metabolic processes. Furthermore, one constituent in lower proportion than the others can limit growth. Only in the very rare situation where two or more ingredients are in exactly equivalent proportions will there be no single growth limiting nutrient. Mixed cultures are more complicated because nutritional requirements are different for the various organisms present. One may be growth-limited by a vitamin, for example, while another is limited by the concentration of magnesium ion. Thus, some types of organisms may have the same growth limitation in a particular medium while other types may have a variety of growth limitations. As relative numbers shift, the importance of certain medium constituents may change. Additional complications arise from some organisms excreting nutrients for others.

There seems to be no mention in the literature of kinetic limitations to growth, but the concept is straightforward and obvious. If all constituents of the medium are in surplus (and none is toxic), there is still a maximum growth rate that must relate to the kinetics of key biochemical steps. Some rate processes such as transport through a membrane, synthesis of a macromolecule, or diffusion of an intermediate from one enzyme to another is going as fast as it can, and no alteration of the composition of the medium produces a higher rate for this step. Perhaps a new ingredient, such as a building block in the key step, could be found and supplemented to the medium to increase the rate, but then another step would become the rate bottleneck. Even in rich media, ingredients are depleted or toxic materials accumulate to limit growth because high concentrations of organisms are supported.

In continuous culture, relatively dilute media generally exhibit a growth-limiting nutrient that operates stoichiometrically. Typical dilution rates for chemostats are well below the maximum specific growth rates of the organisms, thus kinetic limitations do not come into play. Several studies in continuous culture with control of substrate concentration or in chemostats with very high dilution rates have demonstrated that bacteria can double in four to eight minutes (39). This definitely is pushing growth rate toward its maximum, and kinetic limitations are probably important. In fact, an ingredient that was growth-limiting because of its stoichiometry at lower dilution rates may become non-limiting at very high dilution rates. Let us assume that nucleic acid formation is critical for cells growing rapidly, and let's further assume that phosphate for phosphorylation leading to nucleic acids is rate-limiting at a gentle dilution rate, phosphorate might replace it as the limiting ingredient at high dilution rate. Some research on high rate systems will be reviewed subsequently.

V. GROWTH RATE RELATIONSHIPS

Experimental data for growth rates as a function of substrate concentration describe curves as shown in Figure 1. At low concentrations, growth rate rises quite linearly. High concentrations show a constant growth rate or a decline because of inhibition. Although another equation sometimes fits the data well, the most commonly used relationship is the Monod equation

$$\mu = \hat{\mu} \frac{S}{K_s + S} \tag{1}$$

where μ = specific growth rate coefficient
$\hat{\mu}$ = maximum specific growth rate
S = substrate concentration
K_s = a constant

When high concentrations are inhibitory, the Monod equation can be modified arbitrarily to fit the data better (6). There is an analogy between the Monod equation and the Michaelis-Menten equation for enzyme kinetics, but the derivations for either the Monod equation or the proposed modifications are based on quite simplistic assumptions.

The concept of limiting substrate has been misused in formulating growth rate equations for mixed cultures. An example of a common approach for modeling when there are limiting substrates is the work of Miura *et al.* (65). Given two microbes, M_A and M_B, which compete for the same substrate, S_1, the growth rate expression for M_A, was taken as

$$\mu_A = \mu_{AM} \frac{S_1}{K_A + S_1} \tag{2}$$

where the terms are exactly analagous to Equation 1, the Monod equation. The second organism also uses a growth factor, G_a, excreted by M_A, and the assigned growth rate expression was

$$\mu_B = \mu_{BM} \frac{S_1}{K_B + S_1} \quad \frac{G_a}{K_G + G_a} \tag{3}$$

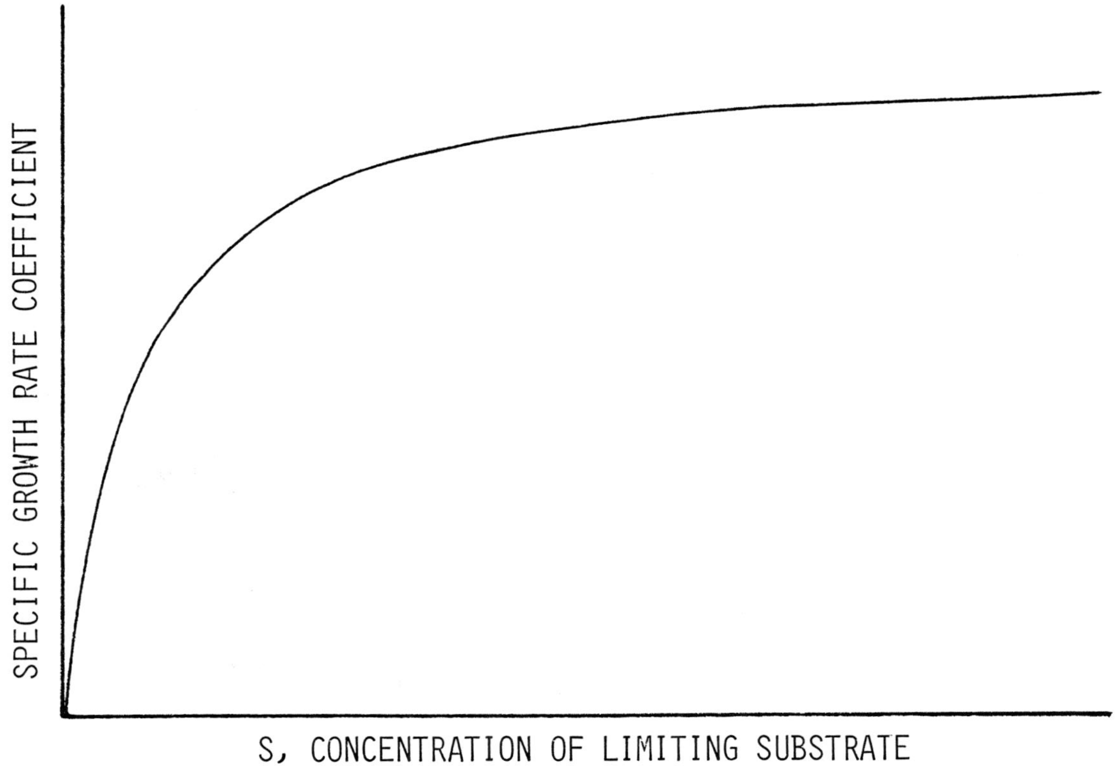

FIGURE 1. Growth Rate versus Substrate Concentration.

where the parentheses enclose a correction term analogous to the Monod equation. There are precedents for selecting this type of equation, and the authors should not be faulted too severely because a true relationship is not known. When μ is a function of two variables, it should be possible to determine from many experiments a three-dimensional plot relating them. Although a computer algorithm can handle such relationships, the techniques are clumsy. It is preferable to have Equation 3, but this equation is contrary to theory. The concept of a limiting ingredient dictates that S_1 *or* G_a should be growth-limiting, and the other should be in excess. Another approach to modeling is to have two separate equations of the Monod type, one for μ as a function of S_1 and the other for μ as a function of G_a (15). Each value of μ would be calculated, and the computer would select the lower (more limited) result for use in the program.

Although the concept of a single limiting substrate has been very useful, real life is seldom that simple. The excellent agreement of real data with the Monod equation probably depends on having a fairly large excess of all but one ingredient. When proportions of ingredients of the medium are closer and shifting because of population changes in mixed cultures, there are very likely switches in limiting ingredients for several types of organisms. Further research to unravel the mysteries of growth rate dependencies would be highly desirable.

The Monod equation is not time-dependent and is valid only for steady state. Some dynamic models incorporate terms with time constants to account for lags in adjustment of μ in responses to changes in S (e.g. 104,108,109). Many other models of mixed culture dynamics ignore these time constants and assume that μ adjusts instantaneously. Mateles *et al.* (61) showed that some adjustment is practically instantaneous if the deviation in S is small, but several hours may be required to reach the final value of μ for larger deviations. The models which ignore these delays may be acceptable when the time periods of interest are several hours or days, and perturbations are gradual. Conclusions concerning short-term stability that are drawn from such models are dubious, however, because, as control engineers are highly aware, delays can seriously impair stability of systems which otherwise appear to have no problem.

VI. ELECTIVE CULTURES

Open systems with fortuitous inoculation of organisms in the fluid feeds or from the atmosphere develop elective cultures which accommodate well and participate in creation of their environment. After a period of time, it becomes unlikely that a dramatic change will result from the random inoculation of a new type of organism, but those already present can exhibit highly dynamic population behavior. Although there has been little research on defining elective cultures, it is worthwhile to review what has been done and to comment on possible approaches.

Natural populations can undergo remarkably abrupt changes. Some examples are the rapid shift from one predominant algal species to another in a lake or pond and sudden shifts in the bacterial composition of a stream. For example, Baker and Herson (9b) found that the diatom *Thallasiosia pseudonanna* stimulated some of its associated bacteria but was also inhibited by them. Cassel *et al.* (18) using pigments as indices of bacterial concentrations in an activated sludge unit observed such pronounced changes. A species that was dominant for several days could fall to undetectable levels and could sometimes reappear much later as a major component of the mixed culture. Such behavior makes it highly tenuous that studies of defined systems with a few components in the laboratory will be acceptable models of natural systems with many biological components.

A qualitative explanation of erratic fluctuations in natural populations can be based on the oscillations of simple systems. In nature it is highly likely that a given organism is participating in many distinct microbial interactions, some of which are oscillatory. Furthermore, these interactions would be expected to be asynchronous. The oscillations in population of a microbial species could reach high numbers when the peaks of several distinct asynchronous interactions are reinforcing, and could fall to very low numbers when the valleys of the distinct interactions overlapped.

A complex interaction explanation is not completely satisfactory because the population decline in a natural population can be too precipitous to be in accord with the observed populations of associated species. In other words, the fall in one type of organism should be accompanied by a rise in numbers of predators or in the natural competitors. Also, it is unlikely that predators would be so specific for one species of prey. It is true that the competitors rise, but not rapidly enough to account for the rate of decline in the other species. Attacks by phage would explain the drastic decline, and this cannot be ruled out because there are no reports of enumerating phage changes in natural populations. However, there should be many different phage in nature to support the general observation that organisms achieving predominance are likely to suffer a severe fall. Ewert and Paynter (31) matched bacteria and virulent phage in an activated sludge process. Of the organism numbers determined by counting with the microscope, less than seven per cent were counted by formation of colonies on culture media. The others may have been dead or may have had special requirements for nutrition or for oxidation-reduction potential. Fastidious anaerobic organisms of intestinal origin would not have grown under the test conditions. Phage increased during activated sludge treatment because effluent counts were higher than those for the inlet. Nevertheless, only three of 48 isolates of dominant bacteria from the process were hosts for the phages present in the system. The hypothesis of phage as the explanation of precipitous drops in natural populations is not supported by this evidence, but it is possible that temperate phage were undetected or that the conditions of the experiment were unfavorable for phage attack on the isolated cultures.

Although there have been no attempts to match up temperate (lysogenic) phage to host species in various ecosystems, there is a great variety of these phages. Dhillon *et al.* (27) noted that only a small percentage of the plaque-forming units obtained from natural habitats, such as sewage treatment plants, were temperate phage. Only phage for *Escherichia coli* were characterized, but at least eighteen different temperate phage were identified. The abundance of both virulent and temperate phages for different strains for this single species would suggest that phages for other organisms are widespread and could be critical factors in population control.

Populations change in response to stress. In most natural ecosystems not already subject to great stress, multitudes of different species are present. A few species may constitute high proportions of the total numbers present while just one or two individuals may represent the least abundant species. Spores or dormant cells may initiate new growth should conditions become favorable. Airborne particles can constantly reinoculate an aqueous system. Applying a stress such as a higher temperature may favor a species that was already active, or perhaps a dormant species, or a variant strain or naturally occurring mutant. It has recently been shown, for example, that planktonic lake organisms exhibit different thermal optima, depending upon isolation conditions (86). Only very heavily stressed environments such as a hot stream with high concentrations of hydrogen sulfide develop pure cultures. Weaker stresses tend to lead to new dynamic population balances, although it is very common to find markedly reduced diversity in a stressed ecosystem compared to its condition before the stress was applied.

An ecosystem rich in species is probably not uniform, and there may be hiding places for the rare species. For example, certain anaerobes are known to have very brief survival times if oxygen is present. Nevertheless, it is possible to culture them from soils that are obviously well aerated. Either there is some physical barrier which protects these organisms from air or there are niches where associated organisms are utilizing oxygen to keep its concentration low.

The ability to accommodate to stress is a selective advantage for natural populations; life persists despite changes in pH, temperature, redox potential, and nutrients. Were there not organisms present to metabolize toxic wastes such as phenol, undesired substances would accumulate. Responses to stress are usually slow; thus shocking a waste treatment plant with phenol can be disastrous, while gradual addition allows those organisms metabolizing the toxic substance to achieve significant numbers. This draws attention to another point: the presence of a few individuals may not be sufficient to allow recovery from stress if the entire population is nearly wiped out. Whereas gradual stress may permit beneficial associations for the needed strains, violent stress may so severely inhibit the associated strains that the desired strains are also lost. Furthermore, gradual stress is accompanied by metabolism of the toxic material so that highly inhibitory concentrations are not reached. Abrupt shocks may exceed the concentrations which inhibit even those organisms which are able to utilize the material.

The persistence of diversity is not in accord with either the theory or experimental data for laboratory systems. For example, in a chemostat with no particular stresses applied, there is a tendency for a mixed population to lead to a pure culture. There may be exceptions because of nutritional interdependencies, reinoculation of an unfavored species which persists by attaching to surfaces, or a sustained interaction such as predation; however, simple, direct competition leads to take over by one type of organism and washout of all others. Analogous reasoning leads to the expectation that in a natural system with flow, a very few well-acclimated species would win out while most others would wash away, but on the contrary, several types tend to be abundant, other are more scarce but significant, and many others persist indefinitely at low numbers. However, natural systems may be far from steady-state, and spatial heterogeneities may invalidate direct extrapolation from well-mixed laboratory reactors.

A powerful competitive stress has been applied to continuous cultures by use of feedback control of substrate concentration (Bungay, *et al.*, 17). Depletion of substrate causes the controller to increase feed rate, and thus residence time in the vessel is reduced. An elective culture with many, many constitutents is quickly converted to a population of just one or two different organisms by selection of fast-growing types which accomodate to increasing dilution rate from control action. The system cannot wash out because accumulation of the substrate would slow the feed rate.

Perhaps nature abhors pure cultures. When field after field is planted with the same crop, insects and microbial diseases specific to that plant may spread rapidly and devastate vast areas. Dense concentration of human population greatly increases the incidence of disease. In like fashion, rich cultures of microorganisms in nature are more subject to predators, parasites, and phage attacks. Stable coexistence of sensitive bacteria and phage is possible when phage are few in number (19). When the product of hosts times phage exceeds a critical value, an outbreak of phage virulence occurs (4). It thus appears that there are adequate obstacles to achieving pure cultures in nature, and only rather severe stresses can overwhelm these obstacles. A natural population may be viewed as highly heterogeneous with the ability to accommodate to almost all gradual stresses. Unfortunately, organic chemists have invented a few industrial compounds that are metabolized slowly, or not at all, by microorganisms, so natural populations can be decimated by human intervention.

The endlessly complicated variations, evolution, and responses of natural ecosystems would seem to defy rational analysis. Laboratory studies of defined mixed cultures are entertaining and of definite value as a foundation for commercial mixed culture fermentations, which are also artificial systems. However, extrapolation of defined systems to natural systems is highly dubious in light of the complexities we have discussed. Explanations for the behavior of mixed cultures in systems such as those for biological waste treatment will require more than laboratory experiments with defined cultures.

VII. COMPETITION

The theory of simple competition in a well-mixed continuous culture is straightforward and is developed in a review by Harder, *et al.* (42) and by Harder (41). If the growth rate relationships are such that one type of organism grows more rapidly than the other under all conditions, it will take over as the other washes out. When the curves for specific growth rate cross as in Fig. 2, the outcome depends upon dilution rate because μ=D at steady state. At the growth rate and dilution rate corresponding to the intersection, both species can persist. At higher or lower dilution rates, the competitor with the higher growth rate will win out. Wilder, *et al.* (104) devised various control algorithms that will maintain both competitors for a case governed by Fig. 2. The biological coefficients were based on a system with *Candida utilis* and *Corynebacterium glutamicum*. One algorithm functioned through detection of total cell density and adjustment of dilution rate. Another control strategy was detection of numbers of each competitor and adjustment of inlet substrate concentration. Still another strategy termed "classical control" manipulated feed rate in proportion to the deviation of one population from a reference state and manipulated feed concentration in proportion to the deviation from reference of the other population. The strategies all produced stable control, but there were differences in slugishness of response or overshoot with decaying oscillations. Stephanopoulus, *et al.* (80) considered the same problem and also developed control systems to insure coexistence. This work is important because it shows that, in special cases, a synergistic mixed population could be maintained in a commercial mixed culture process by a feedback control system.

Trilli and Pirt (88) studied the coexistence of *Klebsiella aerogenes* and *Pseudomonas aeruginosa* in continuous culture. Attempts were made to find different growth-limiting substrates for each organism, but in all cases *Klebsiella* was strongly dominant. There was persistence of *Pseudomonas* at low levels probably as a result of growth on substances excreted by *Klebsiella*. Another competition between organisms with different nutritional requirements was studied by Gottscha, *et al.* (39b). Various thiobacilli were in competition on inorganic or organic limiting nutrients with *Spirillum* present. Smith and Kelly (79) reported a related study. *Thiobacillus neopolitanus* which grows only as a chemotroph and *Thiobacillus A2*, which is a facultative autotroph, were grown in competition for thiosulfate in continuous culture. Although *T.neopolitanus* dominated, there were significant numbers of *T.A2*. Adding glucose to the medium favored *T.A2* with complete washout of *T. neopolitanus*. It was concluded that there was commensalism plus competition for the thiosulfate-limited medium and that facultative organisms may have an advantage over obligate chemotrophs.

The concept of one species completely eliminating others in competition in a continuous culture is not supported by work of Meers and Tempest (64). Binary cultures in chemostats

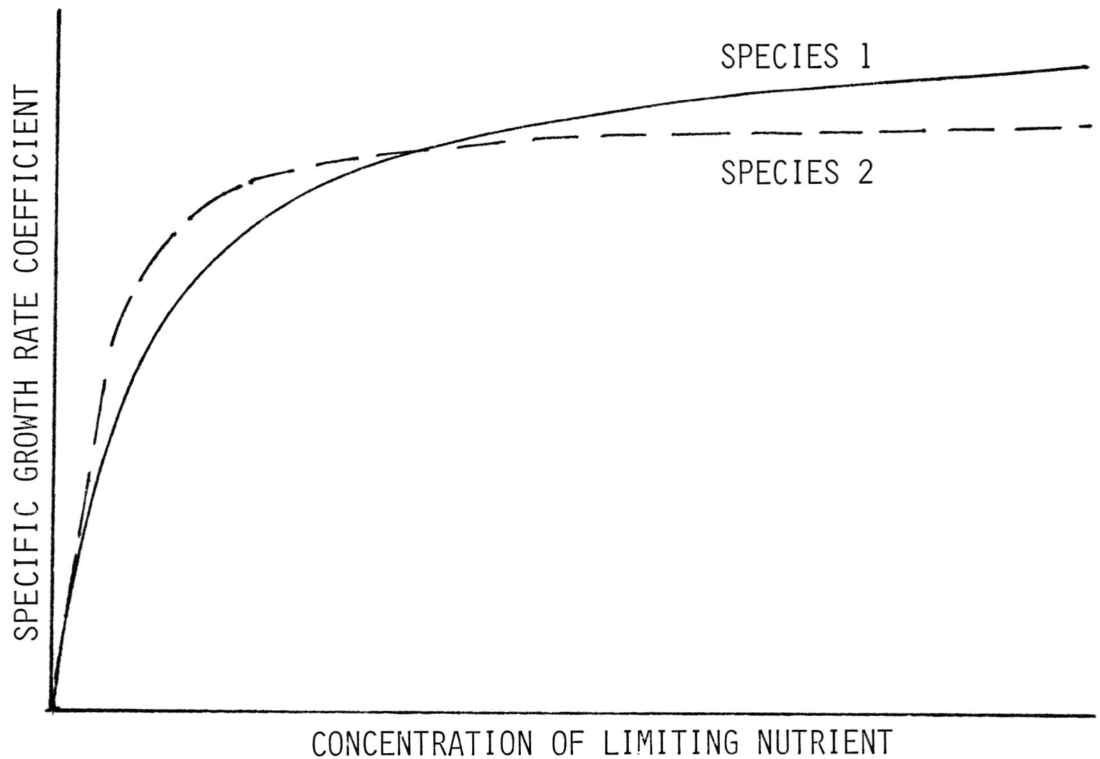

FIGURE 2. Possible Growth Relationship in Mixed Culture.

were limited by the concentration of magnesium ion. Some competitions led to wash out of the less efficient species while others led to coexistence. If one competitor were inoculated at high concentration, wash out of the other could be observed. At low concentrations, both persisted. Each culture was found to excrete growth-promoting substances that worked best for itself but were also effective for the other. Therefore, this association was not a simple competition but a mutualistic situation in which each could benefit over a range of concentrations.

VIII. PREDATION

One of the most interesting reports of predation in continuous culture was by Tsuchiya, *et al.* (89). Typical sustained oscillations were followed by decaying oscillations to apparent steady states that could be followed by resumption of oscillations. These observations were confirmed by Dent, *et al.* (26) who also used *Dictyostelium discoideum* and *E. coli*. When a temporary steady state was established, a change in an experimental variable produced a different steady state. Bazin (11a) and Basin, *et al.* (11b) studied this same system with emphasis on uptake of nutrients from the medium. Several other reports of predation systems are covered in Fredrickson's review (28, 33,52).

Drake and Tsuchiya (30) operated a two-stage continuous culture with *E. coli* in the first stage feeding a second stage which had *Colpoda steinii* as well. The system tended to achieve steady state.

Taylor and Berger (85a,b) have studied predation in batch cultures. Cells of *Aerobacter aerogenes* at various physiological states (exponential phase, stationary phase, etc.) were fed to *Colpodium campylum*. The predators grew most slowly on cells from the late exponential phase. A natural strategy for predators to deal with inhomogeneties in its environment was postulated.

IX. OTHER ASSOCIATIONS

Oberman and Libudzis (68) tested mixtures of *Lactobacillus casei* and a strain of *Streptococcus lactis*, that produced the antibiotic nisin. Both organisms grew in batch culture when started in proportions of ten to one of *L. casei* to *S. lactis*. Similar proportions led to stable mixed continuous cultures, and nisin was not detected. This association illustrates amensalism which is overcome by adjusting relative numbers of organisms.

Levin, *et al.* (56) modeled ecosystems with a limiting nutrient for one prey with one predator or the same plus another prey species. Predictions of the models were tested in the laboratory with *E. coli* and T-2 phage for the one-prey systems and with a sensitive strain and a phage-resistant strain for the two-prey system. Stable states of coexistence were noted, and parameter values for stability were determined. Although such work has implications to a general theory of competition and predation, Bungay (15) has pointed out the problems in formulating the predator growth rate dependency on concentration of mixtures of prey.

The effect of added inert solids was tested by Maigetter and Pfister (59). Pure and mixed cultures of *Chromobacterium lividum* and *Pseudomonas sp.* were grown with and without the clay mineral kaolinite. In all experiments, *C. lividum* adhered to the walls of the culture vessel to complicate interpretation of the results. Coexistence was observed, but the organism which was inoculated first predominated if allowed to become established. Kaolinite reduced numbers of *C. lividum* but had little effect on the *Pseudomonas*. Fluorescent antibody techniques worked very well in resolving the species.

Malhotra and Ghose (60) have reported a symbiotic relationship between *Azotobacter* and yeast.

X. BIOLOGICAL WASTE TREATMENT PROCESSES

There have been several studies of population behavior in biological waste treatment processes (20,21,51). The improvement in treatment efficiency when protozoa were present to graze on the bacterial population is well-documented (24,25, 40). Umorin and Klain (93) have shown that in nitrogen-deficient bacterial cultures with and without the infusorium *Cyclidium glaucoma*, natural death and lysis of bacteria recycled nitrogen less rapidly than did consumption and excretion by the predators. However, Pirt and Bazin (74) found bacterial numbers to be lower and nutrient concentration to be elevated after protozoa were added to laboratory continuous cultures of bacteria. It may be that predators are beneficial only in certain instances of unbalanced nutrition.

Dixit and Patterson (29) investigated the possibility of separating protozoan and bacterial populations in a two-stage modification of the activated sludge process with recycle. A high rate continuous culture was the first stage for dispersed microbial growth. This growth was fed directly to a second stage with dense protozoan populations that grazed on the dispersed bacteria. While soluble substrate removals obtained exceeded those of more conventional processes at equivalent loading rates, the effluent suspended solids were high, and would probably have required additional treatment prior to discharge.

Microbial interactions in photosynthetic systems also improve waste treatment. Swine wastes developed different bacterial populations when the alga *Chlorella vulgaris* was present (4). Some bacteria were favored while others were unaffected or repressed in illuminated systems containing algae. Sawada and Rogers (75) added the photosynthetic bacterium *Rhodopseudomonas capsulata* to batch or continuous cultures of *Klebsiella sp.* on defined media that simulated high BOD wastes. Dissolved oxygen can limit metabolism in rich medium, and photosynthesis improved metabolic rates. Acids produced by *Klebsiella* stimulated the *Rhodospseudomonas.* Aspects of waste treatment when photosynthetic bacteria are present are discussed by Crofts (22).

Unz and Dondero (94) used a micromanipulation technique to separate individual bacteria from activated sludge flocs and trickling filter slimes. A small percentage of the isolated organisms were viable on agar media, and about one-fourth of these were not zoogleal bacteria (those responsible for slime formation). The mixtures of organisms in flocs and slimes are almost certain to be important for high efficiency of metabolism of wastes, and interactions in a gelatinous floc or slime should be different than those in a well-mixed liquid. The mechanisms of interaction in a viscous medium with non-uniform distribution of organisms deserve research.

The activated sludge process depends on retaining most of the culture in the reactor as dense flocs that settle rapidly and can then be recycled easily when collected by sedimentation. Frequently population changes yield less dense cultures which settle poorly and are not collected efficiently in the sedimentation step. This is termed "bulking" and is a common explanation of poor performance by activated sludge. Sykes, *et al.* (83) diverted portions of streams in an actual municipal waste treatment plant to small reactors in which bulking could be studied. The filamentous organisms that competed with the desirable floc-forming organisms and produced bulking were identified. Classification of different bulking systems were:

Sphaerotilus bulking at food to microorganism ratios (F/M) of 1.2 to 1.8 g. of chemical oxygen demand (COD) per g. of mixed liquor suspended solids (MLSS) per day at F/M ratios of about 0.5 day^{-1} and dissolved oxygen below 0.5 mg/liter.

Schizothrix bulking at F/M ratios of 0.45 to 0.7 day^{-1} when acids such as acetic or propionic for substantial amounts of the nutrients.

Bulking of activated sludge in the large plant was caused by *Schizothrix calcicola* a blue-green algae which was growing as a heterotroph because little light penetrates into the reactor. Many older investigations of bulking denote *Sphaerotilus* as the causitive organism but without proof. It is likely the *Schizothrix* was present but not identified. Bulking could probably be prevented or ameliorated by closer control of feed and aeration rates.

In a similar vein, Sezgin *et al.* (75b) concluded from their own experiments and the work of others that the presence of filaments above a critical concentration resulted in a bulking sludge. While no quantitative data were presented, the postulate was that filament-forming organisms could exhibit a higher specific growth rate at lower dissolved oxygen concentrations than could non-filamentous organism. Therefore, it was proposed that the interaction between a high F/M ratio and a low dissolved oxygen concentration within the bulk solution results in a low DO locally within the flox, thus permitting filamentous bulking to occur. A review of bulking by Pipes (73) includes some alternative explanations.

XI. BIOLOGICAL APPROACHES TO POPULATION CONTROL

Venosa (98) has attempted control of undesirable *Sphaerotilus natans* by providing parasitic strains of *Bdellovibrio bacteriovorus*. *Sphaerotilus* is a filamentous organism often associated with bulking of activated sludge or objectionable slimes in piping or in natural waters. The *Bdellovibrio* attacked individual cells of *Sphaerotilis* but were unable to penetrate the sheaths when long filaments had formed. It is unlikely that this control method would be effective after *Sphaerotilus* was established, but individual cells migrating to initiate new filaments might be suppressed. A mathematical model for *Bdellovibrio* infections has been developed (81).

Selection in continuous cultures can have a time span of a few days. When virulent phage are inoculated into a bacterial culture, oscillations of a prey-predator type begin at once but soon become damped and distorted (47,71). Similarly, adding *Bdellovibrio* to sensitive bacteria initiates oscillations which are resistant to the phage or *Bdellovibrio* are greatly favored and increase in numbers relative to the sensitive strains. Varieties or mutants of the parasites have the ability to infect the bacterial strains resistant to the original,

TABLE 2. Processes Using Defined Mixed Cultures

Product	Cultures	Comments	Reference
Amino acids and vitamins	Azotobacter and an activator microorganism		70
Androstane derivatives	Arthrobacter simplex, Nocardia corallina	Much faster in mixed culture than with sequential fermentation	8,69
Aspartic acid	Rhizopus arrhizus, Proteus vulgaris	Changed from fumaric acid fermentation by mixed culture	48a
Copper and nickel salts	Thiobacillus ferroxidans, Beijerinckia lacticogenes	Mutualistic leaching of ores	90
Decomposition of Dalapon	Trichoderma viride (reesei) and 6 bacteria	Herbicide destroyed by stable defined culture	77
Decomposition of polyvinyl alcohol	Neisseria and Pseudomonas	Waste treatment	34,35
Decomposition of phenol	Paramecium caudatum and bacteria	Predators improve nitrogen-deficient process but hurt in rich media	92
Ethanol	Trichoderma viride (reesei) Saccharomyces cerevisiae	Simultaneous hydrolysis and fermentation of cellulose (Many reports in literature of separate addition of enzyme.)	38
Ergot alkaloids	Claviceps purpurea, mixed organisms	Product was mixed alkaloids	53
Food yeast	Candida tropicalis, Saccharomyces cerevisiae, or C. brumptii & C. utilus	Hydrocarbon substrate	62,102

TABLE 2. *(continued)*

Product	*Cultures*	*Comments*	*Reference*
Fuel gas	Clostridium, Acetobacter, Methanobacterium, Methanosarcina, Thiorhodaceae, Athiorhodaceae, Oscillatoria, Anabaena, Microcytsis, Chlorella, Scenedesmus, Euglena, Plormidium, Synechocystis	*Mixtures of hydrogen and methane from manures, using some organisms from list*	*76*
Glutamic acid	Corynebacterium equi, Brevibacterium lacto-fermentum *or* B. saccharolyticum *also* Aspergillus *or* Pseudomonas	*Substrate was hexadecane*	*87*
Guanosine	Corynebacterium glutamicum, Brevibacterium acetylicum	*Synergistic pathways*	*1*
Histidine	Corynebacterium glutamicum, Escherichia coli	*Synergistic pathways*	*7*
Indonesian fermented rice	Amylomyces rouxii *plus one or more yeasts from species Endomycopsis, Candida, and Hansenula*	*Only* E. burtonii *gave the characteristic favor*	*23*
Levorin	Streptomyces levoris *and* Saccharomyces cerevisiae *or* Actinomyces levoris *and* Candida tropicalis, *or other yeast*	*pH optimum shifted when second organism added*	*37,91,106*

Lysine	Brevibacterium, Trichosporon cutaneum	*Product rich in lysine for animal feed*	*12*
Lysine	Pseudomonas brevis, Microbacterium flavum	*Hydrocarbon substrates*	*100*
Lysine	Streptomyces hygroscopicus, Pseudomonas brevis	*Hydrocarbon substrates*	*46*
Malic acid	Rhizopus arrhizus *and* Proteus vulgaris *or* Candida hydrocarbofumarica *and* Mycotorula *or* Sporobolomyces		*84,107*
Single-cell protein	Methylococcus capsulatus, Methylosinus sperium incoloratus, Methylosinus trichosporium rosaceus, Flavobacterium gasotypicum	*Hydrocarbon substrates*	*3*
Single-cell protein	Trichoderma viride *(reesei)*, Candida utilis *or* Saccharomyces cerevisiae	*Substrate was straw*	*72*
Single-cell protein	Candida lipolytica, C. tropicalis	*Percentage of protein higher in mixed than in pure cultures*	*2*
Single-cell protein	Hyphomicrobium sp., Pseudomonas sp.	*Interactions in mixed cultures were well-characterized. Substrate was methane.*	*105*
Succinic acid	Rhizopus arrhizus, Aerobacter aerogenes	*Changed from fumaric acid fermentation by mixed culture*	*48b*
Swiss cheese	Lactobacillus plantarum, Propionibacterium shermanii	*No oscillations in continuous culture*	*55*

Valine	Serratia marcescens, Escherichia coli	*Developed methionine - requiring* E. coli	*101*
Vitamin B_{12}	Rhodopseudomonas spheroides, Corynebacterium	*Hydrocarbon substrates*	*66*
Hydrogen	Azolla *mixed algae*	*Water fern - algae symbiosis*	*67*

predominant parasites. Thus a natural population must be in an evolving complicated state of selection and mutation where the new species could appear identical to the old unless tests for parasites are also evolving and it would be difficult or impossible to obtain and to maintain all of the stock cultures required for characterizing parasite sensitivity of even a few of the various bacteria present.

XII. PRACTICAL PROCESSES

A number of processes employing defined mixed cultures are listed in Table 2. It seems that there are many synergistic combinations of microorganisms that can synthesize substances not produced by pure cultures or that can operate faster or more efficiently than do pure cultures. Searching for new mixed culture processes should be a rewarding investment of research and development resources. Some processes of special interest are not in the table and will be discussed.

Conventional processes for manufacturing paper pulp produce large amounts of wastes containing sulfonated lignin that is highly resistant to microbial attack. Anaerobic cultures do little or nothing to lignin, but it is slowly attacked by aerobes. Ban, *et al.* (10) using a system that was 80 percent bacteria and 20 percent yeast were able to metabolize about half of the calcium lignosulfonate in a wood pulping waste in 24 hours. The culture was developed by inoculation of *Trichosporon cutaneum* and *Trichosporan fermentans* (yeasts) mixed with *Arthrobacter terregens, Arthrobacter citreus, Pseudomonas putida,* and *Chromobacterium violaceum* (bacteria). Most investigators feel that defined cultures would not persist in elective cultures for waste treatment, but the high efficiency of this system might lead to a practical process.

Yogurt is made by the mixed fermentation of fortified milk using mixed strains of *Lactobacillus bulgaricus* and *Streptococcus thermophilis*. A two-stage continuous process can be operated with a well-mixed fermenter developing low pH without casein coagulation followed by a plug-flow reactor for coagulation and further acid production. The first stage seeds the mix with both organisms. MacBean, *et al.* (58) have formulated equations for this process and developed a system based on adjusting feed rate to control pH. Performance and stability were investigated mathematically and experimentally. In milk-based media, lactose is in large excess and nitrogen sources limit growth. As a consequence, when the feed is restricted, acid formation can continue, but at a slower rate, from non-growth-associated metabolism. Buffering by the medium is important to process stability because of this continuing acid production, and the pH set point must be higher than the pH to which the process can drift. The stability of the mixed culture in the actual process was very good although one organism was faster growing than the other. There is a possibility that wall growth was a stabilizing factor that would be of less significance in larger reactors. It was also postulated that nutritional differences mitigated the severity of competition and allowed persistence of the slower-growing organism. In any event, this is a good illustration of a commercial process using feedback control with stable mixed cultures.

Algae are a very promising new source of biomass for energy purposes because of their potentially high growth rates. Vast growth areas are required for plants if biomass is to supply significant amounts of energy on a national basis, and high productivity can reduce the areas and costs dramatically. Efficient inexpensive harvesting of algae is a major problem, however, because concentrations are low and processing large volumes of dilute media is expensive. When large algal species are favored by the culture system, it is possible to harvest by relatively inexpensive straining, but a shift to a smaller species leads to failure at the straining step. Weissman and Benemann (103) have reported on recycling a portion of the large algae to maintain their dominance in the culture. The mathematical analysis and actual experiments showed that recycle affects the outcome of competition. However, recycle reduces yields, and sometimes small algae predominate despite recycle of larger species. Although recycle helps, further research is needed to solve the harvesting problem for algae.

XIII. CONCLUSION

Mixed cultures remain of prime importance for waste treatment despite advances in alternative physiochemical processes. However, much research will be needed to gain adequate understanding of microbial interactions that determine performance and stability of processes based on elective cultures. Many industrial processes based on defined mixed cultures are already important, and progress continues to be rapid. Should ethanol become the main transportation fuel, there will be hundreds of new factories, and mixed culture fermentation of ethanol from biomass could become the major process (99). Harrison and Wren (44) have compared two processes for single-cell protein, one using a pure culture and the other a mixed culture. The mixed culture had superior stability and was more resistant to contamination and foaming. When synergistic cultures yield a product unobtainable with pure cultures, a mixed culture process is dictated, but these other potential advantages suggest that mixed cultures may also replace today's familiar pure culture processes.

The foundations for future mixed culture research are fairly secure. Larsen, *et al.* (54) have presented a system for selecting mixed cultures for commercial processes. Although natural ecosystems and elective culture processes are orders of magnitude more complicated than the models proposed for them, some real understanding and appreciation have been achieved. Investments in mixed culture research and development should be highly rewarding.

REFERENCES

1. Abe, S. and A. Furuya, Japanese Patent 33,393,C.A. 80:119215 (1973).
2. Achremowicz, B., F.V. Kosikowski and K. Masuyama, *Acta Microbiol. Pol. 26,* 265 (1977).
3. All-Union Scientific- Research Institute of Protein Biosynthesis, British Patent 1,481,765, C.A. 87:199180 (1977).
4. Allen, M.D.B. and M.K. Garrett, *J. Appl. Bact. 42,* 27 (1977).
5. Anderson, E.S., *Microbial Ecology,* 7th Symposium of the Society for General Microbiology, Cambridge University Press, Cambridge, England (1957).
6. Andrews, J.F., *Biotechnol. Bioeng. 12,* 679 (1970).
7. Araki, K. and K. Nakayama, *Agric. Biol. Chem. 39,* 127 (1975).
8. Awata, N., K. Dezawa and T. Teramoto, Japanese Patent 00490, C.A. 81:103269 (1974).

9a. Bailey, J.E. and D.F. Ollis, *Biochemical Engineering Fundamentals,* McGraw Hill, New York (1977).

9b. Baker, K.H. and D.S. Herson, *Appl. Environ. Microbiol. 35,* 791 (1978).

10. Ban, S., M. Glanser-Soljan and M. Smailagic, *Biotechnol. Bioeng. 21,* 1917 (1979).

11a. Bazin, M.J., *Ann. Appl. Biol. 89,* 159 (1978).

11b. Bazin, J.J., P.T. Sawndeys, B.A. Owen and D. Kilpatrick, in *Microbial Ecology,* M.W. Loutit, J.A.R. Miles, Eds., Springer, Berlin, Germany (1978).

12. Bekers, M., A. Sakase, M. Kalnina, and U. Viesturs, French Patent 2,244,817, C.A. 83:191356 (1975).
13. Brock, T.D., *Principles of Microbial Ecology,* Prentice Hall, Englewood Cliffs, New Jersey (1966).
14. Bungay, H.R., in *Germentation Technology Today,* G. Terui, Ed., Society for Fermentation Technology, Osaka, Japan (1972) p. 117.

15. Bungay, H.R., in *Modeling Biochemical Processes in Aquatic Ecosystems,* R.P. Canale, Ed., Ann Arbor Science, Ann Arbor, Michigan (1976) p. 377.
16. Bungay, H.R. and M.L. Bungay, *Adv. Appl. Microbiol. 10,* 269 (1968).
17. Bungay, H.R., L.S. Clesceri and N.A. Andrianas, Paper presented at International Fermentation Symposium, London, Ontario (1980).
18. Cassel, E.A., F.T. Sulzer and J.C. Lamb, III, *J. Water Pollut. Control Fed. 38,* 1398 (1966).
19. Chao, L., B.R. Levin and F.M. Stewart, *Ecology, 58,* 369 (1977).
20. Chiu, S.Y., L.T. Fan, I.C. Kao and L.E. Erickson, *Biotechnol. Bioeng. 14,* 179 (1972).
21. Chynoweth, D.P. and R.A. Mah, *J. Water Pollut. Control Fed., 49,* 405 (1977).
22. Crofts, A.R., *Proc. Royal Soc. London,* Ser. B, *179,* 209 (1971).
23. Cronk, T.C., K.H. Steinkraus, L.R. Hacklet and L.R. Mattick, *Appl. Environ. Microbiol. 33,* 1067 (1977).
24. Curds, C.R., *Am. Zoolog. 13,* 161 (1973).
25. Curds, C.R., M.J. Bazin, in *Advances in Aquatic Microbiology,* Vol. 1, M.R. Droop and H.W. Jannasch, Eds., Academic Press, New York (1977) p. 115.
26. Dent, V.E., M.J. Bazin and P.T. Saunders, *Arch. Microbiol. 109,* 187 (1976).
27. Dhillon, E.K.S., T.S. Dhillon, Y.Y. Lam and A.H.C. Tsang, *Appl. Environ. Microbiol. 39,* 1046 (1980).
28. Dive, D., *J. Protozool. 22,* 545 (1975).
29. Dixit, S.N.S. and J.W. Patterson, Paper presented at the 30th Purdue Industrial Waste Conference, Lafayette, Indiana (1975).
30. Drake, J.F. and H.M. Tsuchiya, *Appl. Environ. Microbiol. 34,* 18 (1977).
31. Ewart, D.L. and M.J.B. Paynter, *Appl. Environ. Microbiol. 39,* 576 (1980).
32. Fazel-Madjless, J. and J.E. Bailey, *Biotechnol. Bioeng. 21,* 1995 (1979).
33. Fredrickson, A.G., *Ann. Rev. Microbiol. 31,* 63 (1977).
34. Fukunaga, K., T. Takemasa, K. Ueda, K. Tachibana and S. Sumina, Japanese Patent 70,866, C.A. 85:141339 (1976).
35. Fukunaga, K., K. Ueda, S. Sumina, T. Takemasa and K. Tachibana, Japanese Patent 70868, C.A. 85:141337 (1976).
36. Gall, L.S., *Biotechnol. Bioeng. 12,* 333 (1970).
37. Ganov, V.A., V.P. Namestikova, V.V. Antonova, E.I. Iofina, M.A. Malkov, V.S. Nyn, O.A. Kuznetsova, G.N. Malinovskaya and G.P. Poluparneva, British Patent 1,347,020, C.A. 81:11834, U.S. Patent 3,802,998, C.A. 81:24174 (1974).

38. Gauss, W.F., S. Suzuki and M. Takagi, German Patent 2,609,551, C.A. 86:153959 (1977).
39a. Gitelson, I.I., A.M. Kuznetsov, E.K. Rodicheva, A.M. Fish, R.I. Chumakova and G. Ya. Shcherbakova, *Biotechnol. Bioeng. Symp. 4,* 857 (1974).
40. Gude, H., *Microbial Ecology 5,* 225 (1979).
41. Harder, W., *J. Appl. Chem. Biotechnol. 23,* 707 (1973).
42. Harder, W., J.G. Kupnen and A. Matin, *J. Appl. Bact. 43,* 1 (1977).
43. Harrison, D.E., *Adv. Appl. Microbiol. 24,* 129 (1978).
44. Harrison D.E.F. and S.J. Wren, *Process Biochem. 11,* 31 (1976).
45. Hatch, R.T., C. Wilder and T.W. Cadman, Second International Conference on Computer Applications in Fermentation Technology, Philadelphia, Pennsylvania (1978).
46. Hirakawa, T., K. Takahara, Y. Nakamora, S. Iwasaki and T. Tanaka, Japanese Patent 18,689, C.A. 83:176702 (1975).
47. Horne, M.T., *Science, 168,* 992 (1970).
48a. Hotta, K. and S. Takao, *Hakko Kogaku Zasshi 51,* 12 (1973).
48b. Hotta, K. and S. Takao, *Hakko Kogaku Zasshi 51,* 26 (1973).
49. Jannasch, H.W. and R.I. Mateles, *Adv. Microb. Physiol. 11,* 165 (1974).
50. Jannasch, H.W. and H. Veldkamp, *J. Appl. Chem. Biotechnol. 22,* 105 (1972).
51. Jones, G.L., *Process Biochem. 11,* 3 (1976).
52. Jost, J.L., J.F. Drake, A.G. Frederick and H.M. Tsuchiya, *J. Bacteriol. 113,* 834 (1973).
53. Kobel, H. and J.J. Sanglier, Swiss Patent 577,556, C.A. 86:41798 (1976).
54. Larsen, V.F., R.S. Holdom, M.J. Spivey and M. Todd, *Soc. Appl. Bacteriol. Technical Series 11,* 179 (1978).
55. Lee, I.H., A.G. Fredrick and H.M. Tsuchiya, *Biotech. Bioeng. 18,* 513 (1976).
56. Levin, B.R., F.M. Stewart and L. Chao, *Am. Nat. 111,* 3 (1977).
57. Lovelock, D.W. and R. Davies, *Techniques for the Study of Mixed Populations,* Academic Press, London, New York (1978).
58. MacBean, R.D., R.J. Hall and P.M. Linklater, *Biotechnol. Bioeng. 21,* 1517 (1979).
59. Maigetter, R.Z. and R.M. Pfister, *Can. Jour. Microbiol. 21,* 173 (1975).
60. Malhotra, K.R. and T.K. Ghose, *Eur. J. Appl. Microbiol. 1,* 95 (1975).
61. Mateles, R.I., D.Y. Rye and T. Yasuda, *Nature, 208,* 263 (1965).
62. Matsubayashi, T., T. Shiono, Y. Fukuda and T. Kimura, Japanese Patent 81,582, C.A. 82:56103 (1974).
63. Meers, J.L., *C.R.C. Crit. Rev. Microbiol., 2,* 139 (1973).

64. Meers, J.L. and D.W. Tempest, *J. Gen. Microbiol. 52,* 309 (1968).
65. Miura, Y., H. Tanaka and M. Okazaki, *Biotechnol. Bioeng. 22,* 929 (1980).
66. Nakao, Y., K. Hisano, T. Kanemaru and T. Yamato, Japanese Patent 15,796, C.A. 82:29689 (1974).
67. Newton, J.W., *Science, 191,* 559 (1976).
68. Oberman, H. and Z. Libudzis, *Acta Microbiol. Pol. Ser. B. Microbiol Appl. 5,* 151 (1973).
69. Okamoto, A. and T. Teramoto, Japanese Patent 50,192, C.A. 81:150307 (1974).
70. Panosyan, A.K., G.S. Babayan, V. Nikogosyan, N.M. Sayadyan, D.S. Mkrtchyan, and S.B. Shakhmoradyan, *VOP Mikrobiol. 6,* 145 (1973).
71. Paynter, M.J.B. and H.R. Bungay, in *Germentation Advances,* D. Perlman, Ed., Academic Press, New York (1969).
72. Peiterse, N., *Biotechnol. Bioeng. 17,* 1291 (1975).
73. Pipes, W.O., *Adv. Appl. Microbiol. 24,* 85 (1978).
74. Pirt, S.J. and M.J. Bazin, *Nature 239,* 290 (1972).
75a. Sawada, H. and P.L. Rogers, *J. Ferm. Tech. 55,* 311 (1977).
75b. Sezgin, M., D. Jenkins and D.S. Parker, *J. Water Pollut. Control Fed., 50,* 362 (1978).
76. Sekine, K., Japanese Patent 111,283, C.A. 84:29216 (1975).
77. Senior, E., J.H. Slater and A.T. Bull, *J. Appl. Chem. Biotechnol. 26,* 329 (1976).
78. Sinclair, C.G., W.R. King, D.N. Ryder and H.H. Topiwala, *Biotechnol. Bioeng. 13,* 451 (1971).
79. Smith, A.L. and D.P. Kelly, *J. Gen. Microbiol. 115,* 379 (1979).
80. Stephanopoulos, G., A.G. Fredrickson and R. Aris, *AIChE J. 25,* 863 (1979).
81. Sterkin, V.E. and V.A. Samoilenko, *Mikrobiolog. 41,* 466 (1972).
82. Sudo, R., K. Kobayashi and S. Aiba, *Biotechnol. Bioeng. 17,* 167 (1975).
83. Sykes, R.M., A.F. Rozich and T.A. Tiefert, *J. Water Pollut. Control Fed. 51,* 2829 (1979).
84. Takao, S. and K. Hotta, *Agric. Biol. Chem. 41,* 945 (1977).
85a. Taylor, W.D. and J. Berger, *Can. J. Zool. 54,* 392 (1976).
85b. Taylor, W.D. and J. Berger, *Can. J. Zool. 54,* 1111 (1976).
86. Tison, D.L., D.H. Pope and C.W. Boylen, *Appl. Environ. Microbiol. 39,* 675 (1980).
87. Tomoyuki, I., N. Hiroshi, K. Tanno, K. Kono and N. Miyaji, Japanese Patent 13,111, C.A. 77:32708 (1972).
88. Trilli, A. and S.J. Pirt, *J. Appl. Chem. Biotechnol. 26,* 328 (1976).
89. Tsuchiya, H.M., J.F. Drake, J.L. Jost and A.G. Fredrickson, *J. Bacteriol. 110,* 1147 (1972).

90. Tsuchiya, H.M., N.C. Trivedi and M.L. Schuler, *Biotechnol. Bioeng. 16,* 991 (1974).
91. Tsyganov, V.A., E.P. Yakovleva, V.M. Morozov, E.N. Sokolova, O.S. Kuznetsova and G.M. Ratsun, *Antibiotiki* (Moscow) *18,* 358 (1973).
92. Umorin, P.P., *Tr. Inst. Biol. Vnutr. Vod. Akad. Nau.k. SSSR* No. 24, 201 (1973).
Zh. Obsch. Biol. 35, 119 (1974).
93. Umorin, P.P. and N.P. Klain, *Zh. Obs. Biol. 38,* 573 (1977).
94. Unz, R.F. and N.C. Dondero, *Water Res. 4,* 575 (1970).
95. Van den Ende, P., *Science, 181,* 562 (1973).
96. Varon, M., *Nature, 277,* 386 (1979).
97. Veldkamp, H. and H.W. Jannasch, *J. Appl. Chem. Biotechnol. 22,* 105 (1972).
98. Venosa, A.D., *Appl. Microbiol. 29,* 702 (1975).
99. Wang, D.I.C., I. Biocic, H.-Y. Fang and S.-Y. Wang, 3rd Annual Biomass Energy Systems Conference Proceedings, NTIS, SERI/TP-33-285, p. 61 (1979).
100. Watanabe, K., T. Hirakawa, Y. Nukamuta, S. Iwasaki and T. Tanaka, Japanese Patent 109,584, C.A. 82:137748 (1974).
101. Watanabe, K., T. Tanaka, T. Hirakawa and M. Sasaki, Japanese Patent 33,996, C.A. 86:104462 (1976).
102. Watanabe, K., H. Yano, K. Suzuki, Y. Mekata and H. Kawalarada, Japanese Patent 18,563, C.A. 77:73696 (1970).
103. Weissman, J.C. and J.R. Benemann, *Biotechnol. Bioeng. 21,* 627 (1979).
104. Wilder, C.T., T.W. Cadman and R.T. Hatch, *Biotechnol. Bioeng. 22,* 89 (1980).
105. Wilkinson, T.G., H.H. Topiwala and G. Hammer, *Biotechnol. Bioeng. 16,* 41 (1974).
106. Yakovleva, D.P., *Antibiotiki* (Moscow), *71,* 494 (1976).
107. Yamada, K. and T. Furukawa, Japanese Patent 19,631, C.A. 84:29185 (1975).
108. Young, T.B., D.F. Bruley and H.R. Bungay, *Biotechnol. Bioeng. 12,* 747 (1970).
109. Young, T.B. and H.R. Bungay, *Biotechnol. Bioeng. 15,* 377 (1973).

CHAPTER 2

CONVERTING FOOD PROCESSING WASTES INTO FOOD OR FEED THROUGH MICROBIAL FERMENTATION

M. A. Cousin

Food Sciences Institute
Purdue University
West Lafayette, Indiana

I. WASTE OR RESOURCE?

The changing economic, social, political, and culture values of the world have challenged both industrialized and non-industrialized countries to efficiently use renewable and nonrenewable resources. There have been increasing efforts by industries, businesses, governments, and consumers to recycle traditional waste products. The food industry has not been immune to pressures for adequate resource management. In fact, some scientists believe that no crop or animal or marine commodity should be produced for food if use of only one fraction is expected. Therefore, the planned total use of food raw materials will involve the cooperative efforts of scientists in food and animal sciences, chemistry, agronomy, horticulture, engineering, marine biology, microbiology, genetics, and other related fields.

In addition to the concept of total resource use, various other factors in recent years have focused on the advantages of reusing food wastes. Some of these concerns have resulted from government regulations, increasing world population, reclamation of nutrients, energy usage, and increasing food prices.

In October, 1972, the Federal Water Pollution Control Act Amendment or Public Law 92-500 was enacted to prevent, reduce, and eliminate water pollution in the United States (1). Since that time additional legislation has mandated that a zero discharge of pollutants be achieved by 1985 (2). These changes

ISBN 0-12-040304-8

have environmental impact since they were promulgated to insure that the water quality is suited for recreational activities and maintenance of fish and wildlife (3). Concern over toxicity of waste water to fish has also influenced legislation in the environmental control sector (4). Governmental agencies will continue to monitor the industrial discharges and the effect they have on aquatic life. Food processing wastes are usually toxic to fish because of the presence of organic matter, cleaning and sanitizing compounds, ammonia, and hydrogen sulfide in them that can contribute to the high biochemical oxygen demand (BOD). Therefore, the food industry is faced with selecting other methods for waste treatment in an effort to comply with federal requirements. Some food processors are beginning to adopt waste management programs where they recover and reuse waste products. Other food processors are altering their entire production flow to minimize generation of waste constituents. Since food industries will continually be expected to reduce pollutants in the environment, they must consider conservation, recovery, and reuse as essentials for total resource management.

The increasing world population focuses on the predicted food shortages in various parts of the world (5). Efforts to produce food have been intensified by various means: genetic engineering of plants and animals, exploration of the oceans, chemical or technical fabrication of foods, scavenging for nutrients, and turning food wastes into food products. Recent data indicate that protein is not limiting for much of the world's population. However, certain countries are deficient in protein, but have abundant carbohydrate resources that could be converted to protein if suitable methods were applied. Some of these carbohydrates are classified as waste products that can be converted to human food products or animal feed supplements (6).

Food wastes contain a variety of nutrients that can be recovered for food or feed products. Wastes from fruit and vegetable industries contain cellulose, starches, sugars, pectins, vitamins, minerals, and other nutrients (7). The dairy industry generates a considerable amount of whey which is 80 to 90% of the initial volume of milk and retains around 55% of the milk nutrients. Therefore, whey offers a good source of proteins, lactose, minerals, and vitamins (8). Cereal based products contain a variety of mixed carbohydrates ranging from simple sugars to starch and cellulose plus residual protein, vitamins, and minerals (9,10,11). Thus, the nutrients that can be recovered from wastes fall into two general categories: wastes that can be recovered by chemical or physical means (whey, potato protein, meat and fish lipids or proteins) and wastes that can be biologically converted to new commodities (fermentation of lactose in whey, production of fungi on carbohydrates, and recovery of by-products of microbial metabolism) (5).

Consumption of energy in the food industry has also come under scrutiny as the world envisions depletion of fossil fuels. Considerations of the United States government when it enacted the pollution control legislation took into account the minimum energy consumption necessary to treat wastes and water since sludge waste handling and disposal is an energy user (3). Therefore, various improvements in by-product recovery or other waste management systems are alternatives to saving energy that would otherwise be consumed in the disposal of these waste products. The seafood industry has recently considered the energy used to haul fishery wastes to sea or to landfill sites. Some of the energy conservation has focused on the use of wastes for food or feed products (12). The dairy industry has also considered using whey in an effort to conserve energy that would be used in its disposal (8). Cellulosic wastes have recently been converted to ethanol in an effort to generate energy (10). Hence, the treatment of these various food wastes involves energy consumption that has concerned food processors.

Coupled with the energy, regulation, and population concerns are the rising prices of food products. The increases in the human population as well as the rising affluence of the world seems to place greater demand on plant proteins (12). This increasing demand for plant proteins drives the price of feed grains up substantially, which in turn increases the price of animal proteins. This creates a never ending cycle of inflated food prices. Also, the costs of various regulatory controls in this country for food safety, nutritional content, and waste management have to be paid in some way. In the final analysis, these costs are transferred to the consumers as increased food prices.

Since most generated food wastes are not suitable for human food or animal feed in their natural state, some type of further processing is necessary (13). One processing method for effective conversion of food wastes into food resources is fermentation or use of microorganisms to alter flavor, odor and/or rheological properties of the substrate to produce desirable products. Microorganisms are almost always used in fermentations because they: (a) have greater nutrient absorption than plants or animals, (b) can adapt to many situations and ferment or utilize almost any substance, (c) produce a variety of products, (d) can often carry out more complex syntheses and produce products more easily than synthetically in a laboratory. Therefore, microbial use of waste products holds inherent value for the food industry as a mechanism of producing desirable products for food and feed as well as reducing the BOD of the waste product. Use of microorganisms to alter properties of food is not a new idea. It has been used for thousands of years to produce such foods as: bread, cheeses, alcoholic beverages, soy sauce, and tempeh. Microbial degradation is also used in the traditional waste treatment systems. This review will discuss recent research that is published on the use of microorganisms as a way to convert food processing wastes into food or feed.

II. MICROBIAL FERMENTATION OR USE OF SUGAR FROM FOOD WASTES

One of the easiest substances for microorganisms to either ferment or use is sugar. Most food wastes contain various types of sugars, ranging from simple to complex. These sugars can be fermented to desirable acids, alcohols, aldehydes, ketones, etc. or they can be used to produce biomass or numerous by-products.

A. Animal Industries

The major waste from animal products that has received much attention is whey, a by-product of cheese manufacture. The most economical use of whey is to manufacture a product from the whole whey without further concentration. A potential solution to using unconcentrated whey is to convert it to a nitrogen-rich feedstock for ruminant animals. This can be accomplished by fermenting lactose into lactic acid followed by subsequent neutralization with ammonia (14-19, 21). Usually *Lactobacillus bulgaricus* is used since it grows at 45°C and pH 5-5.8 under anaerobic conditions to produce lactic acid homofermentatively (16). Therefore, the fermentation can be done cheaply and efficiently under nonsterile conditions (18). Reddy *et al.* (19) demonstrated that lactose was metabolized in a batch system within 14-16 h and 95% of the sugar was converted to lactic acid. Continuous trouble-free operations for 42 days have also been reported (16). Various sizes and types of industries could do this fermentation because it can either be operated as a batch or continuous system. When use of deproteinized whey was compared to whole whey, it was concluded that whole whey was a better substrate, but either type of whey could be fermented in a continuous system (14). Fermented ammoniated whey concentrate is an efficient feed supplement for ruminants that is superior to urea but equal to soybean meal (15,19). Juengst (15) concluded that the advantages of such a product were: (a) all whey solids were used, which alleviated further waste treatment; (b) concentration required less energy than spray-drying; (c) the crude protein content was 45%; (d) it was a superior source of nonprotein nitrogen; and (e) this disposal of whey from cheesemaking did not prey on the environment. Erdman *et al.* (21) decided to fractionate the whey-protein-microbial cells from the ammonium lactate for use as nonruminant and ruminant supplements, respectively. The equipment necessary for this separation is normally present in a cheese factory except for filtration systems. However, the additional energy expenditure may off-set the benefits of fractionation. As another way to produce protein, Tzeng *et al.* (20) reported the use of lactose by *Kluyveromyces fragilis*. If the yeast cells were centrifuged and washed, then they could serve as food yeasts. Also by varying the conditions of the fermentation, the yeasts would produce ethanol that could be used for energy and yeast protein, that could be recovered as a by-product.

An interesting approach for use of whey was suggested by Moon *et al.* (22,23), where *Trichosporon cutaneum* and *Candida curvata*, in a two phase fermentation produced cell biomass and oil. Even though the Chemical Oxygen Demand (COD) was reduced by 95%, the yield of oil was very low. They concluded that production of oil by microorganisms will not be as efficient as that made by green plants; however, they suggested that this may be one way to solve the pollution problem of the dairy industry.

In an attempt to use whey for human consumption, production of wine or vinegar from whey has received research interests and holds some promise. Since lactose is not concentrated enough in whey for yeast to produce sufficient alcohol for wine, cottage cheese whey is concentrated to 22-24% lactose, inoculated with *K. fragilis*, and fermented for 15 days at 30°C until 10-14% alcohol is produced (8,24,25). Garvel and Kosikowski (25) reported that the whey was fermented to a crystal clear wine resembling white grape wine with a pleasant wine flavor. Other researchers (8,24) have produced whey wine from concentrated whey with hydrolyzed lactose by inoculating the whey with *Saccharomyces cerevisiae*. Researchers in California and Oregon (26-31) have produced wine from whey supplemented with sucrose by fermenting it with *S. cerevisiae* at room temperature. Residual lactose would serve as a sweetner since *S. cerevisiae* cannot ferment it. Champagne and sherry yeasts also produced desirable wines from supplemented whey. Even though the whey was deproteinated, a clarification step using bentonite was necessary (26,27). This does not create problems for use of the yeasts cells since bentonite increases milk and egg production in cows and chickens, respectively (27). An acceptable wine was produced from the sucrose fortified whey without the need for additional nutrients (31). Efforts to blend whey wine with fruit and berry wines proved successful with berry and citrus wines. However, blending with other fruit wines resulted in off-flavors or synthetic tastes (26,31).

Production of sugar syrups from whey was thought to have food applications. Roland and Alm (32) experimented with the hydrolysis of lactose in whey to sugar syrups for use in wine, beer, and ale manufacture. Preliminary experiments demonstrated that the potential for use existed. However, deterioration characterized by flowery-type aromas and flavors developed in wine within two months of aging. Other research that resulted in alcoholic beverages from whey included a whey champagne in Poland using *Torulopsis* or *Candida* species, Russian whey vodka fermented to 40% alcohol by *Torulopsis*. and whey beer in Poland (24).

There was only one reference in recent years on production of vinegar from whey. As a by-product of single cell protein manufactured by *K. fragilis* in whey, small amounts of ethanol are produced. This ethanol can be further oxidized to acetic acid or vinegar (8).

B. Cereal and Sugar Industries

There are reports in the literature that various cereal and sugar wastes can be used to produce single cell protein or other microbial products. Sugimoto (11) investigated the use of spent solids from production of soy protein isolates for growth of *Candida guilliermondii* and *Debaryomyces kloeckeri*. These substrates contained mainly carbohydrates of low molecular weight. At the end of the 60-h fermentation, 97.2% of total carbohydrate, 97.6% of reducing sugars, and 99.8% of ketohexoses had disappeared from the medium. These carbohydrates could be readily used by the yeasts to produce cell mass. *D. kloeckeri* reduced the BOD by 84%, suggesting that the waste is treated in addition to producing single cell protein. However, *C. guilliermondii* smelled fruity and cells were smaller than those of *D. kloeckeri*, which was easier to separate from the broth and smelled like typical baker's yeast. Kuichi *et al.* (33,34) also experimented with a soybean waste product; however, this was a residue from soybeans used to manufacture miso, and the residue contained carbohydrate and crude protein. The BOD of the residue from the miso process can reach 20,000 to 40,000 mg/l. Various species of *Saccharomyces, Hansenula, Debaryomyces, Pichia, Cryptococcus, Torulopsis, Candida, Kloeckera,* and *Trichosporon* were grown on the soybean waste in an effort to develop flavor compounds which could be characterized. The flavors were described as flower-like, fruit-like, and maple syrup-like. Of the 35 strains studied, only two produced desirable flavors: *Hansenula anomala* produced an ethyl acetate flavor in the early stages, but in later growth stages an unpleasant odor developed which was replaced by a fruit-like flavor and *S. cerevisiae* produced a maple syrup-like flavor that grew stronger and sweeter as growth proceeded. Further studies indicated that *S. cerevisiae* produced mainly ethanol plus acetaldehyde; however, *H. anomola* produced methylaldehyde and ethyl acetate. Therefore, since not all yeasts produce acceptable flavors when grown on soybean wastewater, this process has limitations for commercial use.

Forage (35) experimented with microbial cultivation as a method of treating effluents since this is the principle upon which activated sludge, lagooning, trickling filter, and anaerobic digestion waste treatment systems are based. He suggested that production of single cell protein is a benefit of rather than the primary goal of waste control. Forage (35) reported that confectionery processing effluent that consisted of 70% sucrose, glucose, and other carbohydrates was easily assimilated by *Candida utilis*. The starch had to be hydrolyzed prior to inoculation by *C. utilis* since this yeast cannot degrade it. The BOD from these wastes was reduced by 81% by the growth of this yeast. Hence, he concluded that conversion of confectionery waste to single cell protein was an alternative waste treatment process that required only 80% of the capital investment and used only 0.1% of land area when compared to conventional systems.

Production of baker's yeast from molasses or other by-products of the sugar refining industry has been evaluated for a number of years (36,37,38). Molasses is not only a source of carbon, but also of nitrogen and other essential nutrients that are needed for yeast growth. Saccharose is the major sugar present in molasses, which contains 47 to 52% sugar. Other sugars present in molasses are glucose, fructose, and raffinose. The total nitrogen content, which usually is around 0.82 to 2.2%, is an important nutrient for yeast growth. Certain nutrients affect production of yeast biomass on molasses: (a) magnesium must be present; (b) high calcium content can inhibit yeast growth; (c) microelements such as manganese, molybdate, iron, cobalt, copper and zinc are important; (d) beet molasses contains biotin, pantothenate, and inositol which are essential for good yeast growth; (e) volatile acids (acetic, butyric, formic, etc.) can prevent yeast growth; and (f) sulfrous acid, nitrates, nitrites, color substances, and colloids can also inhibit yeast growth (37). Since *C. utilis* is not a fastidious yeast, it has been selected for use because no additional nutrients need to be added to the molasses (38). The predominant aerobic metabolism of yeast as well as the active degradation of sugars results in a biomass of yeast cells containing about 50% protein, which compares favorably to soybean protein. However, sulfur containing amino acids are limiting, which is common for almost all yeast strains. It has been suggested that yeast cells grown on molasses can be used to increase the protein content of cereal-based foods. Current status of research on molasses indicates that even though it has potential in recycling a waste product, the inherent variations in chemical composition can ultimately prevent its widescale use as a substrate for yeast production.

C. Fruit and Vegetable Industries

Waste generated by the fruit and vegetable industries is usually not salable in the final form and generally cannot be used directly in food or feed. These wastes have traditionally been disposed of by incineration or landfill methods, but these methods have now come under attack in the United States through increased governmental legislation which ultimately strives for a clean environment. Various industries are now searching for ways to recycle these wastes. Some of the research has involved fermentation of these wastes. Moon (39), after screening more than 40 microorganisms, concluded that *Lactobacillus plantarum, L. bulgaricus* and *Streptococcus faecalis* subsp. *liquefaciens* were the best species to ferment sugars in vegetable processing wastes to lactic acid. The vegetable wastes used in the study were green beans, turnip greens, corn, potatoes and black-eyed peas. Since these wastes were not sterilized before inoculation, undesirable microorganisms could dominate the fermentation if sufficient numbers and optimum conditions were present. Turnip greens, which are higher in protein than the other vegetables studied, resulted in slow fermentation rates with minimal acid production and pH reduction. Also, the moisture content affected the fermentation; even though it was optimal at 90%, research has suggested that a moisture content of 70% results in an acceptable fermentation because dissolved solids will not leach from the waste at this moisture level. Since energy would be needed to dry the wastes, fibrous wastes (peanut hulls) were used to absorb moisture to achieve a 70% figure. The results of these experiments suggested that the wastes should be adjusted to contain an optimum protein and carbohydrate content; the best carbon to nitrogen (C:N) ratio was 20:1 to 25:1. Adjusting the C:N ratio may increase the cost of processing wastes by this method; however, this cost may be negligible when compared to those of traditional waste treatment.

Other researchers have considered fungal digestion of processing wastes since this is one step used in traditional waste treatment. Fungi (13,43) are used to treat food processing wastes because they: (a) have a wide range of enzymes that can degrade complex organic compounds, (b) require low concentrations of nutrients to remove BOD, (c) are relatively insensitive to environmental conditions, and (d) can be readily recovered by simple filtration methods. However, various fungal processes can be contaminated by undesirable microorganisms, and thus sterile conditions may be necessary. Some fungi produce mycotoxins so the safety of using these microorganisms is often questioned. Generally, 25 g of fungal protein can be produced from 100 g of carbohydrate. Church *et al.* (40) tried to reduce the organic loading on streams by pea and corn canning wastes through converting the carbohydrate and organic matter to cellular constituents of *Trichoderma viride* and *Gliocladium deliquescens* for use in animal feeds. However, during the midseason of corn processing *Geotricnum* accounted for 70% of the fungi present, but the pea waste was contaminated with *Fusarium*. Since some *Fusarium* spp. produce mycotoxins, this process would result in an unacceptable protein supplement for animal feed. Since the BOD reductions for treating corn and pea wastes were 96 and 95%, respectively, this method could be used to treat vegetable wastes if the system was not contaminated by fungi that produce mycotoxins. An additional bonus of recovering mycelial protein is that it could be sold to offset the waste-treatment costs.

Researchers in New York (41-44) have done various experiments on sauerkraut wastes to produce single cell protein and other by-products. Over 29% of the discarded waste is brine that has a high BOD and contains lactic acid and salt (Table 1).

TABLE 1. Analyses of Sauerkraut Waste Used for Yeast Growth

Component	*41*[a]	*42*[a]	*44*[a]
		mg/l	
BOD	*12,000*	*12,400*	*24,000*
Total Acid, Lactic	*9,800*	*7,400*	*19,900*
NaCl	*14,800*	*18,600*	*26,500*
Nitrogen, Kjeldahl	*657*	*620*	*1,100*
Total Phosphorous	*102*	*81*	*192*

[a]*Number of Reference cited.*

Three yeasts approved by FDA for food use *(S. cerevisiae, K. fragilis,* and *C. utilis)* were grown on the waste. *C. utilis* grew the fastest and reduced the BOD 87% within 48 h. In other experiments Hang *et al.* (42) looked at the production of invertase by *C. utilis* and found 672,000 U invertase/g of dried yeast. Yeasts produced more invertase on sauerkraut waste than on standard peptone sucrose yeast extract broth, possibly because lactic acid is a good carbon source for yeast growth. The BOD was reduced approximately 88.7% in addition to neutralizing the wastes (Table 1). As a result of these experiments, batch processes are being used commercially by sauerkraut manufacturers for daily in-plant treatment of BOD. This system has been so successful that the Environmental Protection Agency has revised the wastewater limitation guidelines for Phelps, New York, which has two processing plants. This process not only reduces the BOD but results in the recovery of a single cell protein and an enzyme which can be sold to various industries. Another industry producing high volumes of salt brine (about 70% of the processing waste) is the pickle industry (45). *C. utilis* grew on lactic acid and produced the greatest cell mass at 30 to 35°C within 24 h after inoculation. Experiments conducted with pasteurized vs. nonpasteurized brine indicated that phosphate, which was precipitated by heat, was essential for good cell production. The BOD was reduced by 90% as a result of the yeast growth in waste containing phosphate.

Experimenters in the fruit industry have studied production of single cell protein as a way to reduce the BOD. Chung and Meyers (46) produced *Pichia spartinae* on banana wastes (Table 2). Both banana pulp and skin liquors contain adequate supplies of usable carbohydrate, nitrogen, and minerals for yeast growth. The major sugars (glucose, fructose, and sucrose) in bananas increase during ripening and are maintained at high levels. Besides *P. spartinae, Saccharomyces* and *Candida* species can grow on banana wastes, producing good

TABLE 2. Proximate Analysis of Dried Banana Wastes (46)

Component	*Banana Pulp Liquor*	*Banana Skin Liquor*
	Percent	
Sugar, Hexoses	*89.5*	*56.5*
Crude Protein	*4.4*	*4.0*
Ash	*5.3*	*37.0*
Fiber	*0.37*	*0.26*
Phosphate	*0.25*	*0.8*

yeast protein. Yeast growth was optimized by addition of only ammonium sulfate. Therefore, this process may have commercial value in tropical countries for using wastes from banana or other fruit processing operations. Another waste product from tropical countries is coconut water, which is a faintly turbid liquid obtained from the processing of coconuts. Tropical countries have large quantities of this material, containing a BOD of approximately 40,000 mg/l. *K. fragilis* used the coconut water, containing glucose, fructose, sucrose and sorbitol to produce single cell protein. Only sorbitol was not utilized by this yeast. Since *K. fragilis* metabolizes mainly through respiration, it produces high cell yields and very little ethanol (47). Supplementation of the substrate with biotin and nicotinic acid increased the biomass yield, but phosphate and magnesium had no effect. Smith and Bull (48), in a subsequent report, concluded that *K. fragilis* when grown on coconut water did not have as high a level of amino acids as when grown on whey. However, fermentation of coconut wastes with *K. fragilis* can be one way for the coconut industry to reduce pollution as well as to recover a protein that can be used as a food or feed supplement.

Use of carbohydrate wastes from date was reported by Kamel (49) and Humphrey (50). Since dates contain approximately 65% sugar as glucose and fructose, these sugars can be obtained from dates by various water extraction methods. Strains of three yeasts *(S. cerevisiae, C. utilis,* and *S. rouxii)* were grown on date juice with the production of single cell protein. Since both *C. utilis* and *S. cerevisiae* have been approved as food-grade yeasts, both yeasts were simultaneously grown on date wastes yielding a higher protein content than when each yeast was grown separately (Table 3). The authors concluded that more work is needed before final evaluations can be made.

TABLE 3. Growth of Saccharomyces *cerevisiae* and Candida utilis *on Date Carbohydrate (49)*

Organism	*% Conversion*[a]	*% Protein*
C. ulitis	*26.4*	*49.4*[b]
S. cerevisiae	*30*	*50.2*[b]
C. utilis + S. cerevisiae	*45.8*	*52.4*[c]

[a] *Medium contained 4% soluble solids at pH 3.5.*

[b] *12 hours.*

[c] *8 hours.*

Labaneiah *et al.* (51) used citrus (grapefruits, lemons, limes, oranges) waste to grow edible mushrooms, *Agaricus bisporus* and *Morchella crassipes*. They concluded that the fungi grew well on the unmodified waste extracts of citrus peels containing basically reducing sugars, crude protein, and ash. The carbon to nitrogen ratio ranged from 8 to 23. Best production of mushrooms occurred on lime wastes, containing higher crude protein than most other wastes whereas poorest production occurred on orange varieties having low percentages of crude protein. If additional nitrogen can be added to the substrate, higher yields of mycelia and protein can be expected. This would be one way to use a waste product to produce a food with which people are familiar.

Durgapal *et al.* (90) briefly mentioned the use of juices extracted from the skin and cores of pineapples and from mango peels as substrates for preparation of vinegar. This process was suggested for cottage-type industries in rural sectors of India for greater use of fruit waste materials.

D. Brewing Industry

Spent grain liquor (Table 4) from brewery wastes is characteristically high in BOD and suspended solids and presents a serious waste treatment dilemma. In an effort to alleviate the waste treatment needs, various researchers have studied production of single cell protein or other by-products on spent grain liquor (52,53,54). Since Hang *et al.* (53) experimented with the growth of four yeasts and four mushrooms on spent grain and achieved only a 20 to 40% reduction in BOD, they evaluated the use of sugar in the spent grain by *Aspergillus niger*. After 72 h the sugar was essentially exhausted, and the BOD was reduced by 96%. This process resulted in production of mycelium with a crude protein content of 29%, and a by-product of citric acid. Additional experiments (52) that monitored production of citric acid by *Asperigillus foetidus* demonstrated that the initial sugar content of the waste dictated the amount of acid produced. High nitrogen contents decreased the amount of citric acid produced even though the cell yield increased. Based on results of these experiments, Hang *et al.* (52) hypothesized that if 1% citric acid were removed from the waste, the BOD would be reduced by 7,500 mg/l, whereas oxidation of citric acid only resulted in removal of 5,200 mg of BOD/l. Spent grain liquor could be used to grow *Aspergillus* species and produce citric acid as a way of lowering BOD values; however, commercial applications must still be assessed. Shannon and Stevenson (54) again experimented with reduction of BOD by growing yeasts and mushrooms. The trub press liquor, a precipitate from hot wort,

TABLE 4. Examples of Brewery Waste Pollution Loads

Sample	*pH*	*BOD*	*Reducing Sugar*	*Nitrogen Kjeldahl*	*Total Phosphorus*
		mg/l			
-----[a]	4.1	22,500	23,000	335	86
A-4[b]	5.0	23,500	24,000	376	87
D-10[b]	5.7	10,800	12,100	280	47
G-17[b]	7.1	8,000	10,900	236	39
GPL[c]	5.96	31,800	30,000	434	--
TPL[c]	5.30	132,000	55,000	764	--
FSL[c]	4.43	67,500	9,900	2,870	--

[a]*Cited in Hang et al. (53).*

[b]*Cited in Hang et al. (52).*

[c]*Gram Press Liquor, Trub Press Liquor, Fermentation Sludge Liquor cited in Shannon and Stevenson (54).*

(Table 4) proved to be the best substrate for all microorganisms studied, with the mushrooms producing the highest cell yields. However, variations noted in the protein content of yeasts and mushrooms were postulated as resulting from the low nitrogen content of the brewery wastes. Both the BOD and COD were reduced from 22 to 56%. As reported previously for culturing mushrooms on citrus wastes, these authors suggested fortifying the wastes with nitrogen before growing the mushrooms. Aguinaldo (91) grew *Volvariella Volvacea* mushrooms on distillery slops containing blackstrap molasses, urea, and phosphate. The waste water had lower BOD after mycelial separation than the treated waste water from the lagoon. However, problems encountered were low yields, bland flavor, and production of nonspherical-shaped mycelium. Tomlinson (55) studied growth of *C. utilis* and *Rhodotorula glutinis* on brewery and distillery wastes in batch and continuous operations. The batch process proved most effective for conversion to yeast protein. It can be concluded from the above reports that the processes have merit in producing fungal proteins or by-products, using waste in a nontraditional manner and thereby lowering treatment costs and requiring low capital and labor investments.

III. MICROBIAL FERMENTATION OR USE OF STARCH FROM FOOD WASTES

Many food commodities, especially cereals and some fruit and vegetable crops, contain large amounts of starch. For wastes that are generated from these commodities to be efficiently used, microorganisms that degrade starch to glucose by amylolytic enzymes must be obtained. Many molds, some yeasts and many bacteria can produce starch-hydrolyzing enzymes (α-amylase, β-amylase, and amyloglucosidase). Industrially, conversion of starch to sugar involves introduction of either microorganisms that produce these amylases or the enzymes themselves into the process. Addition of these starch-containing substances to waste treatment operations increases the BOD loading of the system, which must now be removed from the waste stream.

A. Cereal Industries

Treatment of starch in cereal wastes is important because they increase the biological load of the treatment system, and many starches wastes could be converted to protein, particularly in countries with protein deficits. Much of the research on recycling of starch wastes revolves around fungal conversion of waste materials to food or feed commodities. Gregory *et al.* (6) studied conversion of starch to protein by using thermophilic fungi. Amylolytic fungi were used to eliminate a prehydrolysis step. There was no need to sterilize the system since species of *Aspergillus, Cephalosporium,* and *Rhizopus* grow at pH 3.5 and temperatures of 45 to 50°C. After harvesting, the fungal proteins were fortified with methionine and fed to rats. It was reported that diets containing *Aspergillus fumigatus* protein were only slightly inferior to those containing casein; however, the differences were not statistically significant at the p = 0.05 level. None of the animals showed adverse effects for 90 days and laboratory tests were negative for six common mycotoxins. Even though cassava was used as the substrate, the authors thought that this process could be used for converting starch-containing waste products to protein. Muller (9) studied wastes collected from cereal and flour mills in addition to broken pieces of bakery goods. The starch was enzymatically hydrolyzed before inoculation with *C. utilis*, which was recovered as a nutritional supplement for animal feeds. It was also suggested that *S. cerevisiae* or species of *Hansenula, Lactobacillus,* and *Aerobacter* could be grown on the starch, previously hydrolyzed by acids, enzymes or microorganisms. Church *et al.* (56) grew molds *(Penicillium, Aspergillus, Trichoderma* and *Gliocladium)* on corn syrup, starch, gluten, and corn steep liquor from industrial cereal grain wastes. Growth of these molds in this medium resulted in an 80-95% reduction in BOD with retention times of 6-8 h for sugars and 10-15 h for starches. After the protein was harvested, it was used in animal feeding trials. The weight gain in rats did not significantly differ for either fungal or casein protein, and the appearance of all animals was normal even though the stools of animals on the 70% fungal diet were black. In experiments with chicks, those fed the fungal diet had a greater weight gain than those fed a control diet. Other mice feeding trials that spanned five generations indicated that fungal biomass is a nutritious protein which can serve as the sole protein in the diet. Postweaned pigs showed a greater weight gain when fed fungal protein than when fed soybean meal. The above feeding trials indicate that fungal protein is a good source of feed for animals. Stanton (57) also studied growth of *Aspergillus* on starch residue generated from cassava processing and the manufacture into animal feeds. However, the commercial development of this process has been slow because of the possibility of mycotoxin production by some *Aspergillus* species. Since the above mentioned processes must be done on a large scale for fermentation control and economic feasibility, it is doubtful that they will be used commercially.

Preliminary experiments were conducted on agricultural starch wastes by growing *Pullularia pullulans*, a yeast-like mold, which produces various metabolites besides single cell protein (58). One metabolite, pullulan (a water-soluble polysaccharide), could have many applications such as: (a) material for biodegradable packaging films and food containers, (b) spray-on coating for perishable fruits and vegetables, (c) moisture retention agent for bakery goods, (d) food thickeners, (e) starch substitutes for low calorie foods, and (f) numerous nonfood applications. Complete use of starch waste would minimize pollution problems. Another use of starch waste would be to enzymatically hydrolyze it to glucose-containing syrups (59). These syrups could be used by industries in place of sucrose. Since wastes from potatoes, rice and cassava, which are chiefly starch, could be hydrolyzed without prior isolation of the starch, the economics of this method should be considered for less developed countries that have these wastes.

About 80,000 l of liquid waste are generated daily during processing of baked beans (60). This waste is high in BOD (30,000 mg/l) and produces an objectionable odor when treated by municipal sewage systems. When *A. foetidus* was grown on the waste solids from baked bean processing, a 76% reduction in the BOD was observed. The recoverable mycelium had 40% protein and could be used as an animal feed supplement. The above process had basically four advantages since it: (a) changed an unpleasant odor to an earthy odor, (b) reduced the BOD markedly, (c) resulted in an animal feed supplement, and (d) produced a crude amylase enzyme which could be used as a malt substitute in alcohol production.

B. Fruit and Vegetable Industries

The major starch-containing waste in the vegetable industry originates from potato processing plants. At the present time most of these potato processing wastes are readily recoverable; however, disposal of these wastes has created a financial burden. Since simple disposal of these wastes also results in a loss of potential crude protein, attempts to recover the nutrients are being considered. Because of previous work with recycling lactose in whey, Forney and Reddy (12) studied fermentation of acid-hydrolyzed potato processing wastes by *Lactobacillus delbrueckii, Lactobacillus thermophilus, Lactobacillus bulgaricus, Lactobacillus plantarum* and *Lactobacillus acidophilus* to lactic acid with subsequent neutralization by ammonia. The waste was fortified with minerals, trypticase, yeast extract, and bicarbonate as well as acidified to pH 5.5 because the unsupplemented waste gave poor growth. *L. bulgaricus, L. thermophilus,* and *L. acidophilus* grew consistently better than the other organisms, producing crude protein of around 43%. Since the manufacture of this protein supplement requires hydrolysis, supplementation, and neutralization, it may not be the most economical way to dispose of potato processing wastes.

The Symba process for producing yeast was developed in Sweden to preferentially use two yeasts, *Endomycopsis fibuliger* and *C. utilis*, for growth on starch containing materials (61, 62,66). *E. gibuliger* produces amylases that degrade starch into simple sugars that are then used by *C. utilis* for growth. By symbiotic culture production most of the carbohydrates are used in one step and costly hydrolytic procedures are avoided. The BOD can be reduced by about 90% in this process. The Symba process was first used in Sweden in the early 1970's; the plant produced French fries and potato granules. Both liquids and solids were treated by this process. Since the resulting yeast protein has low nucleic acid and high vitamin contents, it was used to replace skimmilk for feeding baby animals and pets. Studies done at the Swedish University of Agriculture have shown no adverse effects to animals fed diets containing this protein (89). If these yeast are to be used for human food, they can be added to breads, mashed potato powders, and milk-based products. Lemmel *et al.* (62) in Oregon, which produces about 50% of the United States processed potato products, basically looked at a similar process to treat blanch water with *Saccharomycopsis fibuliger* and *C. utilis*. They concluded that single-stage mixed culture was not the best method to use. An alternate two-stage continuous process starting with *S. fibuliger* followed by *C. utilis* produced optimal cell mass. A United States patent was granted for a similar two-stage process to convert raw potato peels, pieces, and water to single cell protein by *E. fibuliger* and *C. utilis* (63).

In another study, processors in Maine suggested that single cell protein be produced from wastewater from potato processing since it has high suspended solids content and BOD value (64). Even though the yeast used was not specified, it was reported that 1 lb of protein is produced for each 2 lbs of dissolved solids. Therefore, the higher the BOD, the greater would be the expected production of protein. It was anticipated that the sale of the protein would help defray operating costs. The process in Maine has been in operation since the mid-1970's with only a 10% downtime because inexperienced operators were chosen from the available labor force. Sewage charges from the municipal system to the Maine processor have been markedly reduced because of the production of this single cell protein. Up to this point, most studies have focused on reducing BOD, and in the process have created single cell proteins. Moreton (65) studied production of *C. utilis* on enzymatically-hydrolyzed potato waste, since previous attempts to grow *E. fibuliger* and *C. utilis* resulted in poor amylase production. Because *C. utilis* produced an acceptable unadulterated yeast protein, it could be used for human food and command a higher price. Therefore, Moreton suggested that the potato industry look at the primary production of single cell protein from waste as a treatment method.

Two studies have reported use of alkaline peelings from fruit and vegetable processing (66,67). These highly alkaline wastes can be fermented and stored in silos for livestock feed. Sistruck *et al.* (66) experimented with conditions for fermentation of potato wastes in these silos for areas that produce low volumes. Aerobic fermentation at 30°C was most efficient in reducing the pH and produced mostly lactic acid microorganisms. *Geotrichum, Aspergillus,* and *Penicillium,* which were identified, were only considered as secondary sources of fermenting organisms. There was little decrease in cellulose content since starches or sugars were preferentially used. *L. bulgaricus, L. delbrueckii, S. faecalis, S. faecium, S. cremoris,* and *S. lactis* were identified as the major organisms in the fermenting wastes; *L. delbrueckii* and *S. faecium* were tentatively identified as most important. If the pH of the waste was greater than 11.0, it could be fermented most rapidly by adding previously fermented waste. Therefore, storage of wastes in covered underground pits could be a successful way to treat waste from the potato industry. Beuchat *et al.* (67) attempted to ferment alkaline wastes by growing *Neurospora sitophila* on the effluents containing high carbohydrate and sodium chloride concentrations. Since growth conditions were not optimized, it may have been possible to obtain more extensive BOD reduction if better operating conditions were achieved. This second process was suggested as a possible way to pretreat alkaline waste discharges before disposal to a municipal system.

IV. MICROBIAL FERMENTATION OR USE OF CELLULOSE FROM FOOD WASTES

Cellulose is the most abundant naturally occurring substance within the ecosystem and can be obtained from various agricultural commodities and food wastes. Since cellulose consists of linear polymeric chains of glucose, it should serve as an attractive substrate for microbial fermentation. However, only a few microorganisms are capable of degrading cellulose. Also, most cellulose is in close contact with lignin, which is not biodegradable. Therefore, use of cellulosic wastes has plagued scientists for a long time. Some workers have suggested that suitable strains of microorganisms, which can grow rapidly on cellulose and produce single cell protein, should be studied with a view to upgrade cellulose for animal feeds. Fungi and some bacteria are major microorganisms which can degrade cellulose, but once the cellulose is hydrolyzed other organisms can use the products of hydrolysis.

A. Marine Industries

In the seafood industry, processors are beginning to consider different waste management systems because of increasing environmental legislation. Processors of shellfish, shrimp, and crab must cope with disposal of chitin waste, which is approximately 75% of the weight of shellfish (12). Chitin is a naturally occurring biopolymer which is very similar to cellulose (the hydroxyl group in the second position in cellulose is replaced by an acetamido group in chitin). Chitin waste in the United States is basically available along the Western Coast and the Gulf of Mexico. A process was developed where a selected microorganism produced chitinase in response to the chitin waste, and this enzyme was used to hydrolyze the waste before single cell protein production (12). In a study designed to isolate chitinase producers, *Bacillus cereus* and a mutant of *Serratia marcescens* were identified as the most active organisms. Since results of this experiment demonstrated that chitin can be hydrolyzed by bacterial enzymes, treatment of this waste could have practical application for seafood industries.

B. Cereal and Sugar Industries

In studies at Louisiana State University (10), experimenters isolated *Cellulomonas uda* from decaying sugarcane stalks and successfully grew it in the laboratory. However, rapid biomass production by this organism is inhibited by accumulation of cellobiose, the disaccharide resulting from cellulose breakdown. Cell production can proceed normally if mixed cultures of *Cellulomonas* and *Alcaligenes faecalis* or *C. guilliermondii* are grown since they produce cellobiase, the enzyme that degrades cellobiose. *Aspergillus fumigatus* and thermophilic actinomycetes can also digest cellulose at rapid rates. Production of single cell protein on cellulose-containing materials may not be feasible since there are inherent technical problems in the process. Another method for reducing the cost of enzymes used in a process is to immobilize the enzymes so that they can be reused. A major obstacle to this method is inactivation of enzymes after short periods because of irreversible adsorption of noncellulosic materials. In an attempt to use waste cellulose, von Hofsten (68) grew thermotolerant basidiomycetes on various carbohydrate sources. One of the fungi that produces enzymes to degrade cellulose was *Sporotrichum pulverulentum*. This fungus can also degrade lignified wood fibers. Even though cellulose can be degraded by certain microorganisms, von Hofsten (68) stated that fibrous plant materials are more suitable for other industrial processes rather than for protein production because of the complexities associated with various interactions during cell growth.

In recent years, cellulose has been considered as a substrate for ethanol production for use in the energy field. However, the by-products of some of these fermentations are not efficiently used at the present time. One process that considered use of waste yeast cells was developed by Savarese and Young (69). This included the enzymatic hydrolysis of cellulose by a cellulase obtained from *T. viride* followed by the fermentation of glucose by *S. cerevisiae*. Ethanol that was produced was used for energy and the yeast cells were recovered as single cell protein. Rosenberg *et al.* (70), on the other hand, looked at the production of fungal protein from oat hulls since they contain approximately 34% cellulose and 30% pentosans. Oat hulls again were partially hydrolyzed by *T. viride*.

They concluded that since the amino acid of profiles of the resulting protein were similar to soybean protein, it would be an acceptable animal feed supplement. In a similar experiment on rice hulls, Thayer *et al.* (71) used a nonspecific mixed culture to produce single cell protein. Rice hulls represent about 15 to 20% of the original rice weight. However, an unusual feature of the composition of rice hulls is ash, containing about 94% SiO_2 (Table 5). Also, because of the high crude fiber and lignin content, rice hulls are poorly digestible. The first step in the process involved treating the hulls with alkali to improve digestibility by microbes and ruminants as well as to remove silica from the ash. The protein that resulted from growth of a mixed culture on the alkali-treated hulls had an amino acid profile that compared favorably with guidelines of the Food and Agricultural Organization/World Health Organization (FAO/WHO). Therefore, cellulose does have potential as a substrate for single cell protein production because of the variety of processing steps that can be used: acid hydrolysis, enzymatic hydrolysis, or direct use of celluloytic microorganisms (72). Presently, the only use of cellulolytic or fibrous plants for food is cultivation of mushrooms on wood, straw or similar materials.

C. Fruit and Vegetable Industries

Production of microbial protein from agricultural wastes has gained attention in recent years because of the need to upgrade the feed value of solid wastes and remove the BOD of liquid waste. Technical problems arise in microbial protein

TABLE 5. Typical Analysis of Ricc Hulls (71)

Component	*Range (%)*
Cellulose	*39.0-42.9*
Crude fiber	*35.7-41.8*
Lignin	*19.8-24.1*
Pentosans	*19.8*
Ash[a]	*16.9-18.4*
Water	*9.0-11.0*
Lipids[b]	*0.8- 1.2*

[a] *Approximately 93.9%* SiO_2.

[b] *Ether extracted.*

production because the chemical nature and physical form of the waste vary greatly. In general, organic waste is mostly carbohydrate of which half is cellulose (73). Fruit, carob pod, date, molasses, corn, and other fibrous wastes have been tested in an effort to produce single cell protein. Processes that were described by Ader and Plaskett (74) involve hydrolysis of cellulose from various agricultural crops to sugar that was fermented to ethanol or converted to single cell protein for animal feed. They postulated that use of wastes from fruit and vegetable industries is difficult because they are in scattered locations and production is seasonal. However, Abou-Zeid *et al.* (75) reported on a process for acid hydrolysis of onion wastes followed by use of this hydrolysate by *S. cerevisiae*. The basic disadvantages of using onion wastes are the presence of volatile sulfur compounds which inhibit growth of *S. cerevisiae* and must be removed and the need for additional nitrogen supplementation. Therefore, this waste would probably be more efficiently used by methods other than fermentation by yeasts.

V. USE OF NITROGEN FROM FOOD WASTES BY MICROORGANISMS

The use of nitrogen from food wastes is important for growth of microorganisms. Since animal and marine products are usually high in protein, microorganisms could use these proteins for growth. However, it may be necessary to supplement some of these food wastes with carbohydrate for optimum biomass production.

A. Animal Industries

There are only a few reports of industries capitalizing on nitrogenous components of waste for microbial growth. In the dairy industry *K. fragilis* has been grown on cottage cheese whey to produce a whey-yeast protein for use as a supplement in bakery products (77). This protein was essentially considered as a nutritious supplement for crackers and cookies. The highest level of protein (20%) could be used in oatmeal cookies because no off-flavors or poor textures were noted. In vanilla-wafer cookies, only 15% of whey yeast protein could be used since higher levels resulted in off-flavors. Cheese-crackers could contain as much as 35% whey-yeast protein without developing bitter flavors. Since both cracker and cookie formulas could be altered to incorporate other flavors, whey-yeast protein could be used as a nutritious supplement for snack items. Mathur and Shahani (76) reported on two other uses of whey proteins; namely for manufacture of Ricotta cheese from blends of whey and milk and for a whey-based culture medium for growing yeasts for bread-making and lactic acid bacteria for dairy use.

The meat processing industry generates collagen which is an insoluble connective tissue protein. This protein accounts for about 30% of the total protein and 6% of animal body weight. At the present time most of the collagen is made into leather, glue, or gelatin. To improve the nutritional value of collagen for feed or food the conversion of this material into single cell protein by *Bacillus megaterium* was investigated (78). Some of the additional nutrients needed to support growth of *B. megaterium* included: phosphate, magnesium, calcium, iron, zinc, copper, and manganese. It can be concluded that a microbial protein with a protein efficiency ratio that is 74% of that of casein could be produced. However, it is doubtful that a fermentation of this type would be extensively used by the industry, since most meat by-products already have established markets.

B. *Marine Industries*

Various nitrogen containing components of fish wastes have been examined for basically two applications: mushroom cultivation (79,80) or production of fish peptones (81,82). Green (80) studied use of fish solubles as the organic nitrogen supplementation in making composts for mushroom cultures and use of fish oils as a stimulant to increase mushroom yield. It was thought that the compost made by fish solubles can help dispose of a fish waste product as well as provide a nitrogen source for the compost. The mushroom grower must increase the nitrogen content of compost by adding chicken manure, dried brewer's grain, cottonseed meal, cocoa bean hulls, or other materials containing organic nitrogen. Since these raw materials become more expensive as the demand for them increases, companies are continually seeking other organic nitrogen sources. Fish solid wastes have a nitrogen content of 5% and could possibly be used in mushroom compost. The nitrogen *per se* is not used by the mushrooms, but is used to support microbial growth in the compost before inoculation with the mushroom spawn. *A. bisporus* is the major species of mushroom commercially produced in the United States. Fish solid wastes replaced either brewers grain or a synthetic nitrogen source in the compost; however, poor growth of mushrooms was observed when brewers grain was replaced in the compost. The presence of fish solids resulted in large-sized mushrooms which have greater consumer acceptance and result in reduced labor of hand-picking. Since previous research indicated that incorporation of polyunsaturated oil into composting stimulated increased yields of mushrooms, Green (80) suggested that the polyunsaturated fish oils could be used in a similar way. Therefore, both fish solids and fish oils could be efficiently used in mushroom cultivation and help the fish industries gain a market for traditional waste products. Other fishery by-products and food processing wastes could possibly be used for mushroom compositing.

Green *et al.* (81,82) studied growth of various microorganisms on fish peptones which are hydrolysates produced by various enzymatic mechanisms. Since these hydrolysates are water soluble and non-coagulable substances, they can be used in place of other peptones in culture media for fermentation, food, clinical, or related industries. Fish peptones were especially well-suited for recovery of fastidious microorganisms, especially lactobacilli, because various amino acids and vitamins were present in them. However, fish peptones were not as effective as Standard Methods agar in recovering organisms from mixed populations. Bough *et al.* (88) reported that microbiological media was produced from peptones derived from shrimp and crab processing wastes. These limited experiments have demonstrated that fish peptones can sustain growth of a number of microorganisms and could be a future source of peptone for microbiological media.

VI. MICROBIAL USE OF LIPIDS FROM FOOD WASTES

Even though lipids are not used by microorganisms as primary nutrients, some lipid-degrading microorganisms can use these materials as carbon sources. Fishery wastes contain little or no carbohydrate that can be used by microorganisms; however, fish oils could serve as metabolizable energy sources for certain microorganisms. It may be necessary to supplement fish oils with some carbohydrates and other nutrients to achieve the most efficient microbial action (79). Production of single cell protein from fish oils may be technically successful, but economically the poor quality protein makes the process undesirable. When Green *et al.* (83) studied stickwater (a by-product of fish meal and oil production that contains protein, nonprotein nitrogen, minerals, oil, and other substances) for growth of lipolytic organisms, *Geotrichum candidum* and *Candida lipolytica* were most effective in using the fish wastes. Since quality and quantity of protein were not determined, there is no evidence to support any recommendations for commercial ventures in this field.

VII. USE OF COMBINED FOOD WASTES BY MICROORGANISMS

Most of the research that has been done on the microbial fermentation or use of food wastes has been concerned with one type of waste. Schmidt *et al.* (84) experimented with production of single cell proteins on a substrate composed of three wastes, collagen, starch, and cottage cheese whey. Since mixed processing wastes contain a variety of substances which may not be readily usable by all microorganisms, a mixed culture may need to be employed to effectively utilize the substrate. Schmidt *et al.* (84) chose species of *Bacillus* and *Lactobacillus* for their process since the mixed culture can use both starch and lactose. The concentration of collagen influenced growth of these organisms but no explanations were given for this observation. Both *B. subtilis* and *B. megaterium* grew abundantly on the substrate because they produced amylase that can hydrolyze starch. Growth of *L. acidophilus* was disappointing; however, the reasons for this were not clearly understood. It was concluded that *B. megaterium* was a better organism for single cell protein production on a mixed substrate than *B. subtilis* because the former produced more amylase and gave greater yields under optimum conditions. The above experiment points out the problems that can be encountered in a multiple fermentation system where one or two microorganisms are degrading a variety of substrates from different industries. It may be necessary to have a mixed culture so that all nutrients are used efficiently or a multiple stage process where microorganisms or enzymes degrade the constituents in a sequential manner. The use of wastes combined from several food processing companies may be needed in the future for plants which generate an insufficient volume of waste to develop a separate system for recovery of food wastes.

VIII. FUTURE DIRECTIONS

There will be a greater need in the future for food processing plants to use wastes in a manner that will expand resources and minimize environmental pollution. Industries shall have to consider other options for use of traditional wastes products. Economically, companies will have to decide whether they wish to make a profit, break even, or maintain a deficit when dealing with their processing wastes. Regardless of the economic conditions, food industries will have to evaluate new business opportunities when auditing the various wastes that they are generating. One example of a complete assessment of waste management is a study done by Kamm *et al.* (85), who surveyed a potato processing industry for the type of waste generated, processing equipment used in the various operations, and amount of energy needed. After considering various treatments that could convert the waste to desirable products, Kamm *et al.* (85) decided to enzymatically hydrolyze the starch waste to produce a dextrose equivalent syrup. Syrup, thus produced, was sufficient to meet most of the internal sugar requirements of the plant. Therefore, the company began to revise the waste treatment facilities of its other potato processing plants to include the recycling of starch-containing wastes. Each food processing company will have to reevaluate its waste treatment approach with the concept of recovery and reuse in mind. In this manner, each organization can expand opportunities to make or save money with resources that would otherwise be a liability on both the environment and company finances.

In the United States, food manufacturers are continually facing problems created by increased government regulations aimed at cleaning up the environment. However, reuse of food waste as a food resource also impinges on legal considerations for compliance with food laws and regulations. These laws and regulations ultimately determine establishment of waste conversion research programs and feasibility of commercial applications. Goldenfield (86) suggested that two areas must be understood when approaching regulatory problems. The first concern is the necessity to survive the critical marketing test that must overcome label and advertising limitations imposed on new foods or ingredients. The second factor to consider is that all new foods must satisfy current regulations regardless of desired economic value. These concerns may cause regulatory agencies to reevaluate the diametrically opposed regulations for cleaning up the environment and for producing food and feed that is tested and retested to meet increasingly smaller tolerance levels of "potential" hazards. The growing need to convert waste to usable food or feed products will undoubtedly cause processors and consumers to adopt "risk-benefit" concepts. In the final analysis, the United States cannot continue to be simultaneously thankful and wasteful when using food, feed, fuel, and fertilizer resources.

When the recent research on waste utilization was compiled, it was obvious that most of it dealt with the microbial production of biomass or single protein (Table 6). Most of the research considered use of wastes to produce single cell protein as a favorable alternative to establishing complicated waste treatment systems which simply dispose of the wastes. Use of single cell protein within our food and feed chains will neither be an easy nor readily acceptable transition. Some factors to be considered are: (a) substrate suitability,

TABLE 6. *Recovery of Food Wastes for Microbial Fermentation or Utilization*

Food Waste	*Major Substrate*	*Microorganism*	*End Product(s)*	*Suggested Use*	*Reference*
Whey	*Lactose*	Lactobacillus bulgaricus	*Ammonium Lactate & SCP*[a]	*Ruminant Feed*	*14, 15, 16, 17, 18, 19, 21*
Whey	*Lactose*	Saccharomyces fragilis	*SCP & Ethanol*	*Feed or Food & Energy*	*20*
Whey	*Lactose*	Trichosporon cutaneum & Candida curvata	*Oil & SCP*	*Food & Feed*	*22, 23*
Whey	*Lactose*	Kluyveromyces fragilis	*Wine, Vinegar & SCP*	*Food*	*8, 24, 25, 77*
Whey	*Added sucrose*	Saccharomyces cerevisiae	*Wine*	*Food*	*26, 27, 28, 29, 30, 31*
Whey	*Lactose*	*Enzyme*	*Sugar Syrup*	*Alcoholic Beverages*	*32*
Whey	*Lactose*	Torulopsis	*Vodka & Champagne*	*Food*	*24*
Soybean Spent Solubles	*Mixed Carbohydrates*	Candida quilliermondii & Debaryomyces kloeckeri	*SCP*	*Feed*	*11*
Soybean Cooking Drain	*Mixed Carbohydrates*	Hansenula anomala & S. cerevisiae	*SCP*	*Feed or Food*	*33, 34*
Confectionary	*Sucrose & glucose*	Candida utilis	*SCP*	*Food or Feed*	*35*

Molasses	*Saccharose, Glucose, Fructose & Raffinose*	S. cerevisiae & C. utilis	*Yeast*	*Food*	*36, 37, 38*
Green beans, turnip greens, corn, potato, peas	*Mixed carbohydrate*	Lactobacillus plantarum, L. bulgaricus Streptococcus faecalis *subsp.* liquefaciens	*Lactic acid*	*Feed*	*39*
Corn & Pea	*Mixed carbohydrate*	Trichoderma viride & Gliocladium deliquescens	*SCP*	*Feed*	*40*
Sauerkraut	*Mixed carbohydrate & Lactic acid*	S. cerevisiae, K. fragilis, & C. utilis	*SCP, invertase*	*Food or Feed*	*41, 42, 43 44*
Pickle	*Mixed carbohydrate & Lactic acid*	C. utilis	*SCP*	*Food or Feed*	*45*
Banana	*Glucose, Fructose & Sucrose*	Pichia spartinae	*SCP*	*Food or Feed*	*46*
Coconut	*Glucose, Fructose, Sucrose & Sorbitol*	K. fragilis	*SCP*	*Food or Feed*	*47, 48*
Date	*Glucose & Fructose*	S. cerevisiae, C. utilis & Saccharomyces rouxii	*SCP*	*Food or Feed*	*49, 50*

TABLE 6. (Continued)

Food Waste	Major Substrate	Microorganism	End Product(s)	Suggested Use	Reference
Citrus	Reducing sugars, crude protein & ash	Agaricus bisporus & Morchella crassipes	Mushrooms	Food	51
Brewery	Reducing sugars	Aspergillus niger	SCP, Citric acid	Food or Feed	42, 53
Brewery	Reducing sugars	C. utilis, S. cerevisiae, A. bisporus, & Morchella esculenta	SCP & Mushrooms	Food	54
Brewery	Reducing sugars	C. utilis & Rhodotorula glutinis	SCP	Food or Feed	55
Cassava	Starch	Aspergillus, Caphalosporium & Rhizopus *Species & Enzymes*	SCP & Glucose syrup	Food or Feed	6, 57, 59
Cereal & Flour	Starch	*Enzymes &*	SCP	Feed	9
Cereal grains	Starch	Penicillium, Aspergillus, Trichoderma & Gliocladium *species*	SCP	Feed	56
Agricultural Wastes	Starch	Pullularia pullulans	Pullulan	Films, Coatings, Thickner	58

Baked Beans	*Starch*	Aspergillus foetidus	*SCP & amylase*	*Feed or Food*	*60*
Potato	*Starch*	L. bulgaricus, L. thermophilus, & L. acidophilus	*Ammonium Lactate & SCP*	*Ruminant Feed*	*12*
Potato	*Starch*	Endomycopsis fibuliger & C. utilis	*SCP*	*Food or Feed*	*61, 62, 65, 66*
Alkaline Fruit & Vegetable Wastes	*Mixed Carbohydrate*	*Mixed lactic* or Neurospora sitophila	*SCP*	*Feed*	*66, 67*
Shellfish	*Chitin*	Bacillus cereus Serratia marcescens	*SCP*	*Feed*	*12*
Sugar Cane Stalks	*Cellulose*	Cellulomonas uda, Alcaligenes faecalis & C. quilliermondii	*SCP*	*Feed*	*10*
Fibrous plants	*Cellulose*	Sporotrichum pulverulentum	*SCP*	*Feed*	*68*
Agricultural Wastes	*Cellulose*	T. viride & S. cerevisiae	*Ethanol & SCP*	*Feed*	*69*
Oat Hulls	*Cellulose*	T. Viride	*SCP*	*Feed*	*70*
Rice Hulls	*Cellulose*	*Mixed Culture*	*SCP*	*Feed*	*71*
Onion	*Cellulose*	*Acid Hydrolysis* & S. cerevisiae	*SCP*	*Feed*	*75*
Whey	*Protein*	——— [b]	*Ricotta Cheese*	*Feed*	*76*
Meat	*Collagen*	Bacillus megaterium	*SCP*	*Feed*	*78*

TABLE 6. (Continued)

Food Waste	*Major Substrate*	*Microorganism*	*End Product(s)*	*Suggested Use*	*Reference*
Fish	*Nitrogen*	A. bisporus	*Mushrooms*	*Food*	*79, 80*
Fish	*Nitrogen*	________[b]	*Peptones*	*Culture Media*	*81 & 82*
Fish	*Lipid*	Geotrichum candidum & Candida lipolytica	*SCP*	*Feed*	*83*

[a] *Single Cell Protein.*

[b] *No Microorganism Identified.*

availability, safety, and cost; (b) microorganisms which can ferment or use the substrate and their safety; (c) nutritional value of final single cell protein (amino acid profile, nucleic acid, vitamin, biological value, etc.); (d) safety aspects in relation to chemical composition, microbiological contaminants, possible carcinogens and toxins, digestibility and possible allergic reactions; (e) regulatory aspects of use; (f) physchological, social, and political aspects if human use is expected and (g) economic feasibility and marketing guidelines (5,43,68,87). Use of single cell protein will cause people to reevaluate the concept of what a food really is, what can be included as a food constituent, what methods will be used to incorporate these proteins into our food supply, what will be printed on the label, and what methods will be used to determine nutritive values and safety. In a sense, some microorganisms have been used for years as single cell proteins in cheese, bread and various other fermented foods. Therefore, the concept of incorporating single cell protein into food should not cause alarm. It may be necessary for food processors to develop new formulas or revise existing formulas to use these proteins (43,68).

Food processors must begin or continue to consider total food resource management. This will cause scientists to generate imaginative ways to effectively use resources that have been labeled as food wastes for far too long. These ideas can be incorporated into recycling by-products in-house or producing a product with market value or establishing new industries to solve old waste problems. Since we shape our lives by what we leave behind, we must leave ideas, processes, and plans for total food resource utilization.

ACKNOWLEDGMENTS

The author thanks Dr. J. V. Chambers and Dr. B. Swaminathan, Purdue University, and Dr. E. H. Marth, University of Wisconsin-Madison, for their helpful comments and suggestions during the preparation of this manuscript.

REFERENCES

1. Cooper, J.L. and Denit, J.D., in "Food Processing Waste Management" (J.H. Green and A. Kramer, eds.) p. 499. AVI Publishing Company, Westport, Conn. (1979).

2. Strolle, E.O., Aceto, N.C., Stabile, R.L., and Turkot, V.A., *Food Technol. 34(2)*, 90 (1980).
3. Denit, J.D. and Forsht, E.H., Proceedings Eighth National Symposium on Food Processing Wastes (EPA-600/2-77-184), 9 (1977).
4. Bissett, D.W., Proceedings of Seventh National Symposium on Food Processing Wastes (EPA-600/2-76-304), 257 (1976).
5. Kapsiotis, G.D., *Food Nutr. 3(2)*, 11 (1977).
6. Gregory, K.F., Reade, A.E., Khor, G.L., Alexander, J.C., Lumsden, J.H., and Losos, G., *Food Technol. 30(3)*, 30 (1976).
7. Anon., *Process Biochem. 9(2)*, 26 (1974).
8. Kosikowski, F.V., *J. Dairy Sci. 62*, 1149 (1979).
9. Müller, H., U.S. Patent 4-056-636 (1977).
10. Srinivasan, M.C., *Chem. Indus. Dev. 11(6)*, 26 (1977).
11. Sugimoto, H., *J. Food Sci. 39*, 934 (1974).
12. Carroad, P.A. and Tom, R.A., *J. Food Sci. 43*, 1158 (1978).
13. Worgan, J.T. in "Food From Wastes" (G.G. Birch, K.J. Parker, and J.T. Worgan, eds.) p. 23. Applied Science Publisher, London (1976).
14. Stieber, R.W. and Gerhardt, P., *J. Dairy Sci. 62*, 1558 (1979).
15. Juengst, F.W., Jr., *J. Dairy Sci. 62*, 106 (1979).
16. Keller, A.K. and Gerhardt, P., *Biotech. Bioeng. 17*, 997 (1975).
17. Meiering, A.G., Azi, F.A., and Gregory, K.F., *Am. Soc. Agric. Engin. Trans. 21*, 586 (1978).
18. Coulman, G.A., Stieber, R.W., and Gerhardt, P., *Appl. Environ. Microbiol. 34(6)*, 725 (1977).
19. Reddy, C.A., Henderson, H.E., and Erdman, M.D., *Appl. Environ. Microbiol. 32*, 769 (1976).
20. Tzeng, C.H., Sisson, D., and Bernstein, S., Proceedings of the Sixth National Symposium on Food Processing Wastes (EPA-600/2-76-224), 118 (1976).
21. Erdman, M.D., Reddy, C.A., and Bergen, W.G., *J. Dairy Sci. 60*, 1509 (1977).
22. Moon, N.J., Hammond, E.G., and Glatz, B.A., *J. Dairy Sci. 61*, 1537 (1978).
23. Moon, N.J. and Hammond, E.G., *J. Am. Oil Chem. Soc. 55(10)*, 683 (1978).
24. Mann, E.J., *Dairy Indus. Int. 45(3)*, 47 (1980).
25. Gawel, J. and Kosikowski, F.V., *J. Food Sci. 43*, 1717 (1978).
26. Anon., *Food Proc. 37(1)*, 58 (1976).
27. Larson, P.K. and Yang, H.Y., *J. Milk Food Technol. 39*, 614 (1976).
28. Palmer, G.M. and Marquardt, R.F., *Food Prod. Dev. 12(1)*, 31 (1978).

29. Carawan, R.E., Chambers, J.V., Zall, R.R., and Wikowske, R.H., Dairy Processing Water and Wastewater Management (Extension Special Report No. AM-18B, North Carolina Agricultural Extension Service, Raleigh), 102 (1979).
30. Palmer, G.M. and Marquardt, R.F., Proceedings Ninth National Symposium on Food Processing Wastes (EPA-600/2-78-188), 61 (1978).
31. Yang, H.Y., Bodyfelt, F.W., Berggren, K.E., and Larson, P.K., Proceedings of Sixth National Symposium on Food Processing Wastes (EPA-600/2-7-224), 180 (1976).
32. Roland, J.F. and Alm, W.L., *Biotech. Bioeng. 17,* 1443 (1975).
33. Kiuchi, K., Suzuki, T., and Ohta, T., *J. Ferm. Technol. 53(6),* 386 (1975).
34. Kiuchi, K., Suzuki, T., and Ohta, T., *J. Ferm. Technol. 53(6),* 393 (1975).
35. Forage, A.J., *Process Biochem. 13(6),* 8 (1978).
36. Lewicki, W., *Process Biochem. 13(6),* 12 (1978).
37. Sobkowicz, G., in "Food From Waste" (G.G. Birch, K.J. Parker, and J.T. Worgan, eds.) p. 42. Applied Science Publishers, London (1976).
38. Lawford, G.R., Kligerman, A., and Williams, T., *Biotech. Bioeng. 21,* 1163 (1979).
39. Moon, N.J., *J. Food Sci. 44,* 1460 (1979).
40. Church, B.O., Erickson, E.E., and Widmer, C.M., *Food Technol. 27(2),* 36 (1973).
41. Hang, Y.D., Splittstoesser, D.F., and Landschoot, R.L., *Appl. Microbiol. 24,* 1007 (1972).
42. Hang, Y.D., Splittstoesser, D.F., and Landschoot, R.L., *Appl. Microbiol. 25,* 501 (1973).
43. Hang, Y.D., in "Food Processing Waste Management" (J.H. Green and A. Kramer, eds.) p. 442. AVI Publishing Company, Westport, Conn. (1979).
44. Rao, M.A. and Hang, Y.D., *J. Milk Food Technol. 38,* 315 (1975).
45. Stevenson, K.E., Black, D.E., and Costilow, R.N., *J. Food Sci. 44,* 181 (1979).
46. Chung, S.L. and Meyers, S.P., *Dev. Indus. Microbiol. 20,* 723 (1979).
47. Smith, M.E. and Bull, A.T., *J. Appl. Bacteriol. 41,* 81 (1976).
48. Smith, M.E. and Bull, A.T., *J. Appl. Bacteriol. 41,* 97 (1976).
49. Kamel, B.S., *Process Biochem. 14(6),* 12 (1979).
50. Humphrey, A.E., *Chem. Eng. 81(26),* 98 (1974).
51. Labaneiah, M.E.O., Abou-Donia, S.A., Mohamed, M.S., and El-Zalaki, E.M., *J. Fd. Technol. 14,* 95 (1979).
52. Hang, Y.D., Splittstoesser, D.F., Woodams, E.E., and Sherman, R.M., *J. Food Sci. 42,* 383 (1977).

53. Hang, Y.D., Splittstoesser, D.F. and Woodams, E.E., *Appl. Microbiol. 30,* 879 (1975).
54. Shannon, L.J. and Stevenson, K.E., *J. Food Sci. 40,* 826 (1975).
55. Tomlinson, E.J., *Water Res. 10,* 367 (1976).
56. Church, B.D., Widmer, C.M., and Espinosa, R., Proceedings of Eighth National Symposium on Food Processing Wastes (EPA-600/2-77-184), 355 (1977).
57. Stanton, W.R., *Process Biochem. 13(12),* 6 (1978).
58. Shipman, R.H. and L.T. Fan, *Process Biochem. 13(3),* 19 (1978).
59. Kearsley, M.W. and Nketsia-Tabiri, J., *Lebensmittel Wissenschaft und Technologie 12(4),* 199 (1979).
60. Hang, Y.D. and Woodams, E.E., *Appl. Environ. Microbiol. 33,* 1293 (1977).
61. Skogman, H., in "Food From Waste" (G.G. Birch, K.J. Parker, and J.T. Worgan, eds.) p. 167. Applied Science Publishers, London (1976).
62. Lemmel, S.A., Heimsch, R.C., and Edwards, L.L., *Appl. Environ. Micro. 37,* 227 (1979).
63. Lines, K.K., US Patent 4-144-132 (1979).
64. Dambois, I., Deeves, R., and Forwalter, J., *Food Proc. 39(5),* 156 (1978).
65. Moreton, R.S., *J. Appl. Bacteriol. 44,* 373 (1978).
66. Sistrunk, W.A., Ismailk, M., and Collins, J.A., *J. Food Sci. 44,* 439 (1979).
67. Beuchat, L.R., Bough, W.A., and Young, C.T., *J. Food Prot. 41,* 24 (1978).
68. von Hofsten, B., in "Food From Waste" (G.G. Birch, K.J. Parker, and J.T. Worgan, eds.) p. 156. Applied Science Publishers, London (1976).
69. Savarese, J.J. and Young, S.D., *Biotech. Bioeng. 20,* 1291 (1978).
70. Rosenberg, H., Obrist, J., and Stohs, S.J., *Economic Botany 32(4),* 413 (1978).
71. Thayer, D.W., Yang, S.P., and Ou, S-Y. L., *Dev. Indus. Microbiol. 19,* 385 (1978).
72. Bellamy, W.D., *Biotech. Bioeng. 16,* 869 (1974).
73. Imrie, F.K.E. and Eighelato, R.C., in "Food From Waste" (G.G. Birch, K.J. Parker, and J.T. Worgan, eds.) p. 79. Applied Science Publishers, London (1976).
74. Ader, G. and Plaskett, L.G., *Food Proc. Ind. 44(521),* 14 (1975).
75. Abou-Zeid, A-S. A., El-Fattah, A.F.A. and Farid, M.A., *Agric. Biol. Chem. 43(9),* 1977 (1979).
76. Mathur, B.N. and Shahani, K.M., *J. Dairy Sci. 62,* 99 (1979).
77. Stewart, C.F. and Gilliland, S.E., *Am. Dairy Rev. 41(4),* 30A (1979).

78. Bough, W.A., Brown, W.L., Porsche, J.D., and Doty, D.M., *Appl. Microbiol. 24,* 226 (1972).
79. Green, J.H. and Mattick, J.F., *J. Food Quality 1(3),* 229 (1977).
80. Green, J.H., *Marine Fishery Rev. 36,* 27 (1974).
81. Green, J.H., Goldmintz, D., Hale, M.B., and Flynn, D.J., *Dev. Indus. Micro. 14,* 310 (1973).
82. Green, J.H., Paskell, S.L., and Goldmintz, D., *J. Food Prot. 40,* 181 (1977).
83. Green, J.H., Paskell, S.L., and Goldmintz, D., *Appl. Environ. Microbiol. 31,* 569 (1976).
84. Schmidt, B.E., Busta, F.F., and McKay, L.L., *J. Food Proc. Pres. 1,* 13 (1977).
85. Kamm, R., Meacham, K., Harrow, L.S., and Monroe, F., *Food Technol. 31(6),* 36 (1977).
86. Goldenfield, I.H., *Food Technol. 32(6),* 34 (1978).
87. Kapsiotis, G.D., *Food Nutr. 4(1-2),* 2 (1978).
88. Bough, W.A., Young, C.T., Stephens, N.L., and Beuchat, L.R., *Food Eng. Int. 2(11),* 74 (1977).
89. Anon., *Food Eng. Int. 1(10),* 32 (1976).
90. Durgapal, B.C., Sharma, D.B., and Ali, M., *Agr. Agro-Indus. J. 11(11),* 3 (1978).
91. Aguinaldo, J.T., *Sugar News 53(9),* 270 (1977).

CHAPTER 3

MICROBIOLOGY OF LIGNIN DEGRADATION

Gamal I. Amer
Stephen W. Drew

Department of Chemical Engineering
Virginia Polytechnic Institute

and

State University
Blacksburg, Virginia

I. INTRODUCTION

Lignocellulosic material comprises about 95% of the earth's land-based biomass. It is composed of approximately 50% cellulose, 25% hemicellulose and 25% lignin, the exact figures varying from species to species. Lignin forms an essential component of the woody stems of arborescent gymnosperms and angiosperms. It is not, however, restricted to arborescent plants, but is found as an integral cell wall constituent of stems, foliage and roots in all vascular plants including herbaceous varieties (1).

In addition to imparting rigidity to plants, lignin acts as a permanent bonding agent between plant cells (8). The fact that lignin is a cell wall constituent allows it to perform multiple functions that are essential to the life of the plant. Lignin decreases water permeation across cell walls in the conducting xylem tissues, and this plays an important role in the internal transport of water, nutrients and metabolites in the plant (1). Lignin also has the important function of protecting plant tissues from invasion by pathogenic microorganisms (1,3). It is most concentrated in tissues where functions such as conduction of solutes (tracheids) and mechanical support (fibers) are specialized.

ISBN 0-12-040304-8

Plant cell structure and function of tracheids and fibers depend on the intimate association of three main features of cellular morphology. Each cell is composed of cellulosic fibrils set in a largely amorphous carbohydrate matrix with a polyphenolic binder. The fibrils are largely microcrystalline with paracrystalline regions surrounding those of higher molecular order.

The carbohydrate matrix is not a homopolymer of a hexose sugar, but rather, is typified by linear and branched pentose sugar polymers (hemicellulose). The major component of the adhesive or encrusting, amorphous, polyphenolic binder is lignin (2).

The three primary features of cellular morphology are combined in different ways to give overall structure to the plant cell. The two most obvious features of overall cellular structure are the primary wall which contains 20-28% cellulose and the thicker secondary wall which contains 45-55% cellulose (2). Lignin is most concentrated in and between the primary walls (middle lamella) of adjacent cells (Figure 1).

Lignin varies in chemical structure from species to species and can be divided into three categories (9): (a) gymnosperms or softwood lignins, (b) angiosperms or hardwood lignins and (c) grass lignins. Homologies in the chemical structure of these lignins are best identified on the basis of nitrobenzene oxidation products (9). Whereas gymnosperm lignins yield mainly vanillin with some *p*-hydrozybenzaldehyde, both syringaldehyde and vanillin are obtained from angiosperm lignins, and significant amounts of all three aldehydes are obtained from grass lignin. In addition, different lignins occur within the same plant (1).

Within the bounds of species and location dependent variability in chemical structure, lignin is a highly polydisperse, polyphenolic macromolecule of nine-carbon phenylpropane units linked in a random fashion by non-labile C-C and C-O-C interunit bonds (4). Polymeric natural lignins arise by enzyme-initiated dehydrogenative polymerization of three primary precursors: *trans*-coniferyl, *trans*-sinapyl and *trans*-p-coumaryl alcohols (1) (Figure 2). Softwood lignin typically contains 80% coniferyl alcohol, 14% *p*-coumaryl alcohol and 6% sinapyl alcohol. Hardwood lignin, on the other hand, contains similar amounts of coniferyl and sinapyl alcohols but a minor amount of *p*-coumaryl alcohol. Bamboo and grass lignin contain a larger amount of *p*-coumaryl alcohol than softwoods or hardwoods (3).

Softwood lignin characteristically contains cinnamyl alcohol and cinnamaldehyde end groups (5). Phenylpropanoid units containing free phenolic hydroxyl groups, common in conifer lignin, may possess an unsubstituted 5-C or aryl-, aryloxy- or alkyl-substituted 5-carbon and are often referred

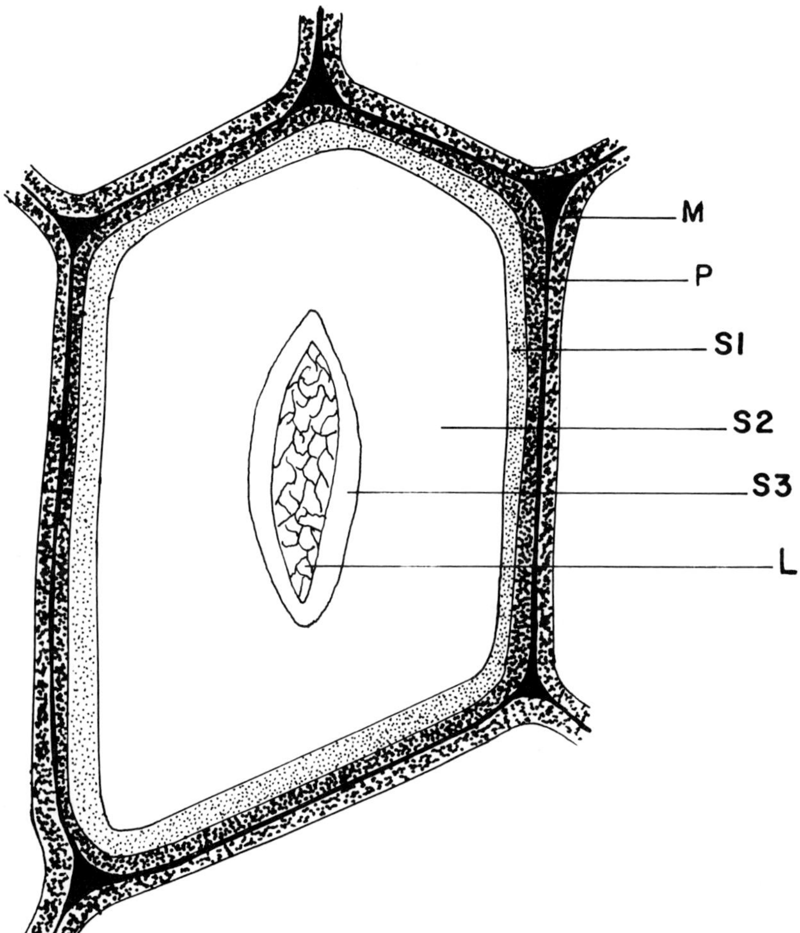

FIGURE 1. A Schematic Representing Typical Vascular Plant Cell Morphology; Middle Lamela (M), Primary Wall (P), First Layer of Secondary Wall (S1), Second Layer of Secondary Wall (S2), Third Layer of Secondary Wall (S3), Lumen (L).

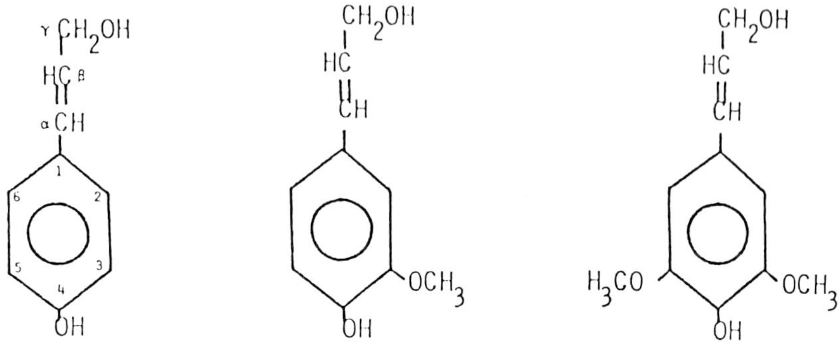

p-COUMARYL ALCOHOL (14 %) CONIFERYL ALCOHOL (80 %) SINAPYL ALCOHOL (6 %)

FIGURE 2. Lignin Building Blocks and Their Relative Abundance. Figures in parentheses represent approximate proportions of these monomers in softwood lignin.

to as "uncondensed" and "condensed" phenolic units respectively. Lignin structures with side-chains connected to the next unit through a β-O-4 linkage are commonly called "arylglycerol-β-ether" structures. These linkages are the most abundant in lignin (5). Other types of structures found in lignin are β-1-lignols, 4-O-1 and 5-1 linkages, β-5 lignols, β-β lignols, etc. A computer-simulated structure for softwood lignin (Figure 3) has recently been proposed (4,6).

In addition to dependence on species and location within the plant, the chemical structure of isolated lignin is very much a function of the method of isolation. Isolation techniques lead to the cleavage of certain bonds (*e.g.* C-C, C-O, arylalkyl ether), and alter the extent of lignin condensation, lignin molecular weight distribution, the content of certain functional groups (*e.g.* OCH_3, COOH, carbonyl), the distribution of phenolic and aliphatic moieties, the elemental analysis of the isolated lignin and the solubility of the resulting lignin in organic solvents and mineral acids. Table 1 summarizes some of the differences in isolated lignins as a function of the method of isolation.

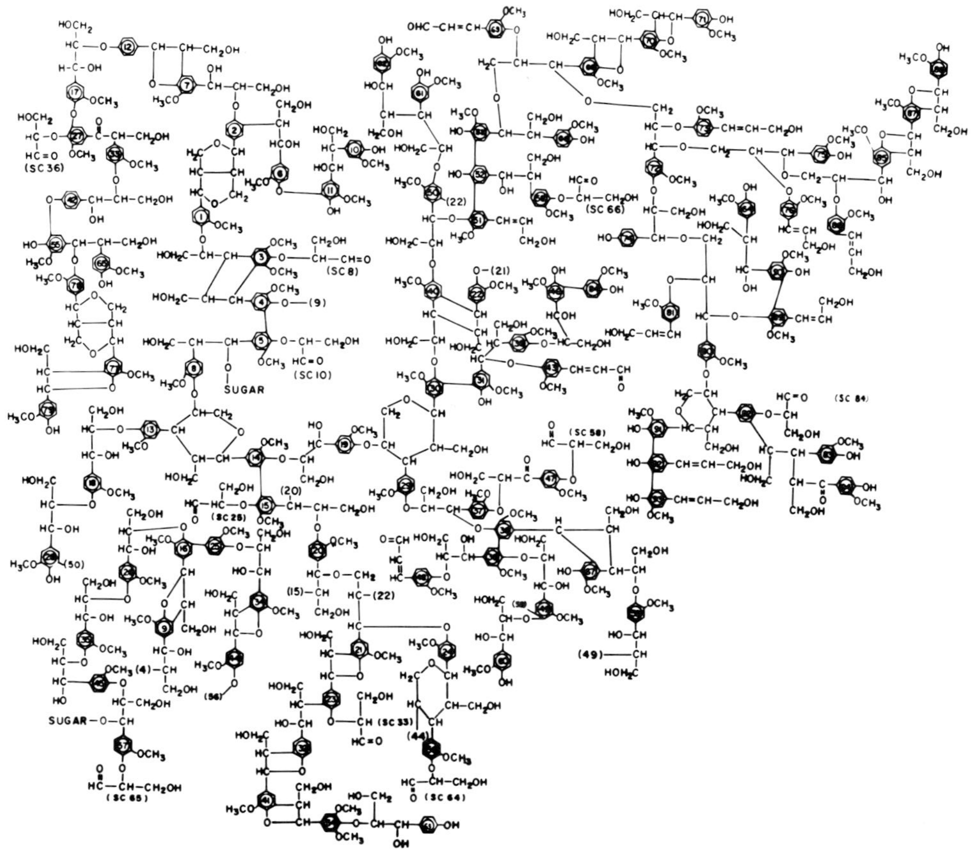

FIGURE 3. Computer-Simulated Structure of Soft-Wood Lignin.

TABLE 1. Characteristics of Differently Isolated Lignins

Method of Isolation	Lignin Recovery (unpurified) % of Total Lignin	Carbohydrate Content, % of Recovered Material	Phenolic Units in Recovered Lignin moles/100 C9 Units	Aromatic Hydroxyl moles/100 C9 Units	Elementary Composition C%	H%	MeO%	Number Average Molecular Weight
(1) Milled wood lignin								
(a) soft wood	40[5] - 50[10]	2 - 8[5]	25 - 35[a]	2.0 - 2.5[a]	64	6	15.7[5]	< 6,000[5]
(b) hard wood			20 - 30[a]		60	6	20	
(2) Brauns lignin								
(a) soft wood	8 - 10[5]	> 2[5]	25 - 35[5]	2.0 - 3.0[5]	64	6	15[5]	850 - 1000[5]
(b) hard wood	< 10[a]		20 - 30[5]		62	5.8	15-18	
(3) Enzymatically liberated lignin								
(a) soft wood	23[10]	present[c]	25 - 35[a]	2.0 - 2.5[a]	64	6	15[5]	850 - 1000[5]
(b) hard wood			20 - 30[a]		62	5.8	15-18	
(4) Kraft lignin								
(a) soft wood	∿ 80[a,b]	0.2[a]			66	6	14[7]	1600[7]
(b) hard wood	∿ 80[a,b]		> 35[7]	> 3[7]	65.5	∿6	19.2	1050[7]
(5) Iotech (Steam Chip explosion) lignin								
(a) soft wood	n.a.	n.a.	n.a.	n.a.		n.a.		n.a.
(b) hard wood	66[a]	< 5[a]	Ca.50[a]	Ca.5[a]	65	7	20[a]	< 1000[a]

n.a. Data not available.

[a] Personal communications with Dr. W. Glasser, Forest Products Department, VPI & SU.

[b] Obtained by adding UV adsorbing and acid soluble fractions.

[c] The amount of carbohydrates depends on the period of incubation.

[d] Numbers in brackets indicate reference quoted.

In addition to dependence on species and location within the plant, the chemical structure of isolated lignin is very much a function of the method of isolation. Isolation techniques lead to the cleavage of certain bonds (*e.g.* C-C, C-O, arylalkyl ether), and alter the extent of lignin condensation, lignin molecular weight distribution, the content of certain functional groups (*e.g.* OCH_3, COOH, carbonyl), the distribution of phenolic and aliphatic moieties, the elemental analysis of the isolated lignin and the solubility of the resulting lignin in organic solvents and mineral acids. Table 1 summarizes some of the differences in isolated lignins as a function of the method of isolation.

The diversity and complexity of lignin chemistry has provided an effective roadblock to developing an understanding of biodegradation of lignin. Rapid, adequate chemical analysis of lignin and lignin degradation products is the single most important limiting factor in this quest for knowledge. In most cases only the most gross estimates of lignolytic activity are available. Although the precise mechanism(s) of lignin biodegradation is still unknown, white-rot and soft-rot fungi are capable of substantially degrading lignin (12-14). On the other hand, brown-rot fungi, which are usually associated with decaying wood (10,15) decompose the cellulose and hemicellulose in wood and cause very little change in the lignin (16). Despite the fact that bacteria have been implicated in the degradation of purified lignin (60,61) there still is much confusion about their role in lignin biodegradation in nature.

II. LIGNIN-DEGRADING MICROORGANISMS

A. *White-rot Fungi*

White rot fungi, wood-decaying Hymenomycetes, are among the most active lignolytic microorganisms known (21) and are capable of degrading all the components of wood including lignin (20,21). Their name arises from the fact that they remove, at least in part, tannin materials from sound wood leaving bleached wood (3). Cowling (17) showed that *Polyporus verisicolor (Coriolus versicolor)* can degrade over 97% of the lignin in sweet gum wood. Early studies (18) found that cultures of *P. versicolor* could be grown on isolated native aspen lignin as the sole carbon source. Van Vilet (19) also found that *P. versicolor* could be cultivated on a medium containing Brauns' lignin as sole organic material. These results are contradictory to more recent results by Eriksson *et al.* (20) and by Drew and Kadam (14) who have concluded that lignin cannot be degraded by white-rot fungi unless a more easily metabolized carbon source is used simultaneously. Since lignin isolation normally results in the presence of some carbohydrate impurities in the lignin, it is conceivable that the amount of carbohydrate present in the isolated lignin in earlier studies was enough to fulfill the requirement for cosubstrate.

That lignin degradation is oxidative is evidenced by the fact that lignin attacked by white-rot fungi contains higher amounts of oxygen than the corresponding sound lignin (4,31). Hydroxyl group content has been reported by some to increase (32,33) and reported by others to decrease (12). Our own observations (4) indicate that hydroxyl content changes very little if at all. Kirk and Chang (12,13) conclude from their extensive data on lignin obtained from white-rotted spruce that substantial side-chain oxidation occurs. They also conclude that, during lignin degradation in their system, oxidative ring-opening occurs while the lignin is still polymerized. This has been confirmed by Chen and Chang (34).

The content of total carbonyl and carboxyl is somewhat higher in degraded lignin than in sound lignin (32,33). The molecular weight of residual biodegraded lignin is usually higher than sound lignin (4). This polymerization may be attributed to the presence of laccase activity found extracellularly in lignin degradation systems (4).

Though it was reported that phenylpropane (C_6-C_3) compounds are intermediates of lignin degradation by white-rot fungi (32,36), only small amounts of such compounds were detected (32). Moreover, the amounts of vanillin and vanillic acid in the fermentation medium were between 0.002% and 0.006% of the lignin (32). Kirk (11) suggests that these compounds are formed from "side groups" or minor structural elements and not from the lignin polymer itself (exo-degradation). A recent study by Haars and Hüttermann (37) supports this suggestion. Based on the degradation of lignosulfonates by *Fomes annosus* (a Basidiomycete), Haars and Hüttermann conclude that lignin biodegradation in nature follows the mechanism of an exo-degradation. They further conclude that the fungus is unable to split the lignin molecule into intermediate-size pieces and that the fungus is most likely attacking side groups during the early stages of the fermentation while leaving the lignin polymer intact.

The fact that lignin is an extensively cross-linked and poly-disperse polymeric structure may preclude direct uptake of lignin by microbial cells; thus extracellular enzymes which facilitate the initial attack on the polymer are plausible. Aromatic ring-opening enzymes are reportedly present in the soluble extracellular proteins in *P. versicolor* culture broth (38). This report on extracellular enzymes and other (39,40) treated later in this review lack confirmation (31). Hall *et al.* (4) have isolated only very small amounts of extracellular proteins from lignin degradation broths. The only potentially lignolytic extracellular enzymes found in lignin fermentation broths are laccases (41), phenoloxidase and quinone-oxidoreductase (4). However, laccase and peroxidase alone are not capable of depolymerizing lignin or even of releasing low molecular weight fragments in detectable amounts (42,43). Furthermore, incubation of lignin with cell free broth from a lignin degrading fermentation does not result in net reduction of lignin molecular weight (4).

In summary, white-rot fungi are able to completely degrade lignin under the right conditions. The lignin undergoes oxidative changes during the early stages of degradation. Although very small amounts of low molecular weight intermediates have been found extracellularly, no extracellular enzymes with lignolytic activity have been isolated from lignin fermentation media. The mechanism of initial lignin breakdown is still a mystery. We will speculate on such a mechanism later in this review.

B. Soft-rot Fungi

Soft-rot fungi are Ascomycetes or Fungi Imperfecti and have been shown to degrade lignin and lignocellulosic materials (14,30,42,48). Hardwoods are more susceptible to soft-rot attack than are softwoods (3,43). Soft-rot fungi, many of which are thermophilic and degrade lignin best when incubated at high temperature (14,44), seem to have an absolute requirement for an easily metabolized carbon source as cosubstrate in lignin degradation (14,30). Eslyn *et al.* (30) have shown that carbohydrates in soft-rotted wood are depleted faster than lignin by four out of six species studied while the two remaining species depleted lignin faster than the carbohydrate. Rosenberg (44) concludes that the moisture content of the solid lignocellulosic substrate influences its degradation by some soft-rot fungi. Soft-rot fungi degrade lignosulfonate (49) as well as whole wood and Kraft lignin (14).

Haider and Trojanowski (45) conclude that soft-rot fungi can attack methoxyl groups, β-ether linkage of the side chain and also the aromatic ring structure of DHP and lignin. Levi and Preston (42) presented evidence that the lignin remaining in soft-rotted beechwood is partially degraded, has a lower methoxyl content and is more acid soluble than the corresponding lignin in sound wood.

It has been suggested (3,46,47) that soft-rot fungi cause limited lignin degradation and that they are less efficient lignin degraders than white-rot fungi. This, however is not always true. Drew and Kadam (14) have found a strain of *Aspergillus fumigatus* (a soft-rot fungus) superior to *C. versicolor* in metabolizing Kraft lignin to CO_2 under identical conditions. From comparison studies of soft-rot and white-rot fungi, Takahashi (48) concludes that soft-rot fungi degrade lignin but that delignification varies not only with the fungal species but also with the type of wood.

C. Bacteria

Examination of early reports pertaining to bacterial degradation of lignin reveals that in many instances weaknesses in experimental technique have led to erroneous or questionable conclusions. Kirk (11) has enumerated these problems. However, reports suggest that bacteria from *Flavobacterium, Micrococcus, Mycobacterium, Pseudomonas* and *Xanthomonas* species are capable of lignin degradation. Ferm (50) reported that both *Acetobacter* and *Pseudomonas* species are capable of degrading the lignin in lignosulfonates. Kleinert and Joyce (51) previously reported this capability in *Pseudomonas*. Pandila (54) concluded that certain soil bacteria can degrade lignin.

Conflicting reports confuse the role of bacteria in lignin degradation. For example, in 1973 Crawford *et al.* (56) claimed that *Nocardia carollina* utilize aromatic compounds but were incapable of degrading lignin. In 1977 Trojanowski *et al.* (57) provided concrete evidence for the ability of *Nocardia* sp. to degrade lignin; and in 1978 Gradziel *et al.* (58) claimed that *Nocardia* sp. was more active in producing $^{14}CO_2$ from ^{14}C-labeled plant lignin and DHP[1] than soft-rot and white-rot fungi tested by Haider and Trojanowski (59).

It has been shown (60) that *Pseudomonas ovalis* degrades a large portion of high molecular weight milled wood lignin (MWL) while increasing the nitrogen content of residual lignin. Because no accumulation of low molecular weight intermediates was observed in the medium, Kawakami (60) concluded that the degradation of lignin by *P. ovalis* leads to a successive breakdown from the terminal ends of the MWL molecule (exo-degradation). Haider *et al.* (61) used ^{14}C-labeled lignin to screen several species of *Nocardia* and *Pseudomonas* and concluded that *Nocardia* sp. were better lignin degraders than *Pseudomonas*.

[1] *DHP; made by dehydrogenative polymerization of coniferyl alcohol with peroxidase and are commonly used in lignin research (45,56).*

The release of p-hydroxybenzoic acid from aspen wood meal by *Pseudomonas* sp., isolated from the intestinal content of a larva living in whole aspen wood, has also been reported (62).

Cartwright and Holdom (55) reported that *Arthrobacter* sp. are capable of utilizing enzymatically released lignin as sole carbon source. The fact that *Arthrobacter* sp. utilized lignin released by the action of brown-rot fungi might indicate that the lignin was partially modified due to the fungal attack, making it more susceptible to bacterial degradation, and contained enough carbohydrate to act as a cosubstrate. The claim that the lignin (implies native lignin) was the sole carbon source in this study would not be appropriate.

Odier and Monties (63) reported that a *Xanthomonas* strain isolated from decomposing wheat straw was able to utilize dioxane lignin as the sole carbon source. They observed that 77% of the initial lignin, in a mineral medium, was degraded in 15 days. They also reported that in the presence of glucose only 23% of the lignin was degraded. Furthermore, they reported that this strain of *Xanthomonas* was capable of degrading wheat dioxane lignin under anaerobic conditions in the presence of NO_3^- and glucose. Under anaerobic conditions small quantities of two phenolic acids were produced. Recently Deschamps *et al.* (94) observed that a strain of *Aeromonas* sp. was capable of utilizing lignin as the sole carbon source. Their results indicate that the *Aeromonas* strain degraded 98% of the lignin present in their fermentation medium (lgm/L) in 5 days of incubation. Crawford (64) provided proof that *Streptomyces* sp. can degrade lignin. *Bacillus megaterium* has also been shown to have lignolytic activity (65).

Reports of the effect of mixed bacterial cultures on the biodegradation of lignin are scarce. Crawford (66) reported that an *Acinetobacter* sp. grew on a lignin model compound and metabolized it to guaiacol, *Nocardia corallina* was then able to utilize the guaiacol. He also found that in the absence of *N. corallina* the *Acinetobacter* sp. died while *N. corallina* did not grow alone on the model compound. Kleinert and Joyce (51) showed that a mixed culture containing two fungal species, *Penicillium* and *Aspergillus,* and *Pseudomonas* was capable of utilizing lignosulfonate as the sole source of carbon. Ban and Soljan (67) showed that a mixed culture consisting of two yeasts, two *Arthrobacter* sp., *Pseudomonas* and *Chromobacterium* genera were able to degrade 50% of the initial lignosulfonate in the medium in 24 hours.

Many of the studies mentioned above indicate the importance of adaptation of the bacterial culture to the substrate prior to observation of lignin degradation (54,55,63). Close contact between the bacteria and the lignin is also reported to be important for extensive degradation (63). Most of the bacterial cultures apparently require the presence of a co-substrate to initiate lignin degradation; *Xanthomonas* 23 (63) and *Aeromonas* L17 (94) are possible exceptions.

Changes occurring in the residual lignin due to bacterial action include a decrease in interunit β-aryl ether bonds, a decrease in molecular weight of MWL and a 70% and 80% decrease in coniferyl aldehyde content of pine and beech MWL due to the action of *P. ovalis* (60). *Nocardia* sp. apparently is capable of removing methoxyl groups and cleaving the side chains and ring structures of DHP (57).

III. MICROBIAL ECOLOGY AND PHYSIOLOGY OF LIGNIN DEGRADATION

The diversity of intermonomer linkages and monomerstructures in lignin, as well as the abundance of various functional groups, provides an opportunity for a given microorganism to cause a variety of limited changes in lignin without necessarily effecting significant degradation. Research to date indicates that the capacity for complete degradation of lignin by a single microorganism may be limited to white-rot fungi and related litter decomposing Basidiomycetes (3,11). It is possible that other fungi (*e.g. Aspergillus fumigatus* (4,14)) and bacteria (*e.g. Xanthomonas* sp/ (63) and *Aeromonas* sp. (94)) can also completely degrade lignin to CO_2 and water.

Sutherland *et al.* (68) has established that *Streptomyces flavovirens* is capable of attacking and destroying the integrity of the cell wall of Douglas-fir phloem. *S. flavovirens* degraded insoluble carbohydrates extensively but did not appreciably degrade the lignin. Their observations over an incubation period of 12 weeks indicated that the nonlignified wall was decomposed first, then the thick, heavily lignified portions of the wall were attacked. Scanning electron micrographs of decayed phloem taken at different time intervals during the incubation period showed the presence of cavities in the cell wall with the actinomycete hyphae penetrating these cavities causing loss of cell wall integrity. It also appears that part of the disruption was caused by the removal of carbohydrates. Crawford and Crawford (70) claimed preliminary evidence indicating that the lignin had been structurally altered as a consequence of the attack on the cell wall by *S. flavovirens* in the experiments of Sutherland *et al.* They (70) also claimed that this was the first report implicating a bacterium in the attack on lignin within intact woody tissues, thus indicating a wood decay role for actinomycetes. The observations of Sutherland *et al.* led Crawford and Crawford (70) to the conclusion that streptomycetes' attack on lignin may follow a pattern similar to white-rot attack on lignin. Preliminary observations, in our laboratory, of micrographs taken at different intervals during lignin degradation by a white-rot fungus also indicate fungal penetration of cavities in isolated particulate lignin exposed to fungal attack.

More recently Phelan *et al.* (69) examined 6 Streptomycete strains to determine their ability to degrade ^{14}C-labeled lignin from Douglas-fir. The substrates were ^{14}C-labeled in side-chains or ring structures of the lignin. Their results clearly indicate that aromatic ring components were cleaved after 1008 hours of incubation and that a substantial amount of the labeled ring carbon was released as $^{14}CO_2$. They also noticed that side-chains were only attacked to a small degree. Phelan *et al.* (69) concluded that Streptomycetes preferentially attack the aromatic rings of the lignin polymer. In contrast *Nocardia* sp. (61) and *Bacillus megaterium* (65) preferentially degrade side-chain components. This may be an indication of the importance of hyphal penetration and disruption of the polymer structure facilitating extensive microbial degradation of lignin.

A. Succession of Microorganisms in Decaying Wood

The biodegradation of lignocellulosic materials is undoubtedly an important step in the process of carbon cycling in nature. Decayed lignified plant tissues are normally attacked on the surface or the upper layers of the soil. The first step of microbial attack on lignocellulosic material, in nature, appears to be mediated by fungi which instigate the decomposition of this intractable substrate. Upon studying the decomposition of lignocellulosic materials by the successive action of two wood destroying fungi, Tichy (71) speculated that fungi provide the primary attack on lignocellulosic materials and cause alterations in the original substrate, other members of the soil microflora will normally amplify the fungal action. Tichy further speculated that in normal succession the organism that will, at a given time, occupy the lignocellulosic material is the organism that can benefit the most from the changes brought about in the substrate by the foregoing partner.

Decomposition of lignocellulosic materials in nature is generally caused by the action of several fungal species and other microorganisms inhabiting the decaying wood. These microorganisms colonize the decaying wood in conjunction and/or in succession. It is reasonable to assume that each organism will, on its own, contribute to the final state of the decomposed substrate. The organism might, for example, suppress other organisms that have attacked the substrate earlier or be influenced by their activity. Tichy (72), who studied fungal succession in the primary attack on lignocellulosic materials, concluded that one important factor determining the colonization of decaying wood by a given organism is substrate availability. He stated that substrate availability is not only determined by the duration and character of the foregoing decomposition, but also by the type of wood being decomposed. Tichy's study (72) indicates a requirement for partial removal of cellulose by cellulose metabolizing fungi in order to make the lignin available for metabolism.

Smith and Ofosu-Asiedu (76) studied the distribution of fungi colonizing wood chips in storage piles. Their observations indicate that the majority of fungi isolated from the wood chip pile were Fungi Imperfecti and Ascomycetes. Furthermore, they observed that certain fungal species suppress other species at different time intervals during the decay of wood.

Recently, Schmidt and Walter (74) studied the succession and activity of microorganisms in stored bagasse. After 6.5 days of storage they were able to isolate 400 different organisms colonizing the bagasse. The flora consisted of bacteria, yeast, fungi and actinomycetes. They observed that microbial succession on the bagasse started with yeast species, followed by bacteria and then by actinomycetes and fungi. They concluded that this succession coincided with the enzymatic activities of the organism during fermentation, noting that residual sugars were consumed at first by the yeasts, followed by bacterial degradation of the pectin, hemicellulose and, in parts, the cellulosic fraction of the substrate. In the later stages the actinomycetes and Fungi Imperfecti attacked the hemicellulose, cellulose and, partially, the lignin fraction of the bagasse fiber. Similar observations concerning microflora colonizing hay during its storage have been reported (75).

These results indicate that wood degradation in nature is the result of the attack of a succession of microorganisms. Brown-rot fungi and bacteria colonize the wood during the early stages of decay, primarily attacking the carbohydrate fractions. Brown-rot fungi cause the removal of cell wall substance beginning in the S2 layer of the secondary wall and continuing through the S1 layer of the secondary wall. The primary wall and the middle lamella are very resistant to degradation by brown rot fungi (77), presumably due to the high lignin content (2). White-rot fungi and actinomycetes amplify the effect of previous microorganisms and initiate significant delignification. In the advanced stages of decay, after most of the polysaccharides are consumed, the cell wall collapses (3). At this point the lignin fraction has undergone physical and limited chemical changes that increase the availability of the lignin fraction and may promote colonization of the decaying wood by other lignin degrading fungi and bacteria. The succession of microbial attack on lignin in nature suggests that mixed or sequential culture techniques may be required for effective bioprocessing of lignin as a means towards its improved utilization.

B. Cosubstrate Requirement in Lignin Biodegradation

The role of bacteria in the degradation of lignocellulosic materials in nature is probably limited, in part, by the lack of mobility of bacteria through woody tissue as compared to the penetration of fungal hyphae. Aerobic degradation of lignin as a sole carbon source by *Xanthomonas* has been documented (63). It is interesting to note, however, that anaerobic degradation of lignin by *Xanthomonas* 23 requires a cosubstrate and NO_3^- as a terminal electron acceptor (63). This specific case may help explain the role of cosubstrates in lignin degradation.

It may be that most lignin degrading organisms require a cosubstrate for lignin metabolism because the energy produced from lignin metabolism is inadequate to support microbial growth. An efficient aerobic electron transfer system utilizing oxygen as a terminal electron accepter can convert the energy inherent in the electrons to high energy phosphate compounds better than a less efficient system utilizing NO_3^- as terminal electron acceptor. Thus in aerobic degradation of lignin by *Xanthomonas* 23, no cosubstrate is required to give enough energy to support microbial growth. In the case of anaerobic degradation of lignin by this culture, less efficient energy conservation may make it impossible for the organism to grow unless an easily metabolizable, high energy cosubstrate is present. Although it seems likely that energy metabolism is closely linked to the ability to degrade lignin it would not be justified to infer that anaerobic metabolism of lignin will occur in nutritionally rich media. The observations on *Xanthomonas* 23 (63) are singular, and the conclusions regarding metabolism in the absence of oxygen are in conflict with other studies (Drew, S.W. and J. Ferry, unpublished data, 73). Nevertheless, the studies with *Xanthomonas* 23 suggest an energy dependent requirement for cosubstrate that may transcend microbial classification.

The importance of energy efficiency in aerobic degradation of lignin may also be indicated by phenomenological observations on the effect of oxygen tension. Increased oxygen concentration allows a higher rate of turnover of the terminal electron acceptor, which results in more efficient energy conservation by the electron transport system. This could explain the observation by many researchers (25,26) that increased oxygen concentrations during lignin fermentations result in lignin decomposition to a greater extent and at a higher rate.

Odier and Monties' (63) observations concerning the ability of *Xanthomonas* 23 to utilize lignin as a sole carbon source suggest that non-cosubstrate requiring microorganisms could be used to selectively remove the lignin fraction of lignocellulosic material while leaving the carbohydrates for other uses. Moreover, since the pulp and paper industry produces industrial wastes containing large amounts of lignin and small concentrations of carbohydrates, such non-cosubstrate requiring organisms might be utilized for treating these wastes without any requirement for additional carbohydrate.

Kadam (52) has shown, based on the evolution of $^{14}CO_2$ from radio-labeled lignin, that the ability of *Coriolus versicolor* to degrade lignin is enhanced by utilizing more slowly metabolized carbohydrates as cosubstrate. In contrast, the presence of an easily metabolized sugar (*e.g.* glucose) in small concentrations has been reported to increase the extent of lignin degradation (20). Another study (21), however, found no effect on the extent of lignin degradation due to the addition of small amounts of glucose. Addition of high concentrations of glucose (35%), however, reduced the extent of lignin degradation by *P. chrysosporium* (21). This observation has been confirmed in our laboratory for *C. versicolor* as well.

Normally carbohydrates used as cosubstrates are metabolized more rapidly and degraded to a higher extent than lignin (23). Santra and Nandi (16), however, claim that *Tremetes scabrosa, Phellinus badius* and *Daedalea flavida* (all of which are white-rot fungi) utilize much of the lignin in rotted mahagony and sapwood and little of the cellulose.

Drew and Kadam (14), in their experiments utilizing *A. gunigatus* and *C. versicolor* to degrade ^{14}C-labelled lignin, found that the ratio of lignin derived carbon dioxide to total carbon dioxide is variable. Based on this observation, they concluded that the lignolytic capacity of their cultures (as defined by metabolism of lignin to CO_2) rises and falls during the fermentation irrespective of total metbaolism (as determined by total CO_2 evaluation). Moreover, they concluded that the metabolic characteristics of *A. fumigatus*, a soft-rot fungus, appear to be very different from those of *C. versicolor*, a white-rot fungus, suggesting a diversity of approaches to cosubstrate and lignin metabolism.

The desirability of noncosubstrate requiring, lignin degrading organisms has lately become the focus of much research (20,78,79). Ander and Eriksson (78) demonstrated that treating wood chips with cellulaseless mutants of white-rot fungi decreases the energy demand for the production of thermomechanical pulp due to lowered lignin content. In another paper (79) they concluded that some of the xylans and low molecular weight sugars in the wood were used as a cosubstrate. In a more recent paper (20) Eriksson *et al.* studied the optimal conditions and the factors influencing the growth of cellulaseless mutants of white-rot fungi during wood rotting. More research is still needed in this field.

C. Nitrogen Limitation and Lignin Biodegradation

A very interesting aspect of lignin degradation is the effect of nitrogen limitation on the breakdown of lignin. Many researchers (20,24,25,28) have concluded that lignin degradation occurs only after the depletion of nutrient nitrogen and the elapse of a subsequent lag phase. Keyser *et al.* (28) showed that adding nutrient nitrogen to the fermentation system delayed the appearance of the lignin degrading system. They also showed that such additions depress existing lignolytic activity. Reid (24) and Kirk and Yang (25) conclude that low nitrogen concentration enhances the rate of lignin biodegradation by *P. chrysosporium*. High nitrogen content, on the other hand, has an inhibitory effect which is relieved upon depletion of excess nitrogen (27,28). Eriksson *et al.* (20) determined the optimal C/N ratio for growth in fermentations containing lignin, for three different white-rot fungi and their cellulase mutants, to be between 160:1 and 400:1 depending on the organism. However, Drew and Kadam (14) concluded from their studies of lignin biodegradation by white-rot fungi that fungal growth and lignolytic activity in the presence of carbohydrates are poorly correlated.

Although the response of lignolytic systems to nitrogen limitation is somewhat variable (14), this response is common in white-rot fungi. To date, no satisfactory explanation has been offered for such a phenomenon. We offer the following possible explanation for the effect of nitrogen limitation on lignin degradation. In the early stages of lignin fermentation the microbial metabolism favors cosubstrate utilization (14). A consequence of carbohydrate metabolization is the production of reducing equivalents (*e.g.* NADH and NADPH). Nitrogen in the medium will be utilized by the microorganisms in conjunction with part of these reducing equivalents for biosynthesis. Once the nutrient nitrogen is depleted, the biosynthetic process will slow or stop. As a result, excess reducing equivalents will accumulate within the cell. Mediation of the electron transport system could alleviate this intolerable condition by increasing electron transfer to the terminal electron acceptor (*i.e.*, oxygen). This would result in the production of an increased concentration of reduced oxygen species (*e.g.*, superoxide radical, hydrogen peroxide... etc.). Such species are very reactive and could conceivably attack the lignin causing its degradation. Evidence for the presence of reduced oxygen species in the medium during lignin fermentation, and the capability of such species to degrade lignin will be discussed in detail in a subsequent section of this paper.

D. Cultivation Parameters and Lignin Degradation

Considerable controversy exists concerning optimum conditions of agitation and aeration. Careful evaluation of the literature has shown that many of the comparisons between static and shaken cultures were performed with conditions that were suboptimal for adequate oxygen transfer. However, the data indicate that conditions which allow morphological differentiation may favor fungal degradation of lignin. Yang *et al.* (21) have shown that degradation of lignin by white-rot fungi in shallow, liquid phase, stationary cultures was ten times as rapid as in agitated cultures. Wilke and co-workers (23) have demonstrated that the extent of lignin biodegradation in damp cultures is higher than in static or shaking submerged cultures. Moreover, lignin degradation by *Planerochaete chrysosporium* was found to be 50% more rapid in submerged, shaken cultures under 100% oxygen than in those under normal air exchange (21). Other studies (23,26) with *P. chrysosporium* indicate a similar increase in the extent of lignin degradation in submerged static cultures under 100% oxygen compared with normal air exchange. In contrast Wilke and colleagues (23) found no difference in the extent of lignin degradation in either submerged shake culture or damp cultures in response to a gas atmosphere of 100% oxygen or normal air. However, a slight increase in the rate of lignin degradation was observed in the case of damp cultures under a 100% oxygen atmosphere as compared with normal air exchange (23).

Another important factor that influences lignin degradation by white-rot fungi is the requirement for close proximity between the fungal hyphae and lignin (22). It has been suggested (23,26) that disrupted contact between lignin and the growing hyphae might be the reason that lignin degradation is inhibited in shaking cultures. Our own observations as well as those of Yang *et al.* (21) indicate that pellets of substrate and mycelia are formed and that the proximity of lignin to cell mass is quite close. Yang and coworkers speculate that interference with reactions requiring oxygen with the macrostructure of the pellets causes the adverse effect on lignin degradation by agitation.

The effect of illumination on wood decay was studied by Duncan (35). Her observations indicate that light from the visible spectrum stimulates the rate of wood decay of three wood-destroying fungi. She found a close correlation between the extent of decay inside blocks of wood and the intensity of light measured in foot-candles. She attributes some of the inconsistencies in results obtained from "presumably well-controlled fermentations" to uncontrolled illumination of such fermentations.

Finally, Fenn and Kirk (29) have demonstrated that o-phthalate (often used as a buffer in lignin fermentations) inhibits lignin degradation. They reported that replacement of o-phthalate with 2,2-dimethyl succinate in *P. chrysosporium* fermentations doubled the rate of lignin degradation. Their observations suggest that lignin metabolism may be particularly sensitive to the presence of nonlignin mononuclear aromatic compounds.

IV. ANALYTICAL TECHNIQUES FOR THE STUDY OF LIGNIN BIODEGRADATION

Analysis of residual lignin is an important step in the study of microbial degradation of lignin. Determination of the extent of lignin degradation as well as the structural changes in the degraded lignin are important factors in assessing the ability of a given microorganism to degrade lignin. Residual lignin analysis, however, is often plagued by interference from other wood components and constituents of the microbial biomass. Phytochemicals of high molecular weight which may be covalently linked to lignin and have phenolic character may be considered, from the analytical point of view, as being part of the lignin (81). However, results from residual lignin analysis will vary from sample to sample and from one method of analysis to another. This is an important first step in the interpretation of patterns of lignin biodegradation. Often the idiosyncrasies of analytical methodology compound errors in interpretation by the uninitiated. Unfortunately no single assay is sufficient to characterize the nature or extent of lignin biodegradation. As a consequence, a thorough understanding of the strength and potential pitfalls of each method of lignin analysis is an essential second step in interpretation of the chemical history of lignin samples.

In the following we describe some of the more common methods used to quantify lignin content in biodegraded samples. Please note that these are not the only methods available, nor is the discussion of each method complete. For more complete discussion of analytical techniques the reader is encouraged to consult references 5 and 82.

A. Sulfuric Acid (Klason's) Method

In this method dried samples (usually 1 gm) of the residual lignin (with or without mycelium) are mixed with 15 ml of 72% sulfuric acid (82). The acid is added slowly while stirring. The mixture is then incubated in a water bath at 18-20°C for 2 to 3 hours with frequent stirring. The mixture is then diluted with distilled water to a sulfuric acid concentration of 3% (82). The weak sulfuric acid solution is boiled for 4 hours under reflux and the insoluble lignin allowed to settle prior to vacuum filtration. The solids are washed free of acid with hot water, dried and weighed.

This technique may underestimate the lignin content of samples which have undergone significant solubilization of lignin or overestimate the lignin content of samples containing large amounts of complexing inorganic salts. Therefore, this assay should be accompanied by others to determine the soluble lignin content and purity of the residual lignin fraction.

B. Spectrophotometric Methods

Lignin determination by ultraviolet absorption is essentially a measure of soluble lignin. In this method the lignin is dissolved in a suitable solvent, such as acetylbromide-acetic acid (81), 80% phosphoric acid (84) or 76% sulfuric acid (85), and the absorbance of the solution measured at 280 nm. Takahashi (48), after determining the insoluble lignin fraction by the Klason method, estimated the acid soluble fraction resulting from the hydrolyzate by measuring its ultraviolet absorption at 205 nm. This method was described first by Schoning and Johansson (87) in 1965.

Dissolved lignosulfonate can be determined by reacting the lignolsulfonate with nitrites in an acidic solution. (Pearl-Benson test) (81). Goldschmidt *et al.* (86) describe a modified method in which the reaction between nitrite and lignosulfonate is carried out in an acetic acid solution. The resulting nitrosolignin derivative is characterized by its intense absorption at 430 nm.

Some of the problems encountered in these methods include the interference of undissolved particles (82) and, in the case of the Pearl-Benson test, interference from other polyphenolic compounds such as humic acids (81). Absorbance measured at 280 nm will not adequately estimate the non-aromatic fraction of lignin samples.

C. Chlorine Number

The chlorine number method is an indirect method of determining the amount of lignin in pulp samples (82). It is based on the fact that lignin is easily oxidized by chlorine. Roe (88), who developed a method for measuring the volume of chlorine gas consumed by pulp samples, defined the chlorine number as the weight in grams of chlorine consumed per gram of pulp under set conditions of time and temperature. Since the chlorine number is proportional to the lignin content in the specific woody material being analyzed, it can be related to lignin content only by an empirically determined factor. TAPPI Standard $T202_{OS}$ is the current standard method for chlorine number determination. Other methods for determining the chlorine number utilize sodium chlorite instead of chlorine gas (81).

Hiroi and Eriksson (89) used the chlorine number method to assay for lignin disappearance during the degradation of milled wood lignin by *Pleurotus ostreatus* (a white-rot fungus). After analyzing their data they concluded that the chlorine number is influenced by the structural changes that occur in the lignin due to enzymatic alterations by the fungus. Furthermore, they calculated that lignin losses measured by this method are 10-15% higher than the true losses after 35 days incubation.

D. Permanganate (Kappa) Number

In unbleached pulps, lignin is more easily oxidized by potassium permanganate than is cellulose (82). The amounts of potassium permanganate consumed by woody materials under standardized conditions give an estimate of lignin content. The Kappa number is usually expressed as the number of ml of 0.1N potassium permanganate solution consumed by 1 gm of woody material. Factors affecting the permanganate number assay include the amount and concentration of permanganate and temperature and reaction time (82) although these later conditions can be easily controlled.

A sample (1 gm) is prepared for Kappa number determination by vacuum drying and subsequent disintegration of the weighed woody material in water (500 mL) until it is free of fiber bundles. The disintegrated material is then reacted with 100 ml of 0.1N potassium permanganate and 100 mL of 4N sulfuric acid while maintaining the temperature of the reaction mixture at 25°C and constantly stirring for exactly 10 minutes. After 10 minutes the reaction is terminated by adding potassium iodide solution (20 mL of 1N). The liberated iodine is then titrated with 0.2N sodium thiosulfate using a few drops of starch indicator (82). Wu and Glasser (90) evaluated the applicability of Kappa number to the determination of lignin in biodegraded lignocellulose and found that the method is affected by the structural changes in the lignin caused by the action of the microorganism. They concluded that the degree of oxygenation of the lignin structure can increase the factor correlating lignin content with Kappa number by 25% or more. Moreover, they suggested that without simultaneous lignin structure analysis, Kappa number alone can lead to overestimation of the ability of microorganisms to delignify lignocellulosic materials.

The methods discussed above are able to quantify polymeric lignin but in some cases these methods are unacceptable for even that task. None of the above methods allow determination of the type or extent of structural change in residual lignin. Table 2 is a summary of the highlights of these methods.

E. Computer Assisted Data Reduction Method

Glasser and Glasser (6,91) have developed an elegant computer assisted data reduction method for elucidating the type and extent of chemical changes in reacted lignins. The method is based on the correlation of several analytical procedures applied to polymeric lignin: elemental composition, OCH_3 content, total OH content, proton NMR spectroscopy of acetylated lignin, ash content and oxidative degradation by alkaline permanganate and hydrogen peroxide before and after oxidative depolymerization with alkaline cupric oxide. The products of this last analysis step are separated by gas and gel permeation chromatography. The results obtained from these analyses are used as input for a simulation computer program (SIMREL Analysis Program) which interprets the data and predicts the chemical behavior and properties of the lignin. Although this approach was developed to simulate the synthesis of natural lignins, its application to analysis of degraded lignin is highly promising.

Understanding the mechanism of lignin biosynthesis and the chemical and analytical behavior of lignin are the basis of SIMREL system (Simulation of Reactions with Lignin by computer). SIMREL can develop a lignin formula of 70 to 100 phenylpropane units by using random coupling reactions between mono- and oligomeric lignin precursors. Data obtained from the literature or from analytical observations can be used as an input for the simulation program. The program itself simulates the expected analytical behavior of the lignin, these predictions can normally be correlated with analytical data obtained experimentally (92,93).

TABLE 2. *Summary of Methods for Residual Lignin Analysis*

Method of Analysis	Basis of the Method	Advantages	Disadvantages	Detectability of Lignin Derived Residuals	Method Used in Ref.
1. Sulfuric acid (Klason)	Insolubility of lignin in 72% sulfuric acid	Works well with soft wood lignin	·Does not work well with hardwood lignin that contain large fractions of acid soluble lignin ·Can introduce positive error due to the condensation of carbohydrate degradation products which become acid insoluble	Quantification possible if acid insoluble Characterization not possible	20,21,22 83
2. Spectrophotometric	UV adsorbing properties of solubilized lignin	·Good precision ·Simple ·Fast	·Interference from insoluble particles	Quantification possible is soluble. Some characterization possible	48,49,95
3. Chlorine number	Selective response of lignin to Chlorination	Good for non-microbially degraded lignin	·Affected by structural changes in lignin caused by enzymatic attack	Quantification as well as characterization not possible	80,89
4. Permanganate (Kappa) number	Selective response of lignin to oxidation by potassium permanganate	Good for non-degraded pulps	·Affected by structural changes in lignin caused by enzymatic attack	Quantification as well as characterization not possible	25,90

This computer assisted data reduction method is by no means complete. The technique is still in an evolutionary state and is currently incapable of accurately determining the phenolic OH content, carboxyl group content, and the structural side chain features of the lignin polymer. Dr. W.G. Glasser of the Department of Forestry and Forest Products at Virginia Polytechnic Institute and State University and his group are working continuously at complementing and filling the gaps in this promising technique for lignin analysis. An example of data generated by this analytical approach applied to microbially degraded lignin can be found in Reference 4.

F. Radioiosotopic Tracer Studies

Rapid, sensitive analysis of subtle changes in residual lignin structure and functionality still elude the research community. Although great strides have been made in sensitivity and breadth of analysis, virtually all of these procedures are tedious or require sophisticated instrumentation that may not always be available. On the other hand, a rapid and sensitive technique for assessing the overall extent of lignin metabolism and, in some cases, for pinpointing the functionality involved in sequential breakdown has been developed. This technique is based on the degradation of ^{14}C-labeled natural lignin (45,46) and extractive free lignocelluloses which contain ^{14}C only in their lignin component (45,98,99).

The evolution of $^{14}CO_2$ by cultures growing on ^{14}C-labeled lignin has given important clues to the mode of action and extent of lignin degradation. The presence of low molecular weight intermediates in lignin degradation can be detected by monitoring water soluble ^{14}C-labeled compounds in the fermentation broth. Moreover, the study of sequential attack on the lignin molecule has been facilitated with lignins that are ^{14}C-labeled only in side-chains, methoxyl groups, or ring structures (45,57).

The methods for ^{14}C-labeling of the lignin component in lignocellulosic materials have been described by many researchers (45,98,99). An aqueous solution of L-*[*$U^{14}C$*]* phenylalanine or other suitable precursor is fed to live saplings through their cut stems (97). It is shown that the resulting lignocellulosic materials contain at least 90% of their ^{14}C in non-acid-hydrolyzable components (98). In the preparation of specifically ^{14}C-labeled DHP's, specifically labeled coniferyl alcohols (^{14}C in aromatic rings, in propyl side-chains, or in methoxyls) are oxidatively polymerized *in vitro* using a peroxidase/H_2O_2 system as a catalyst (46).

Radioisotopic methods for the study of lignin biodegradation have recently been reviewed by Crawford *et al.* (97,100).

A cautionary note about the utilization of DHP in lignin studies has been sounded by Glasser in his recently published, excellent review article (81). Glasser has enumerated the main shortcomings of utilizing DHP in lignin research. These short comings are: (1) the preparation of the polymer ignores the role of carbohydrates in the lignification mechanism, (2) the polymer structure is greatly affected by polymerization conditions and (3) the polymer structure usually deviates from the structure of isolated lignin. These observations could explain the differences between the character of ^{14}C-labeled DHP and ^{14}C-labeled lignin degradation observed by Trojanowski *et al.* (57) and Gradziel *et al.* (58).

V. THE CHEMISTRY OF LIGNIN BIODEGRADATION

It is perhaps presumptuous to speak of the chemistry of a phenomenon that is so poorly understood. Nevertheless, it is possible to glean, at least, a feeling for the sequence of events involved in the metabolism of lignin.

A. *Low Molecular Weight Intermediates and Lignin Biodegradation*

The question of whether low molecular weight intermediates are formed as a consequence of lignin biodegradation is an important one. The ability to produce low molecular weight compounds from lignin by microbial degradation might allow commercial utilization of lignin, and their analysis could lead to a better understanding of the initial steps in lignin breakdown. Many researchers have addressed this subject (4, 32,36,52,62,63,70,95) but no general consensus regarding the extent or type of low molecular weight species involved in lignin breakdown has been reached.

Phenylpropanoid (C_6-C_3) compounds were implicated in early studies (32,36) as being intermediates in lignin degradation by white-rot fungi. However, only small amounts of such compounds have been reported to accumulate (32,36). Kirk (11) suggests that such small amounts could have been formed from minor structural elements instead of being formed as a consequence of the breakdown of the bulk of the polymer itself.

In a recent study Kawakami (60), who examined the degradation of MWL by *P. ovalis*, found no accumulation of low molecular weight intermediates in the culture medium. Meanwhile,

Odier and Monties (95) were also unable to detect low molecular weight intermediates in the fermentation media during the degradation of dioxane lignin from wheat straw by *Pseudomonas, Flavobacterium* and *Aeromonas* species. However, in a later study (63) they reported detecting small quantities of two phenolic acids that appeared rapidly after innoculation; they did not observe significant accumulation of those compounds.

Kadam (52) utilized a dialysis fermentation in an effort to detect the presence of low molecular weight intermediates during biodegradation of ^{14}C-labeled Kraft lignin by a variety of white- and soft-rot fungi but was unable to detect accumulation of dialyzable intermediates. His results indicate that no more than 2% of the original radioactivity in the lignin was initially dialyzable and that biodegradation rapidly lowers their concentration. The organism probably competes with autorepolymerization for available dialyzable species. Perhaps low molecular weight lignin breakdown products not only fail to accumulate but also gradually disappear in lignin fermentations because of an imbalance in reaction rates. For example, the rate of microbial uptake of low molecular weight intermediates may be more rapid than their generation, or autorepolymerization may be so rapid that low molecular weight breakdown products are removed as fast as they are released. Quite probably a combination of these possibilities occurs.

Low molecular weight intermediates undergo fast repolymerization, mediated by phenoloxidases which are normally found extracellularly during lignin fermentations. Haars and Hüttermann (37) used a homogeneous lignosulfonate preparation and thioglycolic acid, an inhibitor of polyphenol oxidase activity, in the study of lignosulfonate degradation by the Basidiomycete *Fomes annosus*. Their data, over a 20-day incubation period, indicate the accumulation of phenols and phenolic ether compounds in the fermentation medium.

B. Sequential Attack on the Lignin Molecule

The literature indicates that residual lignins have been biologically attacked in a sequential manner with the order and sites of attack dependent on the microbial species. Haider and Trojanowski (45) studied the biodegradation of specifically labeled ^{14}C- DHP's and found that soft-rot fungi attack the side-chains and aromatic rings in the early stages of degradation. Ultimately, radioactive carbon dioxide derived from methoxyl groups was released at a higher rate than radioactive carbon dioxide derived from side-chains or aromatic rings.

In studying the degradation of ^{14}C-labeled lignin from maize straw by a *Nocardia* sp., Trojanowski *et al.* (57) showed that $^{14}CO_2$ was released first from methoxyl groups followed immediately by the release of $^{14}CO_2$ from the side-chains. Twenty-four hours after inoculation, $^{14}CO_2$ from the aromatic rings was released; their data show that the highest rate of $^{14}CO_2$ released came from side-chains followed by $^{14}CO_2$ released from methoxyl groups.

Kirk and Lundquist (96) found that *P. versicolor* degraded sweetgum lignin by sequential attack starting with demethylation and proceeding to more extensive breakdown of the phenolic structure. Kirk *et al.* (73) offered the following scenario to describe the sequential degradation of lignin by white-rot fungi: first, the degradation proceeds by a demethylation of phenolic units in different parts of the lignin polymer to produce diphenolic units. This first step is thought by these workers to be the rate limiting step. Diphenolic units are then cleaved to produce aliphatic and carboxyl-rich residues. The residues formed are then further degraded by both hydrolytic and oxidative reactions. They speculated that while the attack on aromatic regions is progressing, some terminal side-chains are being oxidized to aromatic carboxyl residues. Presumably the low molecular weight aliphatic products are readily taken up by the fungal hyphae.

These studies suggest that the degradation of lignin follows an exodegradation pattern. That is, microbial attack proceeds from the outer macromolecular limits of the lignin and progresses towards the inside of the lignin molecule. It also appears that lignin degrading microorganisms are unable to split the lignin molecule into polyphenolic moieties of intermediate size during the early stages of degradation. This suggestion appears to have been partially confirmed by the recent work of Haars and Hüttermann (37).

VI. MECHANISM OF INITIAL ATTACK ON LIGNIN BY MICROORGANISMS

Complete microbial degradation of lignocellulosic material requires coordinated breakdown of both the carbohydrate fraction and the encasing lignin fraction. Delignification not only facilitate mircobial penetration but also allows higher rates of enzymatic attack on the cellulose fraction. *In situ* depolymerization of cellulose by the action of extracellular microbial enzymes is well established, but the microbial breakdown of lignin is not well understood. Although lignolytic cultures do elaborate enzymes that modify lignin structure, none of these have the independent ability to depolymerize lignin, nor are whole culture filtrates able to reduce the number averaged molecular weight of residual lignin (4).

Although the plant synthesizes the precursors of lignin via enzymatic pathways, the ultimate formation of lignin is a random free-radical, one-electron oxidative polymerization of intermediates catalyzed by the action of plant peroxidases (53). The resultant polymer is randomly linked and polydisperse with no identifiable repeating subunits. In other words, nature has provided an extremely efficient mechanism for the synthesis of a nearly ideal multifunctional structural material. It follows that nature would also allow the development of highly efficient means for the breakdown of this important carbon source. Although there is some risk in second-guessing Mother Nature, one might expect that such a system for lignin breakdown and carbon recycling might not have a plethora of extracellular enzymes, each specific for a different inter- and intra-unit bond. Rather, it seems likely that activities of broad specificity would develop that could lead to simpler substrates for classical catabolic pathways. Such a system for the initial stage of lignin breakdown might involve cell-bound activities and require intimate contact between cell membrane and macromolecular lignin. Alternatively, there may be, unidentified, diffusible agents, enzymatic or non-enzymatic, that could catalyze the initial step(s) of lignin breakdown. Support for these possibilities exists in the literature, but in each case the data still leave inconsistencies.

Crawford *et al.* (101,102) studied the degradation of veratryl-glycerol-β-(o-methoxyphenyl) ether, a lignin model compound, by *Pseudomonas acidovorans*. This model compound contains the arylglycerol-β-aryl ether linkage which is quantitatively the most important intermonomeric linkage in lignin (5). Their results indicate that *P. acidovorans* was capable of degrading the ring and propanoid side-chain of this model compound. Furthermore, they observed that an early step in its degradation was demethylation (101). This led them to speculate that the enzyme effecting such a demethylation is a mixed-function oxidase requiring Fe^{2+}. They also speculated, that further degradation of such a compound was effected by a dehydrogenase requiring NAD^+ as a cofactor. It is interesting to note that this strain of *P. acidovorans* was found to be incapable of effectively degrading ^{14}C-labeled lignin from spruce. This led Crawford and Crawford (70) to conclude that the correlation between the degradation of model compounds and the degradation of lignin is that lignin degrading microorganisms can degrade lignin models while model-degrading microorganisms are not necessarily able to degrade lignin.

Kirk and Chang (12,13) have concluded from their analytical data of lignin obtained from rotted spruce that oxidative ring-opening degradation of lignin has occurred while the aromatic rings are still part of the lignin polymer. This observation has been supported in the recent literature (34). These results as well as others (73,105) seem to indicate that enzymes responsible for initial breakdown of lignin are of the mono- and dioxygenases. It is important to note that most of these enzymes require redox cofactors (*e.g.* NADH and NADPH) (31), thus making the possibility of extracellular attack by mono- and dioxgenases somewhat unlikely. Similarly, the suggestion that membrane bound mono- and dioxygenases cause the initial breakdown of lignin (13,70), requires that the lignin penetrate the cell wall to reach the membrane activities, a mechanically and chemically arduous task. The fact that lignin is a large cross-linked polymer appears to preclude such a penetration.

The early claim (38) that aromatic ring-opening enzymes were found extracellularly in *P. versicolor* culture broths has not been confirmed (31). Other findings (103,104) which claim the presence of hydrolytic enzymes that can cleave β-0-4 linkages in lignin in cultures of several white-rot fungi still lack confirmation.

Higuchi and his coworkers (39,40) have proposed that oxidative cleavage between C_α and C_β of the phenylpropanoid side-chains of dehydrodiconiferyl alcohol suggests a mechanism for C=C double bond cleavage in lignin. Enzymes which can effect such a bond cleavage in lignin or lignin model compounds have not yet been found (31).

Kirk *et al.* (106) and Ishihara (41) have suggested that phenol oxidizing enzymes (*e.g.* laccase and peroxidase) play an important role in lignin degradation. However, the studies by Gierer and Opara (107), as well as other studies (108,109), indicate that neither peroxidase nor laccase alone can cause the depolymerization of lignin or the release of low molecular weight fractions.

To date, no enzymes capable of carrying out depolymerization of lignin have been shown to exist in lignin degrading systems. Hall (31), upon comparing the chemistry of biologically degraded lignin with the chemistry of reactions between reduced oxygen species and lignin model compounds has suggested that reduced oxygen species such as superoxide radical, hydroxyl radical and singlet oxygen may be the attacking agents that cause the initial step in depolymerixation of lignin. The possibility that reduced oxygen species excreted by microbial cells could initiate the breakdown of lignin can be supported by the following facts:

(1) The diversity of intermonomer linkages and monomer structures makes it highly unlikely that one or several extracellular enzymes can cause lignin polymer cleavage.

(2) The nature of lignin biodegradation is oxidative as indicated by residual fermentation lignin which contains significantly more oxygen than the corresponding sound lignin (4,11).

(3) The extent of lignin biodegradation is increased by increasing the oxygen tension in fermentation media (21,23).

(4) Close proximity between microbial cell and lignin is required to achieve degradation of the lignin (4,23). Wilke *et al.* (23) suggest that attacking agents needed for the initial breakdown of the lignin polymer by white-rot fungi are unstable or inactive at all but very small distances from the growing hyphae. The property of instability applies to reduced oxygen species.

Preliminary results in our laboratory indicate the presence of significant amounts of extracellular superoxide radical and hydrogen peroxide in the fermentation medium during lignin degradation by white-rot fungus. The presence of extracellular hydrogen peroxide formed by wood rotting fungi has been observed by Koenings (110,111). Based on his observations, Koenings (111) proposes that brown-rot fungi cause the partial decay of wood via a hydrogen peroxide - Fe^{2+} system. Despite these observations, our preliminary results do not indicate whether such reduced oxygen species are involved actively in the degradation of the lignin polymer or whether they are produced as a consequence of lignin degradation. This question is currently under investigation in our laboratory.

Another aspect of lignin biodegradation that could indicate that extracellular enzymes have a limited role in lignin degradation is the fact that lignin adsorbs proteins to its surface. Our own observations indicate that enzymes added to fermentations containing lignin are readily adsorbed to the lignin with considerable loss of activity. Moreover, the fact that reduced oxygen species cannot be produced microbially in the absence of oxygen during fermentation might explain why anaerobic biodegradation of lignin is limited.

Another aspect of lignin degradation that has not yet been extensively studied is the antibiotic effect of lignin components on microorganisms. Zemek *et al.* (112), in a recent paper, have studied the effect of lignin model compounds with guaiacyl and syringyl structures on a variety of yeasts, bacteria and fungi. Their observations indicate that such compounds exhibit an inhibitory effect on the growth of the studied microorganisms.

VII. SUMMARY

Lignin degradation, is quite common in nature and is an important link in the natural carbon cycle. A large variety of microorganisms are known to degrade lignin in nature as well as in contrived fermentation systems. White-rot and soft-rot fungi, as well as Actinomycetes, are apparently the most active lignin degraders in nature.

The large, cross-linked, polymeric structure of the lignin macromolecule makes its direct uptake during the initial stages of its degradation by microbial cells improbable. Moreover, the fact that the lignin macromolecule is composed of different monomeric units linked by a large variety of non-hydrolyzable intermonomeric bonds precludes hydrolytic cleavage of the biopolymer. Despite the fact that many extracellular and membrane-bound enzymes have been suspected in the initial breakdown of lignin, such activities have not yet been found. A close review of the literature indicates that the initial breakdown of the lignin macromolecule may be non-enzymatic. In addition, the degradation of the lignin polymer appears to follow an exo-degradation mechanism. That is, many lignin degrading microorganism are apparently incapable of splitting the lignin molecule into intermediate molecular weight polyphenolic moieties which are further degraded; instead, they attack the periphery of the macromolecule.

Studies of the biodegradation of monomeric lignin model compounds do not shed light on the initial step(s) involved in the breakdown of the lignin polymer. Such studies assume that microbial breakdown of lignin model compounds is similar to microbial breakdown of lignin an assumption that may not be correct. It is true that degradation of monomeric lignin model compounds can conceivably elucidate the mode of degradation of low molecular weight moieties resulting from initial breakdown of the lignin macromolecule. However, the chemical identities of these low molecular weight intermediates are not yet known. The efficacy of studies using aromatic, monomeric lignin model compounds in attempts to identify intracellular pathways for metabolism of lignin depends on the assumption that lignin breakdown products are indeed mononuclear phenolic materials.

Careful analysis of soluble and insoluble residual lignin resulting from lignin fermentations is a critical step in assessing the lignolytic ability of microorganisms. Furthermore, such analyses are essential in understanding the steps involved in lignin metabolism by microorganisms. To date the methods for residual lignin analyses are complex, time-consuming and error-prone. There is an urgent need to develop a quick and simple method for residual lignin analysis that will yield accurate and reproducible results capable of elucidating structural changes in residual, biodegraded lignin. The development of such an analysis technique will undoubtedly lead to a better understanding of the complex problem of lignin biodegradation.

ACKNOWLEDGMENTS

The work in our lab has been supported by NSF grant number AER 76-11050 A02. We thank Dr. W. Glasser from the Forest Products Department of the Virginia Polytechnic Institute and State University and Mr. C. Edwards from the Department of Chemical Engineering for their invaluable assistance in writing this paper.

REFERENCES

1. Sarkanen, K.V. and C.H. Ludwig, Definition and nomenclature, p. 1, in "Lignins: Occurrence, Formation, Structure and Reactions" K.V. Sarkanen and C.H. Ludwig eds., Wiley-Interscience, N.Y. (1971).
2. Wardrop, A.B., Occurrence and Formation in Plants, p. 19 in "Lignins: Occurrence, Formation, Structure and Reactions" K.V. Sarkanen and C.H. Ludwig eds., Wiley-Interscience, N.Y. (1971).
3. Ander, P. and K.E. Eriksson in "Progress in Industrial Microbiology", Vol. 14, M.J. Bull. ed., Elsevier Sci. Pub., p. 1, N.Y. (1978).
4. Hall, P.L., S.W. Drew, and W.G. Glasser in "Lignin Biodegradation: Microbiology, Chemistry, and Potential Applications" T.K. Kirk, T. Higuchi and H.M. Chang eds., CRC Press, West Palm Beach, FL. (1979).
5. Lai, Y.Z. and K.V. Sarkanen, Isolation and Structural Studies, p. 165, in "Lignins: Occurrence, Formation, Structure and Reactions", K.V. Sarkanen and C.H. Ludwig eds., Wiley-Interscience, N.Y. (1971).
6. Glasser, W.G. and H.M. Glasser, *Holzforschung, 28,* 5, (1974).
7. Marton, J., Reactions in Alkaline Pulping, in "Lignins: Occurrence, Formation, Structure and Reactions" K.V. Sarkanen and C.H. Ludwig eds. Wiley-Interscience, N.Y. (1971).
8. Higuchi, T., *J. Jap. Forest. Soc., 35,* 77 (1953).

9. Sarkanen, K.V. and H.L. Hergert, Classification and Distribution, p. 43 in "Lignins: Occurrence, Formation, Structure and Reactions" K.V. Sarkanen and C.H. Ludwig eds., Wiley-Interscience, N.Y. (1971).
10. Schubert, W.D. "Lignin Biochemistry", Academic Press, N.Y. (1965).
11. Kirk, T.K. in "Annual Review of Phytopathology", J.G. Horsfall ed., Vol. 9, p. 185 (1971).
12. Kirk, T.K. and H.M. Chang, *Holzforschung, 28,* 217 (1974).
13. Kirk, T.K. and H.M. Chang, *Holzforschung, 29,* 56 (1975).
14. Drew, S.W. and K.L. Kadam, *Development in Industrial Microbiology, Vol. 20,* 153 (1979).
15. Kirk, T.K. and D.D. Nicholas eds., Wood Deterioration and its Prevention by Preservative Treatments, Vol. 1, Syracuse University Press (1973).
16. Santra, S. and B. Nandi, *Wood and Fiber, 8,* (4), 249, (1977).
17. Cowling, E.B., USDA Tech. Bull. 1258, 79, (1961).
18. Dion, W.M., *Can. J. Botany, 30,* 9 (1952).
19. Van Vliet, W.F., *Biochim. Biophys. Acta, 15,* 211 (1954).
20. Eriksson, D.E., A. Grunewald and L. Vallander, *Biotech. Bioeng., 22,* (2), 363 (1980).
21. Yang, H.H., M.J. Effland and T.K. Kirk, *Biotech. Bioeng., 22,* (1), 65 (1980).
22. Rosenberg, S.L. "Patterns of Diffusibility of Lignin and Carbohydrate Degrading Systems in Wood-Rotting Fungi" Presented at the ACS meeting, Washington, D.C., September 9-14, (1979).
23. Wilke, C.R. and S.L. Rosenberg, Progress Report for NSF, Department of Chemical Eng. and Lawrence Berkely Lab., University of California, Berkeley, Cal., March 20 (1978).
24. Reid, I.D., *Can. J. Bot., 57,* 2050 (1979).
25. Kirk, T.K. and H.H. Yang, *Biotech. Letters, 1,* (9), 347 (1979).
26. Rosenberg, S.L. And C.R. Wilke, in "Lignin Biodegradation: Microbiology, Chemistry and Potential Applications" T.K. Kirk, T. Higuchi and H.M. Chang eds., CRC Press, West Palm Beach, FL (1979).
27. Kirk, T.K., E. Schultz, W.J. Connors, L.F. Lorenz and J.G. Zeikus, *Arch. Microbiol., 117,* 277 (1978).
28. Keyser, P., T.K. Kirk and J.G. Zeikus, *J. Bacteriol., 135,* 790 (1978).
29. Fenn, P. and T.K. Kirk, *Arch. Microbiol., 123,* 307 (1979).
30. Eslyn, W.E., T.K. Kirk and M.J. Effland, *Phytopathology, 65,* 473 (1975).
31. Hall, P.L., *Enzyme and Microbial Technology,* in Press.
32. Ishikawa, H., W.J. Shubert, and F.F. Nord, *Arch. Biochem. Biophys., 100,* 131 (1963).

33. Hata, K., *Holtzforschung, 20,* 142 (1966).
34. Chen, C.L. and H. Chang, ACS/CSJ Chemical Congress, Abstract Cellulose, Paper and Textile Division, No. 132, Honolulu (1979).
35. Duncan, G.C., *Phytopathology, 57,* 1121 (1967).
36. Fukuzumi, T., *Bull. Agr. Chem. Soc. Jap., 24,* 728 (1960).
37. Haars, A. and A. Hüttermann, *Naturwissenschaften, 67,* (1), 39 (1980).
38. Flaig, W. and K. Haider, *Arch. Microbiol., 40,* 212 (1961).
39. Higuchi, T., in "Lignin Biodegradation: Microbiology, Chemistry and Potential Application" T.K. Kirk, T. Higuchi and H. Chang eds., CRC Press, West Palm Beach, FL (1979).
40. Ohta, M., T. Higuchi and S. Iwahara, *Arch. Microbiol., 121,* 23 (1979).
41. Ishihara, T., in "Lignin Biodegradation: Microbiology, Chemistry and Potential Applications" T.K. Kirk, T. Higuchi and H. Chang eds., CRC Press, West Palm Beach, FL (1979).
42. Levi, M.P. and R.D. Preston, *Holzforschung, 19,* 183 (1965).
43. Preston, R.D., *Wood Sci. Technol., 13,* (3), 155 (1979).
44. Rosenberg, S.L., *Mycologia, 70,* (1), 1 (1978).
45. Haider, K. and J. Trojanowski, *Arch. Microbiol., 105,* 33 (1975).
46. Kirk, T.K., W.J. Connors, R.D. Bleam, W.F. Hackett and J.G. Zeikus, Proc. Nat. Acad. Sci. USA, 72, 2515 (1975).
47. Seifert, K., *Holz Roh-Werkstoff, 24,* 185 (1966).
48. Takahashi, M., Wood Research, Bulletin of the Wood Research Institute, Kyoto University, *61,* 1 (1976).
49. Suzuki, F., J.E. Zojic and N. Kosaric, Proc. 9th Canadian Symp.: Water Poll. Res. Can., 142 (1974).
50. Ferm, R. and A.C. Nilsson, *Svensk Papperstid., 72,* 531, (1969).
51. Kleinert, T.N. and C.S. Joyce, *Svensk Papperstid., 62,* 37 (1959).
52. Kadam, K.L., Ph.D. Dissertation, Department of Chem. Eng., Virginia Polytechnic Institute and State University, Blacksburg, VA (1979).
53. Freudenberg, K. and A.C. Neish, Constitution and Biosynthesis of Lignin, Springer, Berline (1968).
54. Pandila, M.M., *Pulp and Paper Magazine of Canada, 74,* (3), 80 (1973).
55. Cartwright, N.J. and K.S. Holdom, *Microbios., 8,* 7 (1973).
56. Crawford, R.L., E.M. McCoy, J.M. Harkin, T.K. Kirk and J.R. Obst, *Appl. Microbiol., 26,* 176 (1973).
57. Trojanowski, J., K. Haider and V. Sundman, *Arch. Microbiol., 114,* 149 (1977).

58. Grandziel, K., K. Haider, J. Kochmanska, E. Malarczyk and J. Trojanowski, *Arch. Microbiol. Pol., 27,* 103 (1978).
59. Haider, K. and J. Trojanowski, *Arch. Microbiol., 105,* 33 (1975).
60. Kawakami, H., *Mokuzai Gakkaishi, 22,* 252 (1976).
61. Haider, K., J. Trojanowski and V. Sundman, *Arch. Microbiol., 119,* 103 (1978).
62. Danilewicz, K. and M. Tomazewski, *Acta. Microbiol. Polon., 4,* 37 (1972).
63. Odier, E. and B. Monties, *Ann Microbiol. (Inst. Pasteur), 129A,* 361 (1978).
64. Crawford, D.L., *Appl. Environ. Microbiol., 35,* 1041 (1978).
65. Robinson, L.E. and R.L. Crawford, *FEMS Microbiol., letters, 4,* 301 (1978).
66. Crawford, R.L., *Can. J. Microbiol., 21,* 1654 (1975).
67. Ban, S. and M.G. Solgan, *Bioetch. Bioeng., 21,* 1917 (1979).
68. Sutherland, J.B., R.A. Blanchette, D.L. Crawford and A.L. Pometto III, *Current Microbiol., 2,* (2), 123 (1979).
69. Phelan, M.B., D.L. Crawford and A.L. Pometto III, *Can. J. Microbiol., 25,* 1270 (1979).
70. Crawford, D.L. and R.L. Crawford, *Enzyme Microb. Technol., 2,* (1), 11 (1980).
71. Tichy, V., Folia Fac. Sci. Nat. Univ. Purkynianae Brunensis, *Biologia 27, 11,* 81 (1970).
72. Tichy, V., *Ceska Mykologie, 29,* 90 (1975).
73. Kirk, T.K., W.J. Connors and J.G. Zeikus, Recent Advances in Phytochemistry, Vol. 11, 369 (1977).
74. Schmidt, O. and K. Walter, *Europ. J. Appl. Microbiol. Biotech., 5,* (1), 69 (1978).
75. Gregory, P.H., M.E. Lacey, G.N. Festenstein and F.A. Skinner, *J. Gen. Microbiol., 33,* 147 (1963).
76. Smith, R.S. and A. Ofosu-Asiedu, *Can. J. For. Res., 2,* 16 (1972).
77. Wilcox, W.W., *Botan. Rev., 36,* 1 (1970).
78. Ander, P. and K.E. Eriksson, *Svensk Papperstid., 78,* 641 (1975).
79. Ander, P. and K.E. Eriksson, *Svensk Papperstid., 78,* 643 (1975).
80. Ander, P. and K.E. Eriksson, *Arch. Microbiol., 109,* 1 (1976).
81. Glasser, W.G., Pulp and Paper, 3rd ed., Vol. 1, p. 39, J.P. Casey ed., John-Wiley and Sons Inc., N.Y. (1980).
82. Browning, B.L. "Methods of Wood Analysis", Vol. 2, Interscience, N.Y. (1967).
83. Rosenberg, S.L., *Devel. in Ind. Microbiol., 20,* 133 (1979).

84. Bethge, P.O., G. Gran, and K.E. Ohlsson, *Svensk Papperstid., 55,* 44 (1952).
85. Loschbrandt, F., *Norsk Skogind, 4,* 119 (1950); *10,* 402 (1956).
86. Goldschmid, O. and L.F. Maranville, *Anal. Chem., 31,* 370 (1959).
87. Schoning, A.G. and G. Johansson, *Svensk Papperstid., 68,* 607 (1965).
88. Roe, R.B., *Ind. Eng. Chem., 16,* 808 (1924).
89. Hiroi, T. and K.E. Eriksson, *Svensk Papperstod., 79,* 157 (1976).
90. Wu, L.C.F. and W.G. Glasser, *Biotech. Bioeng., 21,* (9), 1679 (1979).
91. Glasser, W.G. and H.R. Glasser, *Macromol., 7,* (1), 17 (1974).
92. Glasser, W.G. and H.R. Glasser, *Cellulose Chem. Technol., 10,* (1), 23 (1976).
93. Glasser, W.G. and H.R. Glasser, *Cellulose Chem. Technol., 10,* (1), 39 (1976).
94. Deschamps, A.M., G. Mahoudeau and J.M. Lebeault, *Eur. J. Appl. Microbiol. Biotechnol., 9,* (1), 45 (1980).
95. Odier, E. and Monties presented by M.A.R. Prevot, C.R. Acad. Sci. Paris, t 284, Series D-21752178 (June 6, 1977).
96. Kirk, T.K. and K. Lundquist, *Svensk Papperstid., 73,* 294 (1970).
97. Crawford, R.L. and D.L. Crawford, *Devel. Ind. Microbiol., 19,* 35 (1978).
98. Crawford, D.L. and R.L. Crawford, *Appl. Envir. Microbiol., 31,* 714 (1976).
99. Crawford, D.L., R.L. Crawford and A.L. Pometto III, *Appl. Envir. Microbiol., 33,* 1247 (1977).
100. Crawford, R.L., L.E. Robinson and A.M. Cheh in "Lignin Biodegradation: Microbiology, Chemistry and Potential Applications" T.K. Kirk, T. Higuchi and H.M. Chang eds., CRC Press, West Palm Each, FL (1979).
101. Crawford, R.L., T.K. Kirk and E. McCoy, *Can. J. Microbiol., 21,* 577 (1975).
102. Crawford, R.L., J.M. Harkin, T.K. Kirk and T.K. McCoy, *Appl. Microbiol., 25,* 322 (1973).
103. Ishikawa, H. and T. Oki, *Mokuzai Gakkaishi, 12,* 101 (1966).
104. Fukuzumi, T., H. Tatatuka and K. Minami, *Arch. Biochem. Biophys., 129,* 396 (1969).
105. Haider, K. and K. Grabbe, *Zentralbl. Bakteriol. Parasitenk. Infeklionskr. Hyg., 205,* 91 (1967).
106. Kirk, T.K., J.M. Harkin and E.B. Cowling, *Biochem. Biophys. Acta, 156,* 145 (1968).

107. Gierer, J. and A.E. Opara, *Acta Chem. Scand., 27,* 2909 (1973).
108. Ferm, R.K.P. Kringstad and E.B. Cowling, *Svensk Papperstid., 75,* 859 (1972).
109. Hartenstein, R., E.F. Newhauser and R.M. Mulligan, *Phytochem., 16,* 1855 (1977).
110. Koenigs, J.W., *Phytopathology, 62,* 100 (1972).
111. Koenigs, J.W., *Wood and Fiber, 6,* (1), 66 (1974).
112. Zemek, J., B. Kosvikova, J. Augustin and D. Joniak, *Folia Microbiol., 24,* 483 (1979).

CHAPTER 4

IMMOBILIZED MICROBES

R. A. Messing

Research and Development Laboratories
Sullivan Science Park, Corning Glass Works,
Corning, NY

I. INTRODUCTION

Microbe or cell immobilization is a natural phenomenon. Stones in a body of water are generally slippery. Those that are in a position to receive light and are within a few feet of the water surface contain a film of photosynthetic organisms (algae, bacteria). Those stones that are in deeper water and receive little or no light contain a film of microbes that are not photosynthetic. These immobilized microbes aid in the reduction of BOD within the body of water. We can almost be certain that immobilized microbes came into existence when the first cells evolved. As a matter of conjecture, the very earliest cells may have been formed on the surface of inorganic materials.

Immobilized microbe technology is not of recent vintage. The Schuetzenbach "quick vinegar" process which employed microbes immobilized as a thin film on wood shavings for the conversion of ethanol to an acetic acid was invented in the early 1820's. Thin films of microbes on stones have been employed in the trickling filter waste treatment processes for many years. The tenderization of meat by aging it employes immobilized microbes and has been practiced since early man. What then is new about immobilized microbes?

The past thirty years has witnessed the generation of knowledge concerning the metabolism of microbes on a molecular level and the understanding of fermentation processes. This knowledge offered the opportunity to employ these microbes in a much more efficient manner. Some of these basic studies

ISBN 0-12-040304-8

have led to the understanding of the interaction of cells with their environment. Armed with this information, both metabolic cycles and cellular reproduction could be controlled and optimized. Now, with genetic manipulation, an opportunity may be available to select a gene responsible for a particular metabolic cycle in one organism for transfer to a preferred microbe which may either reproduce more rapidly or be less noxious than the source microbe.

Now let us turn our attention to microbes as a tool of production. Under these circumstances, the feed or nutrients (carbon, nitrogen, minerals), the gas (oxygen, carbon dioxide) and energy must be used as efficiently as possible. If a product is produced by a single enzyme within the microbe, it would not be efficient to expend carbon, nitrogen, minerals and gas for reproduction and maintenance of cells to utilize this individual enzyme. Under these circumstances, it would be appropriate to find a condition that kills the cell without affecting the enzyme activity.

If a product is produced only with a viable cell in either the stationery phase or the death phase, but not in the logarithmic growth phase, then the nutrients would be wasted if they were utilized for growth. Under these circumstances, a confinement of the cell by various immobilization approaches, such as encapsulation or entrapment, prevent reproduction, and most of the nutrients may be utilized for production of the product.

There are products which are either most efficiently produced or always produced under logarithmic growth conditions. Representative examples of this are the reproduction of cells for either single cell proteins or yeast production. In order to optimize growth and cell reproduction, effective delivery of nutrients at a rapid rate and the subsequent efficient removal of metabolic wastes are required. A continuous reactor, such as a chemostat, fulfills some of these requirements; however, if the flow rates are too high, the cells are washed out. Immobilization technology offers an opportunity for the retention of cells under the normal wash out conditions employed in high flow-rate continuous reactor processes. The immobilization practiced for this application is markedly different than that described for the previous cases. One cannot confine the cells by either encapsulating or entrapment since that would minimize the cell reproduction. Therefore, immobilization approaches that allow for growth and reproduction are required.

II. DEFINITION OF IMMOBILIZED MICROBE

An immobilized microbe, as used in this discussion, is a microbe either alive or dead, that is fixed to a carrier or to another cell by either covalent bonding or by adsorptive forces. This definition is intended to include a cell fixed within a membrane which completely surrounds it (encapsulation, etc.).

III. SURVEY OF REVIEW LITERATURE

The immobilization of microbes (whole cells) was the subject of four major reviews during the period from 1976 through 1978. The interest in this blossoming technology was stimulated by the successes in immobilized enzyme applications which were commercialized in the late 1960's through the late 1970's. The focus of the Chibata and Tosa (1) reivew was the industrial applications of immobilized enzymes and immobilized microbial cells. The immobilized cells of this review were, in fact, dead and could be viewed solely as a form of stabilized and immobilized enzymes. Generally, this review is versed towards the utilization of but a single enzyme within the dead cell.

Another fairly comprehensive review of immobilized cells is that of Abbott (2). The major emphasis of this review is production of pharmaceutical compounds by immobilized cells. Immobilization techniques and methods are briefly summarized but more comprehensively discussed under specific applications. The author's main emphasis is the immobilization of cells for the purpose of using a single enzyme within a particular cell. Most of his discussions concern dead cells and the immobilization of cell-free enzymes. Extensive tables have been compiled which summarize specific examples of immobilized enzymes and immobilized cells, the technique for immobilization, and the company or institution at which these efforts were performed. Abbott does mention a few applications in which the cells were immobilized, not only for the enzyme present, but also for the cofactor required by the enzyme and for sequential multi-enzyme reactions. One example was cited in this review of an immobilized microbe that apparently exhibited growth after immobilization and, thus, this is the single concrete reference to a living immobilized microbe in this article.

Jack and Zajic (3) devote themselves solely to the immobilization of whole cells. In addition to the immobilization of microbial cells, these authors very briefly discuss the immobilization of animal tissue cells. A very pertinent contribution of this review is the emphasis placed by the authors on the viability of the cells. They make clear distinctions between dead cells which have been immobilized solely for the activity of the single enzyme, cells that retain viability but are not reproducing in which multi-enzyme reactions occur and cofactors are available, and, finally, cells that are rapidly reproducing with full viability. These authors further differentiate between viable cells that are reproducing normally and viable cells that have been somewhat modified after an immobilization process. They clearly present the general advantages and disadvantages of immobilizing cells. Their approach to this topic is that of the specific techniques of immobilization. More than half of this review is devoted to entrapment techniques. Their discussion of immobilization by adsorption is rather brief and some of the conclusions may be considered arbitrary. The discussion of covalent or coordinate bonding of cells to supports was simply a citation of three papers.

The most comprehensive treatment of microbial cell immobilization is that of Durand and Navarro (4). These authors have surveyed modern immobilized microbe technology and have included both the positive and the negative aspects. The salient points concerning the immobilization and use of either non-growing or dead cells for a single enzyme reaction that is non-cofactor-linked are thoroughly but succinctly discussed and evaluated. The immobilizations of cells without supports and with supports by adsorption, by entrapment, by encapsulation and by covalent binding are concisely surveyed and analyzed. Perhaps the most important contributions of these authors are their rigorous examination of immobilization techniques with respect to the application and their emphasis of the value of immobilized viable cells as a microbial culture system. In fact, they indicate, in this review, that the cells found in the effluent from an immobilized microbe reactor divided synchronously and that this synchronized growth appeared to reduce generation time.

IV. IMMOBILIZED CELLS FOR A SINGLE ENZYME REACTION

This approach is an alternative to the immobilization of an enzyme. The factors that favor the cell rather than the enzyme immobilization are:

(1) Isolation and purification are unnecessary.

(2) In most instances, the enzyme is more stable in the cell environment than it is after isolation.

The disadvantages encountered when the cells are immobilized rather than the isolated enzyme are:

(1) An additional barrier to diffusion is encountered at the cell wall and/or membrane.

(2) Less enzyme or enzyme activity is immobilized per unit surface area or volume of reactor which, in turn, implies larger reactors for a given activity.

(3) The cells may contain unwanted extraneous reactions which may affect the product.

A. Immobilized Microbes for Glucose Isomerization

The immobilized biological system that has had the greatest economic impact upon the market is that which contains glucose isomerase. Glucose isomerase is an intracellular enzyme that catalyzes the conversion of less than one-half of the glucose in solution to fructose. When this enzyme is applied to glucose derived from corn, the resulting product is known as "high fructose corn syrup". This syrup is used as a substitute for sucrose in confectionary, food and beverage applications. The commercial implications of this enzyme have induced the studies of a wide variety of immobilization techniques as applied to the cells containing this enzyme in order to improve the economics of this conversion.

Although the initial efforts of Takasaki and Kanbayashi (5) and Tsumura (6) did not truly involve immobilized cells, these researchers established that an enzyme could be fixed within the cell such that it would not leak out during the employment of those cells for enzyme conversions. Takasaki and Kanbayashi fixed the glucose isomerase within a *Streptomyces sp.* by heating the cells to between 60° and 85° for ten minutes. Tsumura fixed the isomerase within *Streptomyces phaeochromogenes* cells by β-ray irradiation. Tsumura and coworkers did actually immobilize cells of *Streptomyces phaeochromogenes* containing glucose isomerase by forming a complex with chitosan (7,8).

Vieth *et al.* have immobilized whole cells of both *Streptomyces phaeochromogenes* and *Streptomyces venezuelae* containing glucose isomerase in collagen which were subsequently cross-linked with either formaldehyde or glutaraldehyde (9,10,11). These immobilized cells were employed in a column reactor at 70°C for 40 days in which the isomerization of glucose was accomplished on a continuous basis.

Chibata, Tosa and Sato (12) chose polyacrylamide entrapment to immobilize cells of *Streptomyces griseus* which contained glucose isomerase. They found that this system required magnesium ions for activation and that cobalt ions helped stabilize the enzyme activity of these immobilized cells.

Poulsen and Zittan (13) describe a relatively simple procedure for immobilizing *Bacillus coagulans* cells that contain glucose isomerase by simply crosslinking the cells for 60 minutes at 25°C with glutaraldehyde.

B. Other Immobilized Cells with Single Enzyme Reactions of Commercial Significance

The Japanese researchers were the pioneers in innovations leading to the commercial exploitation of immobilized microbes for single enzyme reactions. One of the early efforts was that of the group at Tanabe Seiyaku Company to develop an immobilized system to convert fumaric acid to aspartic acid (14, 15,16). The enzyme, aspartase, derived from *Escherichia coli*, mediates the addition of ammonia to the double bond in fumaric acid to form L-aspartic acid. The initial endeavors (17) of this group to obtain a continuous system by entrapping the purified enzyme in polyacrylamide led to the conclusion that the operational stability of this system was not satisfactory for industrial application. These researchers, however, were rewarded when they successfully immobilized whole cells of *E. coli* in polyacrylamide gels by polymerizing the acrylamide monomer in the presence of N, N'-methylene-bisacrylamide, β-dimethylaminopropionitrile and potassium persulfate for between 10 and 30 minutes at 40°C. These investigators further noted that when the entrapped *E. coli* cells were suspended in a solution containing 1 M ammonium fumarate and one mM Mg^{++}, pH 8.5, and incubated at 37°C, the activity increased 10-fold. This activation appeared to be due to an increase in membrane permeability for the substrate and/or products due to the autolysis of the cells within the gel. Even with this lysis, the enzyme did not leak from the entrapped preparation. This immobilized cell system for the production of L-aspartic acid has been in industrial operation at Tanabe Seiyaku Company, Ltd. since 1973. The overall production costs of the immobilized system is about 60% of the conventional batch process using intact cells. Recently this group has evolved an entrapment procedure which employs kappa-carrageenan gelation of the *E. coli* by cooling to 10°C, and soaking in 0.3 M KCl and, finally, crosslinking with glutaraldehyde and hexamethylenediamine (18). The carrageenan immobilized cells proved to be far more productive than the cells immobilized in polyacrylamide.

Another system of immobilized cells which is of significance to the medical industry, the conversion L-arginine to L-citrulline, was developed by the same group at Tanabe Seiyaku Company, Ltd. (19). The enzyme, L-arginine deiminase, that mediates the conversion of the imine in arginine to the carbonyl in citrulline is found in *Pseudomonas putida*. These investigators entrapped the *Pseudomonas putida* cells in polyacrylamide gel by a method similar to that described for the *E. coli*. Although the addition of a surfactant such as cetyltrimethylammonium bromide is required for a high rate of L-citrulline formation in the free cells, no surfactant is required for the immobilized microbes. This indicates that there may have been a change in the cell walls during the immobilization process which readily allowed the penetration of the arginine to the active site of the enzyme or the removal of citrulline from that same active site.

L-Malic acid is an antidote for hyperammoniemia and is also used as a component of amino acid infusion. Malic acid is produced by the addition of water to the double bond of fumaric acid which is catalyzed by the enzyme fumarase. The enzyme, fumarase, is found both in *Brevibacterium ammoniagenes* and in *Brevibacterium flavum*. The Tanabe Seiyaku group initiated a program to convert this batch process into a continuous reactor process (20,21,22). *B. ammoniagenes* cells having high fumarase activity were entrapped by the polyacrylamide gel procedure. Initially it was found that these immobilized cells formed succinic acid as a by-product. The separation of this acid from malic acid was very difficult. The production of succinic acid could be effectively suppressed by treatment of the immobilized cells with deozycholic acid, bile acid or bile extract. In addition, these detergents markedly enhance the formation of L-malic acid by the immobilized cells. Subsequent efforts (22) indicated that *B. flavum* had a high fumarase activity and high operational stability after immobilization. This organism was chosen for immobilization by a more recent procedure (23) which involved entrapment in carrageenan gels. It was found that the carrageenan entrapped *B. flavum* was 5.2 times more productive than the polyacrylamide entrapped *B. ammoniagenes*.

Urocanic acid is a sun-screening agent which is produced from L-histidine through the catalytic removal of the amine and the formation of the double bond adjacent to the carboxyl group by the enzyme L-histidine ammonia-lyase. Yamamoto, Sato, Tosa and Chibata decided to convert the batch process which employed either the extracted enzyme or the microbial broth of *Achromobacter liqudium* to a continuous packed-bed process (24). This organism, however, has an undesired enzyme activity, urocanase, which converts urocanic acid, the product, to imidazolone propionic acid. The detrimental urocanase activity can be eliminated without affecting the desired, histidine ammonia-lyase, enzyme by heat treating the cells at 70°C for thirty minutes prior to immobilization (25). The heat treated cells are then immobilized in polyacrylamide gel. The half-life of the continuous reactor column packed with the immobilized cells and in the presence of magnesium ions at 37°C was estimated to be approximately 180 days.

C. Miscellaneous Information Concerning Immobilized Microbes for Single Enzyme Reactions

L-sorbosone is an intermediate for the synthesis of vitamin C. Martin and Perlman (26) entrapped *Gluconobacter melanogenus*, which contains the enzyme L-sorbose dehydrogenase that is capable of oxidizing sorbose to sorbosone, in polyacrylamide gel. A major problem encountered in the application of this immobilized system was the delivery of oxygen to the entrapped cells. If pure oxygen was delivered to the entrapped cells, the enzymatic activity was rapidly destroyed.

The pharmaceutical industry requires 6-aminopenicillanic acid (6-APA) for the production of synthetic penicillan. The enzyme penicillan amidase found in *Escherichia coli* is capable of deacylating penicillan to produce 6-APA. Sato *et al.* (27) immobilized *E. coli* in polyacrylamide gel. Although the immobilized cells contained high penicillan amidase activity, they also contain penicillinase activity which decomposes both penicillan and 6-APA. The specific inactivation of penicillinase was a difficult task. However, a compensating factor appeared to be that the penicillinase activity was much lower than the penicillan amidase activity. In a case of this type, it may be advantageous to employ the purified enzyme rather than the immobilized cell.

The immobilization of cells by entrapment in polyacrylamide gel was first performed by Mosbach and Mosbach in 1966. They originally applied this technique to the immobilization of the lichen, *Umbilicaria pustulata*. They demonstrated that the immobilized cells retained both esterase and decarboxylase activity with periodic testing over a three month period at 20°C. Mosbach and Larsson (29) subsequently immobilized *Curvularia lunata* by this technique. The cells contain the enzyme, 11-β-hydroxylase, which convert Reichstein compound S to cortisol. Although the immobilized cells lose part of their enzyme activity with storage, the activity may be restored by incubating the entrapped cells in a nutrient medium containing cortisone. This reactivation may be one of the very early demonstrations of viable growth after entrapment since the reactivation may be due to the growth of cells.

A British Patent Application, published recently, describes an immobilization process which involves the cross-linking of cells containing glutaraldehyde sensitive intracellular enzymes by glutaraldehyde in the presence of polyamine (30). *Bacillus pasteurii* containing urease, *Kluyveromyces fragilis* containing lactase, and *Leuconostoc oenos* containing malolactic enzyme were successfully immobilized by this procedure.

V. IMMOBILIZED CELLS FOR MULTI-SEQUENTIAL ENZYME REACTIONS AND COFACTORS

The continuous production of coenzyme A involves panthothenic acid, L-cysteine and ATP. The reaction carried out by *Brevibacterium ammoniagenes* involves five sequential steps: Panthothenic acid → phosphopantothenic acid → phosphopanthothenoylcysteine → phosphopantetheine → dephosphocoenzyme A → coenzyme A. The microbial cells were entrapped in polyacrylamide gel. The addition of sodium lauryl sulfate to the reaction mixture containing the entrapped cells enhanced the productivity. In a continuous column reactor the half-life was observed to be 5 days at 37°C (31).

The reduction of nitrate to nitrite and then to ammonium can be performed by *Micrococus denitrificans*. Mohan and Li (32,33) encapsulated this microbe in a liquid-surfactant membrane and demonstrated the disappearance of nitrate and nitrite substrate. The continuous reduction without an additional supply of cofactors demonstrates the viability of this multi-enzyme system and the liquid membrane encapsulation approach.

Corynebacterium glutamicum converts glucose to glutamic acid. This conversion is a multi-sequential enzyme reaction. Glowinski and Charm (34) entrapped this mircobe in polyacrylamide gel. The immobilized cells were then suspended in a medium containing glucose and incubated at 30°C with stirring. The quantity of glutamic acid produced after 144 hours was determined. It was found that the productivity of the immobilized cells was greater than that of the free cells. However, when this immobilized system was employed in continuous column reactors, difficulties were encountered in supplying oxygen to the entrapped cells.

Saccharomyces cerevisiae has been immobilized in calcium alginate gel (35). When a glucose solution was continuously fed to a column packed with the entrapped cells, it was estimated that the efficiency of the conversion of glucose to ethanol was 90% of theoretical yield. The half-life of this system is approximately 10 days. However, it has been used for a total of 24 days for the continuous production of ethanol.

Bennet and Weetall (36) demonstrated that the photobacterium, *Rhodospirillium rubrum* immobilized in agar was capable of producing hydrogen from malate and water in the presence of light.

VI. IMMOBILIZED VIABLE CELLS FOR STATIONARY PHASE PRODUCTS

The classical examples of stationary phase products are antibiotics. In order to produce antibiotics, cells must be viable. Suzuki and Karube (37) set for themselves the goal of producing antibiotics from immobilized microbes. They evaluted collagen, alginate and polyacrylamide as entrapment gels for the immobilization of *Penicillium chrysogenum* for the production of penicillin. Although the immobilized cells in alginate proved to contain the highest activity, this preparation was too fragile in the presence of phosphate ions to permit its use in a reactor. The collagen immobilized mycelium was relatively inactive. The authors speculate that this may be due to the tanning reagent, glutaraldehyde. Polyacrylamide gel entrapment of the mycelium was selected for further comparative studies of the immobilized mycelia versus the free mycelia. The free mycelium decreased in production with repeated use while the immobilized mycelium increased initially in production but decreased gradually thereafter. The overall productivity of the immobilized cells was substantially greater than that of the free microbe. The estimated half-life of the immobilized mycelium was 6 times longer than that of the free mycelium. An interesting point to note was that the oxygen uptake of the immobilized mycelium was about 30% of that of the free mycelium. The penicillin production by the immobilized cell under air and under nitrogen were compared. Only a small amount of penicillin was produced by the immobilized cells under anaerobic conditions. Therefore, oxygen was required for the system synthesizing penicillin. Further, the authors note that upon grinding the immobilized cells viable mycelium could be recovered.

Bacitracin is a peptide antibiotic produced by *Bacillus* species. Suzuki and Karube (37) immobilized *Bacillus sp.* in polyacrylamide gel. The productivity of these immobilized cells was found to be only 20 to 25% that of the washed free cells. It was suggested that the lower rate of bacitracin production is mainly caused by inactivation of enzymes with polymerizing reagents and also may be partly due to hindered diffusion of substrates and/or products through the gel. Early experiments with the immobilized cells indicated that the bacitracin production was markedly reduced upon reuse when a fermentation medium was employed. It was further noted that bacteria were released from the immobilized preparation and growth was noted in the fermentation medium. The effect of medium composition on bacitracin productivity was then examined. It was found that bacitracin productivity in a medium containing meat extract or peptone was greater than that in a medium containing carbohydrate. In addition, the productivity upon reuse was retained when the peptone or meat extract was employed. The tendency to maintain a constant level of bacitracin production with reuse of free cells was also noted when peptone medium was employed. The productivity of bacitracin in peptone medium from free cells was compared to that of the immobilized cells with repeated use in batch production. The production by the free cells decreased gradually after the third reuse and precipitously after the fourth reuse. On the other hand, the productivity of the immobilized cells increased with successive utilization in peptone medium until about the seventh reuse. After about the seventh reuse the productivity appeared to remain constant through about the 14th run. The authors attribute this increased activity of the immobilized preparation to the synthesis of additional cells within the gel.

A study of *Clostridium butyricum* entrapped in polyacrylamide gel for sensing biochemical oxygen demand (BOD) in wastewater indicates that these immobilized microbes remain active for long periods of time under certain conditions. The study also indicates that BOD may be rapidly estimated (38).

VII. IMMOBILIZED VIABLE MICROBES UNDER GROWTH CONDITIONS

The discussion to this point has primarily been focused upon cells that were either dead or were not reproducing or the reproduction was unintentional or at a very low rate. Perhaps intermediate between those studies and the rapid growth of cells are the efforts of Kennedy *et al.* (39) in which they entrapped cells in hydrated titania or zirconia gels. The cells, *Saccharomyces cerevisiae* and *E. coli*, were examined for viability by oxygen uptake. They found that the rate of oxygen uptake of the immobilized cells was about 30% of that of the same number of free cells. It was suggested that the reduced rate of oxygen uptake may have been caused by the restriction of the metal hydroxides to access of aerated buffer and a decrease of cell surface area available for oxygen uptake. An additional study employing this technique for immobilization of *Serratia marcescens* indicated growth of the microbe associated with the matrix when fresh culture was added. The prime indication for this growth was the increase of the red pigment characteristic of this microbe.

An excellent discussion of the application of immobilized microbes to microbial culture systems or continuous fermentation is offered by Atkinson and Knights (40). This discussion focuses upon film fermentors and offers some insight into their present and future applications.

McCarty *et al.* (41,42) were the early proponents of the anaerobic filter for waste treatment. This system is a continuous fermentation under anaerobic conditions which employ immobilized microbes. Essentially, both facultative and anaerobic microbes are adsorbed on the surface of stones. This immobilized film of cells produce progeny which are elaborated into the flowing system. Essentially, the anaerobic filter is a continuous packed-bed reactor. The process time for reducing the BOD and producing methane is markedly reduced as a result of the elaboration of the progeny cells from those that are immobilized in the form of a film on the surface of the stones.

Perhaps the most revealing studies in this area are those produced as a result of the efforts of Durand and coworkers at the Laboratoire de Génie Biochimique de l'Institut National des Sciences Appliquées (INSA) in Toulouse, France. These researchers investigated the relative retention of microorganisms (4,43) on various supports such as wood, P.V.C., pouzzolan (a finely, ground burnt clay or shale), resins and silica. In one comparative study of the yeast, *Saccharomyces carlsbergensis*, they found that on a milligram dry weight of cells per gram of support basis, wood chips retain between 6 and 120 times more cells than any other of the previously described supports. In this study they also revealed the adverse effect of glutaraldehyde on these cells in a comparison of untreated silica and silica treated with glutaraldehyde. Their conclusion is that retention simultaneously depends on the type of adsorbents and the type of microorganisms.

The foundation for the studies at INSA was prepared by the work of Harttori and Furusaka (44,45) which indicated that *E. coli* cells immobilized by adsorption on ion-exchange resins displayed a periodicity in the release of free cells, and it was found that these cells would divide synchronously in growth media. The release of progeny by cell division from the resin surface proceeded according to the age distribution of the adsorbed cells. Helmstetter and Cooper (46,47) further indicated that *E. coli* cells which were adsorbed on a filter membrane and subjected to a flowing nutrient medium exhibited shorter doubling time than freely suspended microorganisms.

These observations were further expanded by Hattori *et al.* (48,49,50) when they noted that both batch studies and continuous cultivation of adsorbed cells demonstrated diminished generation times.

The INSA group has constructed a pilot plant for producing fermented beverages (51) based upon the retention of microbes on a solid support. The solid supports that have been tested are P.V.C. and porous bricks. The equipment consists of two columns, the first of which is used for changing wort into beer and the second one, which is termed the storage column, is employed for the maturation of the beer. The equilibrium state is obtained very rapidly and the reactor, once stopped, may be started again without any lag time. Wort is fermented with a residence time of 90 minutes. The lifetime of the support appears to be unlimited and fermentations have been performed without the addition of yeast over an eight month period. No plugging of the reactor has been noted because the microbes are washed out after multiplication by the flowing liquid.

Navarro and Durand (52) studied the generation time of the yeast, *Saccharomyces uvarum*, in the non-immobilized state and compared it to those that were immobilized by adsorption on the γ-aminopropyltrimethoxysilane treated porous silica, covalently coupled through glutaraldehyde to the γ-aminopropyltrimethoxysilane treated porous silica and by adsorption to granules of porous brick. These studies were performed in a column reactor. The generation time for the free cells was found to be 8 hours. The generation time for the cells adsorbed on the γ-aminopropyltrimethoxysilane treated porous silica was found to be about 4 hours at 22°C, which was the same temperature at which the free cell generation time was determined. The generation time of the cells attached to the porous brick was found to be 2 hours and 45 minutes, while that of the cells covalently coupled through glutaraldehyde to the silane treated porous silica was found to be 1 hour and 45 minutes at 22°C. In addition, these authors found that not only did the immobilized cells reproduce synchronously with reduced generation times, but also, their daughter cells did when liberated from the reactor and incubated. In the free state a new synchronous culture of cells appears with a generation time that is equal to the immobilized cells. As a matter of fact, this synchronous growth has been noted for two to three generations after liberation from the immobilized state. Synchronous growth has been noted for four to five generations after immobilization by adsorption on silane-treated silica, five to seven generations with the brick, and ten to twelve generations after immobilization by covalently coupled yeasts to silica surface. The authors attribute this synchronous growth to the fact that the daughter cells, in the form of the buds, are released from the surface while the parent cell is retained on the support.

VIII. OPTIMIZATION OF SUPPORT STRUCTURES FOR BIOMASS ACCUMULATION AND GROWTH

Relationships between the accumulation of stable and viable microbes and the pore morphology of a dimensionally stable inorganic support were determined (53,54,55). That relationship is dependent upon the cell or spore dimensions, the mode of reproduction and the pore diameter of the porous support. In order to achieve high accumulation of microbes that reproduce by fission, at least 70% of the pores of an inorganic support should have pore diameters in the range of 1 times the smallest major dimension through 5 times the largest major dimension of the cell. To achieve the highest accumulation of microbes that reproduce by budding, at least 70% of the pores should have pore diameters in the range of 1 times the smallest dimension of the cell and less than 4 times the largest cell dimension. In the case of microbes that form spores and exhibit mycelial growth in order to achieve high accumulation of biomass, at least 70% of the pores of the inorganic support should have pore diameters in the range of 1 times the smallest dimension of the spore but less than about 16 times the largest dimension of that spore. These relationships were established by varying the physical parameters of the carriers as well as their chemical composition. In addition, it was demonstrated that the accumulation of biomass was independent of the mode of attachment but appeared to be more dependent upon the surface contributions of the support material. This latter point was demonstrated in comparative studies of *Streptomyces olivochromogenes* and *Penicillium chrysogenum*. The initial growth of *S. olivochromogenes* was substantially greater on the more negative surfaces of glass than on cordierite. To the contrary, the growth of *P. chrysogenum* was greater on the more positively charged cordierite than on the glass.

IV. PROSPECTS FOR THE FUTURE (CONCLUSION)

Immobilized microbe technology will not only impact the production of fine chemicals and pharmaceuticals in the form of single enzyme reactions and multi-sequential enzyme reactions to deliver products that are relatively free of cells and debris, but also, it will impact, to perhaps a greater degree, continuous fermentation processes. The continuous fermentation processes that will probably be most affected by the reduced generation times of immobilized cells and optimized immobilization surfaces are those that involve energy production processes, such as methane and alcohol and anaerobic waste treatment or conversion processes.

The major limitation of this technology appears to be the problems encountered in delivering gases such as oxygen and the removal of waste gases such as carbon dioxide.

Perhaps the most exciting event of the future will be the joining of the genetically optimized microbe, immobilization technology and the continuous reactor.

REFERENCES

1. Chibata, I., Tosa, T., *Appl. Biochem. Bioeng. 1,* 329 (1976).
2. Abbott, B.J., *Adv. Appl. Microbiol. 20,* 203 (1976).
3. Jack, T.R., Zajic, J.E., *Adv. Biochem. Eng. 5,* 125 (1977).
4. Durand, G., Navarro, J.M., *Process Biochemistry 13,* 14 (1978).
5. Takasaki, Y., Kanbayashi, A., *Kogyo Gijutsuin Biseibutsu Kogyp Gijutsu Kenkyusko Kenkyu Kokoku 31,* 31 (1969).
6. Tsumura, N., Annu. Meet. Soc. Ferment. Technol., Jpn., 21st, (1969), p. 81.
7. Tsumura, N., Kasumi, T., 5th Int. Ferment. Symp., Berlin, (1976) p. 561.
8. Tsumura, N., Kasumi, T., Ishikawa, M., *Rep. Nat'l. Food. Res. Inst.* (Tokyo) *31,* 75 (1976).
9. Vieth, W.R., Wang, S.S., Saini, R., *Biotechnol. Bioeng. 15,* 565 (1973).
10. Saini, R., Vieth, W.R., *J. Appl. Chem. Biotechnol. 25,* 115 (1975).
11. Venkatsubramanian, K., Saini, R., Vieth, W.R., *J. Ferment. Technol., 52,* 268 (1974).
12. Chibata, I., Tosa, T., Sato, T., Japanese Patent Kokai, 74/132,290 (1974).
13. Poulsen, P.B., Zittan, L., "Methods in Enzymoloty", (Mosbach, K., ed.), Vol 44, p. 809, Academic Press, NY (1976).
14. Chibata, I., Tosa, T., Sato, T., *Appl. Microbiol. 27,* 878 (1974).
15. Tosa, T., Sato, T., Mori, T., Chibata, I., *Appl. Microbiol., 27,* 886 (1974).

16. Chibata, I., Tosa, T., Sato, T., Mori, T., Yamamoto, K., "Enzyme Engineering", (Pye, E.K., Wingard, L.B., Jr., Eds), Vol. 2, p. 303, Plenum, NY (1974).
17. Tosa, T., Sato, T., Mori, T., Matuo, Y., Chibata, I., *Biotechnol. Bioeng. 15,* 69 (1973).
18. Nishida, Y., Sato, T., Tosa, T., Chibata, I., *Enzyme Microb. Technol. 1,* 95 (1979).
19. Yamamoto, K., Sato, T., Tosa, T., Chibata, I., *Biotechnol. Bioeng. 16,* 1589 (1974).
20. Chibata, I., Tosa, T., Yamamoto, K., *Enzyme Eng. 3,* 463 (1975).
21. Yamamoto, K., Tosa, T., Yamashita, K., Chibata, I., *Eur. J. Appl. Microbiol. 3,* 169 (1976).
22. Takata, I., Yamamoto, K., Tosa, T., Chibata, I., *Eur. J. Appl. Microbiol. 7,* 161 (1976).
23. Takata, I., Yamamoto, K., Tosa, T., Chibata, I., *Enzyme Microb. Technol. 2,* 30 (1980).
24. Yamamoto, K., Sato, T., Tosa, T., Chibata, I., *Biotechnol. Bioeng. 16,* 1601 (1974).
25. Shibatani, T., Nishimura, N., Nabe, K., Kakimoto, T., Chibata, I., *Appl. Microbiol. 27,* 688 (1974).
26. Martin, C.K.A., Perlman, D., *Biotechnol. Bioeng. 18,* 217 (1976).
27. Sato, T., Tosa, T., Chibata, I., *Eur. J. Appl. Microbiol. 2,* 153 (1976).
28. Mosbach, K., Mosbach, R., *Acta Chem. Scand. 20,* 2807 (1966).
29. Mosbach, K., Larsson, P.O., *Biotechnol. Bioeng. 12,* 19 (1970).
30. Gestrelius, S.M., U.K. Patent Application GB 2019410A (Oct. 31, 1979).
31. Shimizu, S., Morioka, H., Tani, Y., Ogata, K., *J. Ferment. Technol. 53,* 77 (1975).
32. Mohan, R.R., Li, N.N., *Biotechnol. Bioeng. 17,* 1137 (1975).
33. Mohan, R.R., Li, N.N., *Biotechnol. Bioeng. 16,* 513 (1974).
34. Slowinski, W., Charm, S.E., *Biotechnol. Bioeng. 15,* 973 (1973).
35. Kierstan, M., Bucke, C., *Biotechnol. Bioeng. 19,* 387 (1977).
36. Bennet, M.A., Weetall, H.H., *J. Solid-Phase Biochem. 1,* 137 (1976).
37. Suzuki, S., Karube, I., "Immobilized Microbial Cells" (Venkatsubramanian, K., Ed.) p. 59, ACS Symposium Series 106, Washington, D.C. (1979).
38. Karube, I., Tadashi, M., Suzuki, S., *J. Solid-Phase Biochem. 2,* 97 (1977).
39. Kennedy, J.F., Barker, S.A., Humphreys, J.D., *Nature 261,* 242 (1976).

40. Atkinson, B., Knights, A.J., *Biotechnol. Bioeng. 17,* 1245 (1975).
41. McCarty, P.L., *Pub. Works 95,* 107 (1964).
42. Young, J.C., McCarty, P.L., *Jour. Water Poll. Control Fed. 41,* R160 (1969).
43. Maignan, C., Navarro, J.M., Durand, G., *Qecol. Plant. 9,* 365 (1974).
44. Hattori, T., Furusaka, C., *Biochem.* (Tokyo) *48,* 831 (1960).
45. Hattori, T., Furusaka, C., *Biochem.* (Tokyo) *50,* 312 (1961).
46. Helmstetter, C.E., *J. Mol. Biol. 24,* 417 (1967).
47. Helmstetter, C.E., Cooper, S., *J. Mol. Biol. 31,* 507 (1968).
48. Hattori, R., Hattori, T., Furusaka, C., *J. Gen. Appl. Microbiol. 48,* 271 (1972).
49. Hattori, R., Hattori, T., Furusaka, C., *J. Gen. Appl. Microbiol. 18,* 285 (1972).
50. Hattori, R., *J. Gen. Appl. Microbiol. 18,* 319 (1972).
51. Corrieu, G., Blachere, H., Ramirez, A., Navarro, J.M., Durand, G., Duteurtre, B., Moll, M., Proc. 5th Int. Ferment. Symp. (1976) p. 294.
52. Navarro, J.M., Durand, G., C. R. Acad. Sc. Paris, t. 290, (11 février 1980) Serie D, p. 453.
53. Messing, R.A., Oppermann, R.A., *Biotechnol Bioeng. 21,* 49 (1979).
54. Messing, R.A., Oppermann, R.A., Kolot, F.B., *Biotechnol. Bioeng. 21,* 59 (1979).
55. Messing, R.A., Oppermann, R.A., Kolot, F.B., "Immobilized Microbial Cells" (Venkatsubramanian, K., Ed.) p. 13, ACS Symposium Series 106, Washington, D.C. (1979).

ANNUAL REPORTS ON FERMENTATION PROCESSES, VOL. 4

CHAPTER 5

PEPTIDES

L. C. Vining

Biology Department
Dalhousie University
Halifax, Nova Scotia

New peptides continue to be discovered at a steady rate in the search for biologically active and useful substances from natural sources. A high proportion of them exhibit antibiotic activity, but peptides are among the most versatile of metabolites and the manner in which they inhibit the growth of microorganisms is quite varied. This chapter reviews the peptides discovered as microbial products since 1975. Those isolated earlier are referred to only where they relate to more recent discoveries, or where new information modifies or extends previously accepted ideas about them. As much as possible, substances of like function are grouped together. Frequently this also brings together chemically related compounds and peptides of uncertain function are grouped broadly according to chemical structure. The full range of amino acid-derived secondary metabolites has been surveyed in several earlier articles (1-4). Certain groups have also received more recent coverage. These include the polymyxins (5), the virginiamycins (6), the β-lactams (7-9) and antibiotics produced by *Bacillus* species (10,11). Up-to-date reviews of the biosynthesis of peptides have also appeared (12,13). Where such current information is available, the topic will be dealt with only briefly here.

ISBN 0-12-040304-8

I. INHIBITORS OF CELL WALL SYNTHESIS

A. β-*Lactams*

The isolation of 7-methoxycephalosporins from *Streptomyces* (14) has been followed in recent years by the discovery of other and more unexpected β-lactams in the actinomycetes. This latter development owes much to innovative new screening procedures. The noncardicins were uncovered by mutant test organisms supersensitive to β-lactam antibiotics (15); clavulanic acid, the olivanic acids and metabolites MC 696 SY2-A and -B emerged from searches for β-lactamase inhibitors (16-18); the thienamycins were detected as inhibitors of peptidoglycan synthesis (19,20). Thienamycins, epithienamycins (21), olivanic acids, antibiotic PS-5 and the MC 696 SY2 metabolites form a structurally related group with the 7-oxo-1-azabicyclo-(3.2.0)help-2-ene ring system (22-26). They combine a high level of antibiotic activity with inhibition of β-lactamases. Clavulanic acid, on the other hand, is primarily a β-lactamase inhibitor (27).

The variety of naturally occurring β-lactames now also includes the bleomycin group of antibiotics (28), certain *Pseudomonas* toxins (29), a glutamine antimetabolite (30) and the pachystermine group of plant alkaloids (31). Most of these new metabolites are clearly different in biosynthetic origin from the classical penam and cephem derivatives. An investigation of clavulanic acid formation in *Streptomyces clavuligerus* with ^{13}C-labeled precursors indicates that the β-lactam-ring carbons are derived from glycerol (32). The remainder of the molecule comes from α-ketoglutarate, probably via glutamate. The carbon skeleton of the thienamycin-olivanic acid group can be plausibly dissected into acetoacetyl and glutamate components. Both clavulanic acid and the thienamycins are co-produced with penicillin N, cephamycin C and related β-lactams known to be formed by the adipylcysteinylvaline route and it has been suggested that some enzymes, possibly those catalysing ring closure, may be common to the different pathways (32).

δ-(L-α-Aminoadipyl)-L-cysteinyl-D-valine, recognized as a probable intermediate in the pathway to penam and cephem antibiotics since its discovery in *Penicillium chrysogenum* (33), has been identified in *Caphalosporium acremonium* (34), and *S. clavuligerus* mycelium (35). *Paecilomyces persinicus*, a fungus that produces the cephalosporin rather than the penicillin type of β-lactam, has been found to form, in addition to the tripeptide, two tetrapeptides. These are glycyl-δ-(α-aminoadipyl) cysteinyl-valine and the corresponding glycyl-δ-(α-aminoadipyl)-cysteinyl-β-hydroxyvaline (36). Certain mutants of *Cephalosporium acremonium* which have lost the capability to produce cephalosporins accumulate the thiomethyl derivative as well as the disulfide dimer of δ-(L-α-aminoadipyl)-L-cysteinyl-D-valine (37). These products were found extracellularly in about the same total amount as the β-lactam metabolites produced by the parent strain, suggesting that they were shunt metabolites.

B. Glycopeptides

Several new members have been added to the vancomycin group of cell-wall inhibitors. The mannopeptins, isolated from *Steptomyces platensis*, give only mannose on mild acid hydrolysis and also differ from vancomycin in not giving aspartic acid and N-methyl-D-leucine on vigorous hydrolysis. Mannopeptin A contains β-methylphenylalanine whereas mannopeptin B contains phenylalanine (38). Production conditions are unusual in that a high (1.8 mM) concentration of ferrous salt is required in the medium. Also, unlike previously known glycopeptides, the mannopeptins are active against some Gram-negative bacteria. Another pair of glycopeptides, azureomycins A and B, were found in *Pseudonocardia azurea* by Omura's group, during screening for inhibitors of cell wall biosynthesis (39). Azureomycin A causes accumulation of UDP-N-acetylmuramyl-L-alanyl-D-glutamyl-*meso*-diaminopimelyl-D-alanyl-D-alanine within the cytoplasm, indicating that it blocks the assembly of peptidoglycan (40).

Ristomycin A, a purified component of the ristomycin complex produced by *Proactinomyces fructifera*, contains the branched tetrasaccharide, O-β-D-arabinopyranosyl-(1→2)-O-α-D-mannopyranosyl-(1→2)-O-(α-L-rhamnopyranosyl-(1→6))-D-glucopyranose, together with mannose and ristosamine (3-amino-2,3,6-trideoxy-L-ribohexopyranose) linked to an unusual peptide. The structure of the tetrasaccharide was established by chemical and ^{13}C nuclear magnetic resonance studies (41) and the peptide has recently been shown (42) to consist of three components, ristomycinic acid (I), actinoidinic acid (II) and the acid-and alkali-sensitive *(bis)*-dechlorovancomycinic acid (III). These three peptide components account for the entire aglycone of ristomycin A but the linkage sequence is not yet known. All three components are present in ristocetin A which may, in fact, be identical to ristomycin A (43). Actinodinic acid and vancomycinic acid (IV) are components of the aglycone in vancomycin (44).

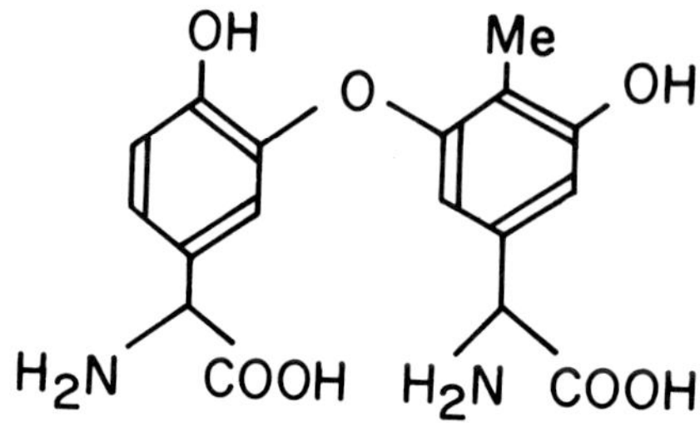

I

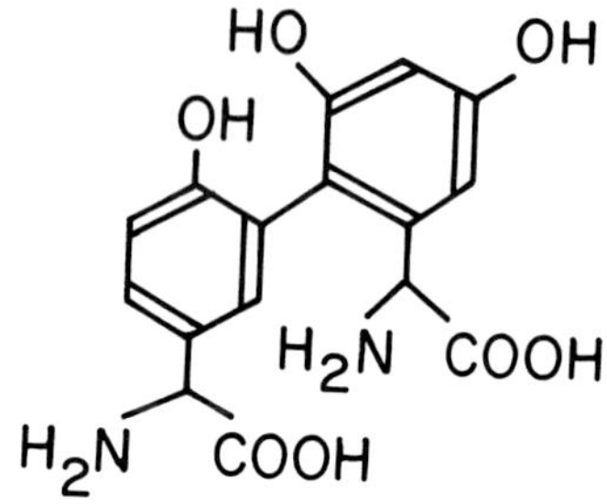

II

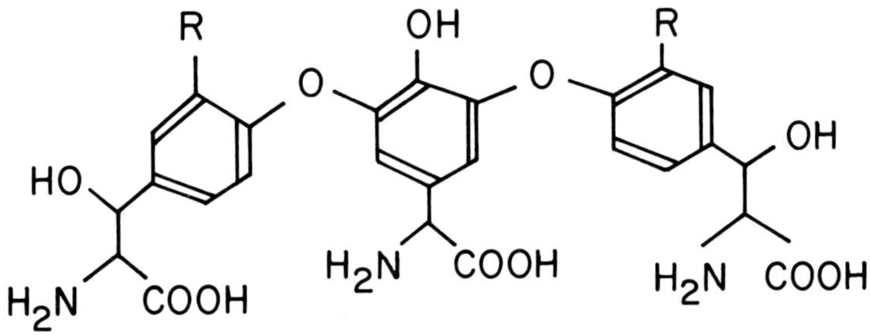

III , R = H

IV , R = Cl

Avoparcin (antibiotic LL-AV290) and antibiotic AB-65, produced by *Streptomyces candidus* (45) and *Saccharomonospora viride* (46), respectively, are chlorinated glycopeptides. Although they may belong in the vancomycinristocetin group, the inforamtion is not yet decisive. Avoparcin contains p-hydroxyphenylglycine and its chloro-derivative, which are also present in janiemycin (47) and enduracidin (48). The latter inhibits peptidoglycan synthesis (49) but is not closely related in structure to vancomycin.

C. Acyl Peptides

Stushimycin, aspartocin and amphomycin (glumamycin) are linear acyl peptide antibiotics with similar structures (50) and probably, also, similar modes of action. Amphomycin causes accumulation of UDP-N-acetylmuramyl pentapeptide in *Bacillus megaterium*. Since neither disaccharide nor lipid-bound intermediates are formed, the antibiotic is believed to inhibit attachment of UDP-N-acetylmuramyl pentapeptide to undecaprenol phosphate (51). Gardimycin, a linear peptide from *Actinoplanes liguriae* with two unidentified sulfur-containing amino acids, falls outside this group and inhibits peptidoglycan biosynthesis by blocking the transfer of lipid-bound intermediates to cell wall acceptors (52). Its action is thus very similar to that of vancomycin, although it is not known whether the similarity extends to specific binding with the terminal peptide sequence of cell wall biosynthetic intermediates (43). Gardimycin is of particular interest because it shows activity against Gram-positive bacteria, Gram-negative cocci and obligate anaerobes. It is also of low toxicity and is active *in vivo*.

D. Gram-Negative-Specific Inhibitors

Bicyclomycin (V), isolated from *Streptomyces sapporoensis*, binds to proteins on the inner membrane of *E. coli* (53). The binding is specific; it does not occur with *Bacillus subtilis* cells, the sites differ from those at which penicillin binds and peptidoglycan synthesis is not affected. Reaction probably occurs between thiol groups in the protein and the $\Delta^{4,5}$ exocyclic double bond since thioether adducts form *in vitro* when bicyclomycin is exposed to mercaptoethanol (54). Tests with a series of semi-synthetic derivatives have shown that the 3'-hydroxyl group is also essential and that the overall structure is necessary for activity (55). Bicyclomycin gives rise to both spheroplast and filamentous forms of *E. coli* and microscopic examination of treated cells shows blebs forming on the cell surface in the stages before lysis (56,57). Filament formation suggests that bicyclomycin has affected membrane proteins functioning in cell division. These must be additional to those that bind penicillin.

Electrophoretic examination of envelope proteins from bicyclomycin-treated *E. coli* showed a decrease in free lipoprotein and even stronger depletion of the peptidoglycan-bound form. A likely response to bicyclomycin was considered to be synthesis at the ribosome of an incomplete or altered form of free lipoprotein that it is incapable of binding to the peptidoglycan. However, a mutant *E. coli* strain lacking peptidoglycan-bound lipoprotein grows as well as the wild type under most conditions and is equally sensitive to bicyclomycin. Thus lipoprotein biosynthesis may not be the primary site of action (53).

Several *Streptomyces* species produce peptidolactone antibiotics that exhibit specific action against Gram-negative bacteria (58,59). Globomycin (VI) and antibiotic SF-1902 A5 (VII), which differ only in their β-hydroxy acid component, cause spheroplast formation in growing *E. coli* cells. The effect on wall synthesis has been traced to inhibition of the prolipoprotein-processing enzyme. The observed result is a marked decrease in the amount of bound lipoprotein formed, as seen also with bicyclomycin. This coincides with accumulation of a lipoprotein fraction that can be converted to the normal free form by incubation with cells in the absence of globomycin. Inukai and his colleagues (60) have pointed out that globomycin bears a resemblance in its amino acid sequence to the sequence near the processing site of prolipoprotein.

E. Fungus-Specific Inhibitors

Aspergillus aculeatus produces a group of peptides, the aculeacins, active specifically against fungi (61). Exposure to aculeacin A causes lysis of *Saccharomyces cerevisiae* and electron micrographs show material being released at bud tips. Since the antibiotic inhibits glucan synthesis strongly and its ability to disrupt yeast protoplasts is much weaker than that shown by amphotericin B, the primary effect was considered to be on cell-wall synthesis (62). All six aculeacins contain threonine, hydroxyproline and palmitic acid. The remaining hydrolytic products vary within the group but include two imino acids and a basic amino acid (63). In several respects the aculeacins resemble echinocandin B, an antifungal agent isolated from *Aspergillus nidulans* (64) and, as antibiotic SL 7810, from *Aspergillus rugulosus* (65). However, echinocandin B contains linolenic acid.

II. ANTIBIOTICS AFFECTING MEMBRANE PERMEABILITY

In his survey of recent work on peptide antibiotics, from the genus *Bacillus*, Shoji, puts the number of known substances at well above one hundred (11). They fall into three groups--linear peptides, cyclic peptides and peptidolactones. Where their mode of action has been examined, most have been shown to affect membrane permeability.

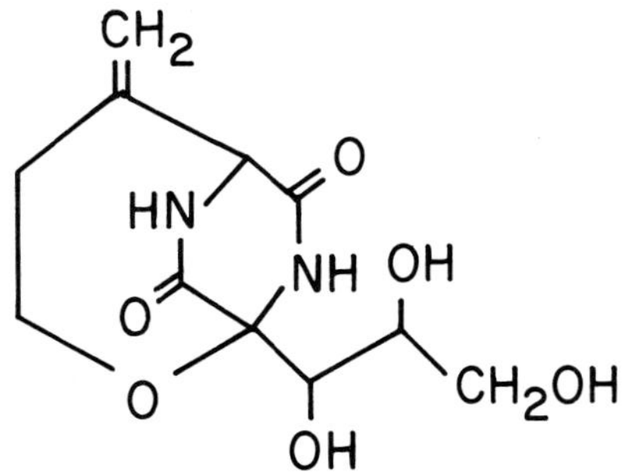

V

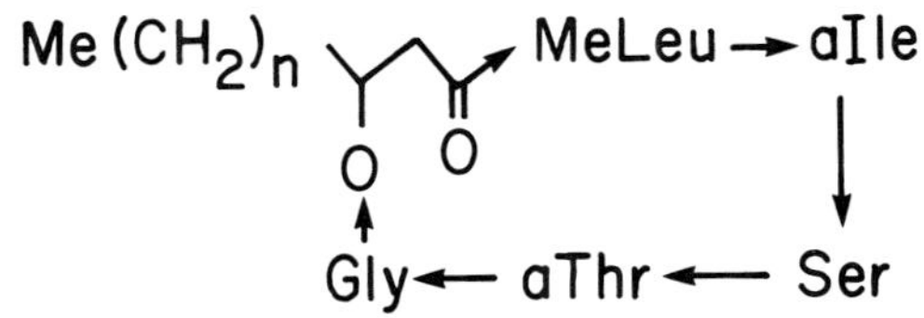

VI, n = 5

VII, n = 7

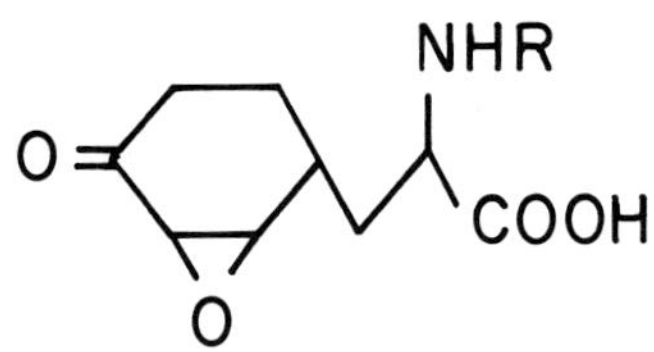

VIII, R = Ala

IX, R = H

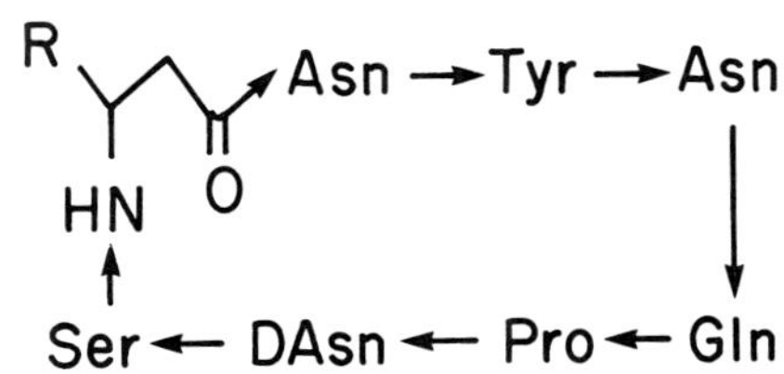

X, R = iso or anteiso $C_{11}H_{23}$

A. *Linear Peptides*

Two new types of linear peptides, the cerexins (66) and tridecaptins (67) have been added in the past few years to the well-known gramicidins and edeines. Bacilysin (VIII) has been found to be identical to tetaine and bacillin (68). Contained within its dipeptide structure is anticapsin (IX), the unusual amino acid antibiotic of *Streptomyces griseoplanus* (69).

B. *Cyclic Peptides*

This group also contains new additions, both to the established family of polymyxins and through recognition of a subgroup, the octapeptins, with eight instead of ten amino acids. The size difference is confined to the linear section (70). The octapeptins disrupt the outer membranes of Gram-negative bacteria. Their effects at minimal inhibitory concentrations resembly those of uncouplers of oxidative phosphorylation and are ascribed to interference with selective ion permeability (5). As a result, octapeptins inhibit the generation of membrane potential (71). Although polymyxins also affect the Gram-negative outer membrane, the mechanism is probably different. Initially, they are believed to form complexes with lipopolysa-charide (72), thereby activating enzymes that disrupt phospholipid layers and destroying the barrier function of the outer membrane (73). While the lethal result may be due to changes in selective ion permeability, polymyxins do not behave live uncouplers of oxidative phosphorylation (72).

A survey of 38 *B. subtilis* strains exhibiting antifungal activity and reexamination of many antibiotics previously reported from this species, has reduced the number of distinctly different compounds to three - iturin A, bacillomycin L and mycostubtilin (74). In the absence of convincing evidence to the contrary, all previously described preparations can be assumed identical to one of these compounds which, moreover, represent a homogeneous peptide class. The unique features, exemplified in iturin A (X), are the presence of a lipophilic β-amino acid and seven or eight D- or L-amino acids in a cyclic structure. Certain sterols and membrane lipid components counteract the inhibitory action of these antibiotics on the growth of *Saccharomyces cerevisiae*, suggesting interference with membrane function as the likely mode of action (75).

C. Peptidolactones

The peptidolactone group of *Bacillus* antibiotics includes two structural types, differing in the manner of forming the lactone ring. In the esperins and surfactin the peptide is cyclized by amide and ester bonds to a C_{13}-C_{15} β-hydroxy acid (11). To this group the polypeptins of *Bacillus circulans* can now be added, though here the hydroxy acid is of shorter chain length. Polypeptin A (XI), the main component, has a C_7 anteiso acid (76). Closely related to the polypeptins are the permetins of *B. circulans*. Polypeptin A and permetin A (XII) differ only in the exchange of serine for threonine.

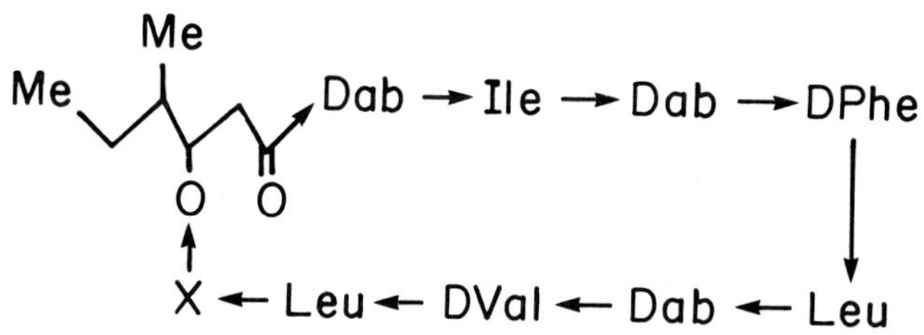

XI , X = Thr

XII , X = Ser

Similar relationships may hold for a number of less well characterized products from *B. circulans,* such as antibiotics 4205 and B-43 (77). All of these substances show activity against both Gram-positive and Gram-negative bacteria. Permetin A inhibits some anaerobic species as well (78).

In the brevistins (XIII) from *Bacillus brevis* (79) and antibiotic TL-119 (antibiotic A-3302-B) from *B. subtilis* (80),

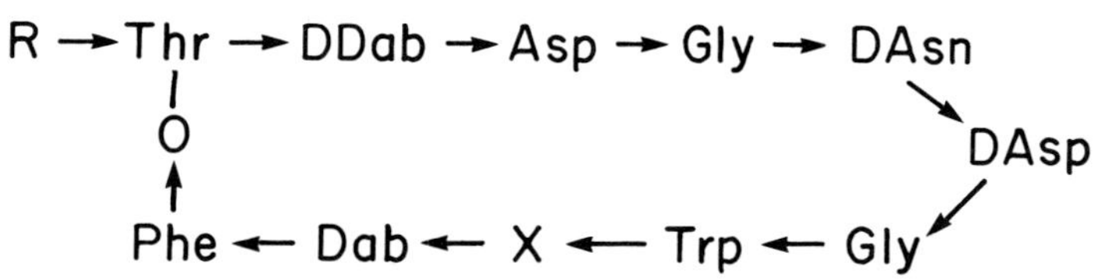

XIII , R = Fatty acid;
X = Val or Ile

the lactone ring structure is generated by an ester bond between the C-terminal amino acid and the hydroxyl group of threonine. The N-terminus is masked by acylation. Brevistins and antibiotic TL-119 inhibit primarily Gram-positive bacteria.

D. Peptaibophols

Reinvestigation of the structure of alamethicin, a membrane-active antibiotic from *Trichoderma viride*, by gas-chromatography-mass spectroscopy and high resolution field-desorption mass spectroscopy has shown that it is a linear peptide masked at the N-terminus by an acetyl group and at the C-terminus by phenylalaninol (81). Alamethicin (XIV) is a mixture of two components, I and II, which differ only in the exchange of an alanine for an α-aminoisobutyric acid residue, respectively. Related closely to the alamethicins in both structure and biological effects are suzukacillins A and B, produced by a different strain of *Trichoderm viride* (82). The two components have been separated by preparative thin-layer chromatography and a tenative structure has been assigned to suzukacillin A from gas chromatographic-mass spectrometric data. Conformational analysis indicated that the N-terminal sequence forms a lipophilic α-helix. The N-terminal region of alamethicin is similar in character but shorter, accounting for differences in behaviour between the two substances in mixed hydrophilic-hydrophobic systems. However, both antibiotics generate pores in cell membranes by aggregating initially in the lipid surface as dimers with their hydrophilic C-termini adjacent (83). Such pores are capable of conducting cations at rates proportional to the potential or chemical gradient across the membrane (84). The action of these antibiotics thus resembles that of the linear gramicidins (85) rather than the ionophorous desipeptide antibiotics that ferry ions through the bilayer (86).

The presence of phenylalaninol and a number of α-aminoisobutyric acid residues is characteristic of a number of peptide antibiotics which Pandey and colleagues (87) suggested by called peptaibophols. Using the techniques successfully applied to the alamethicins, they have established the structures of several additional members. In each of these the amino acids have the L-configuration. The antiamoebins, discovered in *Emericellopsis poonensis* because of their activity against protozoa and helminths and separated by counter-current distribution into fractions I and II, are assigned the revised structures XV and XVI (88). Antiamoebin I has the same amino acid content and chromatographic behaviour as stilbellin, isolated from a *Stilbella species* (89). Very similar peptide antibiotics are produced by *Emericellopsis microspora* when the

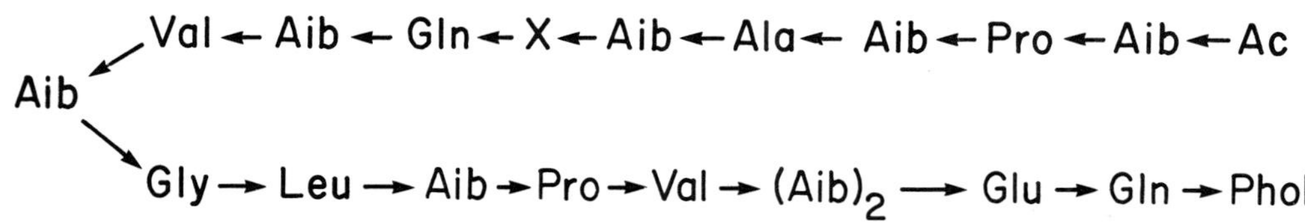

XIV , X = Ala or Aib

Ac → Phe → $(Aib)_3$ → X → Gly → Leu → $(Aib)_2$ → Y → Gln → Iva → Hyp → Z → Phol

	X	Y	Z
XV ,	Iva	Hyp	Aib → Pro
XVI ,	Iva	Pro	Aib → Pro
XVII ,	Val	Hyp	Ala
XVIII ,	Val	Hyp	Aib

medium is supplemented with propylproline (89,90). Emerimicins III (XVII) and IV (XVIII) contain proline as an additional residue adjacent to the C-terminus. The full structure of emerimicin II, which contains several amino acids not present in its congeners, has not been determined.

A third *Emericellopsis* species, *E. salmosynnemata*, produces zervamicins I and II without addition of propylproline to the medium (91). Amino acid analyses indicate that these are also peptaibophols but their structures have not been elucidated. Interestingly, the bioinactive N,O-diacetyl-L-phenylalanyl-L-phenylalaninol is the main product in *E. salmosynnemata* cultures (92). The N-benzoyl-O-acetyl analogue, asperglaucide has been isolated from *Aspergillus glaucus* (93). An additional peptaibophol, samarosporin, has been obtained from a *Samarospora* species. It contains phenylalaninol and six α-aminoisobutyric acid residues, (94).

III. INHIBITORS OF PROTEIN SYNTHESIS

A. *Polythiazole Group*

Thiostrepton has been found in cultures of *Streptomyces laurentii* (95) as well as *Streptomyces azureus* and *Streptomyces sioyaensis* (96) and the thiopeptins from *Streptomyces tateyamensis* (97). The antibiotic complex A 7413 produced by an Actinoplanes probably belongs to this group as well (98). The core structure of thiostrepton, initially determined by X-ray crystallography, has been extended by ^{13}C nmr studies which place an additional dehydroalanine at the amide end of the open-chain segment. From these investigations, also, the relationship between thiostrepton and siomycin A was clarified (96). Additional work using ^{15}N nmr and 270 MHz ^{1}H nmr has confirmed these structures and elucidated that of thiopeptin B_a (99). Other chemical (100) and spectroscopic (101) investigations directed at the thiopeptins have established the structures of eight components in this complex. A striking feature common to all thiopeptins is the presence of a thioamide linkage.

The conformation of these antibiotics in solution is similar to that deduced for thiostrepton from crystallographic analysis. There is considerable structural congruence between them and another polythiazole-containing antibiotic, nosiheptide, from *Streptomyces actuosus* (102). Nosihephide has also been isolated as multhiomycin from *Streptomyces antibioticus* (103). Extensive investigation of this compound by chemical (104), ^{13}C and ^{15}N nmr spectroscopic (105,106) and X-ray crystallographic (107) methods has provided full details of its structure and conformation.

There is also a noteworthy resemblance between these *Streptomyces* antibiotics and a group of polythiazoles in Gram-positive eubacteria. The complete structure of micrococcin P from *Bacillus pumilis* has now been deduced from ^{13}C nmr spectroscopy and extensive chemical investigations (108). The *B. pumilis* product is not homogeneous but the main component is probably the same as the original micrococcin isolated by Su from a *Micrococcus* species (109). Micrococcins M_1 and M_3 have been reported as products of *Micrococcus* and *Staphylococcus* species (110) and three compounds related to but different from micrococcin P have been found in strains of *Bacillus cereus, Bacillus megaterium* and *Bacillus badius*. In view of this rather broad distribution, Shoji and colleagues have proposed the group name thiocillins for these compounds (111).

Althiomycin is a less complex representative of the polythiazole class. Isolated from *Streptomyces chartreusis, Streptomyces althioticus* and *Streptomyces matensis* (as matamycin) its structure was determined independently by three groups (112-114) as (XIX). Bycroft and Pinchin (114) have suggested that it may be a highly modified derivative of the pentapeptide XX.

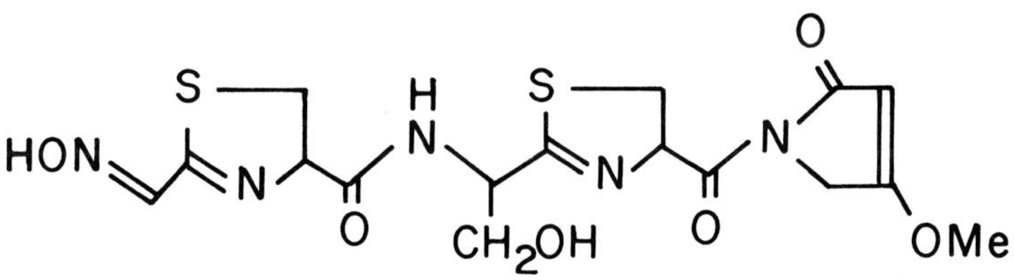

XIX

H-Gly Cys Ser Cys Gly-OH

XX

Berninamycins have probably only a single thiazole ring which forms part of the berninamycinic acid moiety (XXI). The parent antibiotic has been separated into A and B components (115) and the products obtained by acid hydrolysis of A were shown to be berninamycinic acid, threonine, β-hydroxyvaline, pyruvic acid and ammonia. Examination of berninamycin A by nmr identified six component parts of the molecule but could not determine the manner of their assembly (116). The sulfomycins appear to be closely related to the berninamycins. Abe and coworkers (117) have identified berninamycinic acid, threonine and pyruvic acid among the hydrolysis products.

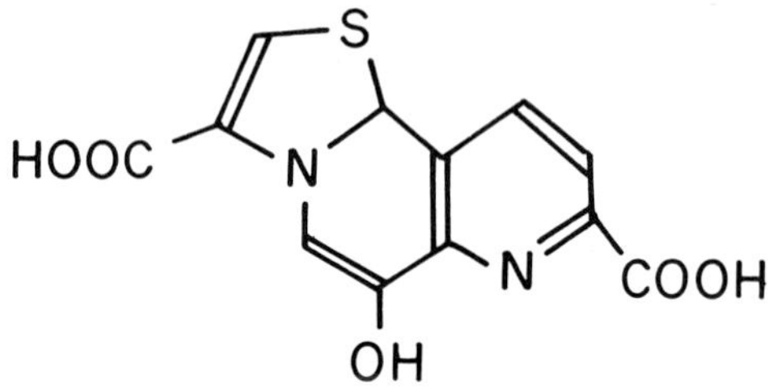

XXI

B. Negamycins

The negamycins form a group of simple dipeptide-like antibiotics in which the C-terminal component is N-methylhydrazinoacetic acid. In common with aminoglycosides, they cause miscoding during protein biosynthesis and they also prevent termination in translation systems directed by phage f2 RNA. However, they bind to different ribosomal sites from streptomycin and show no cross-resistance with aminoglycosides (118). Interference with termination occurs at a hitherto unrecognized step that is specifically blocked by negamycin (119). Negamycin (XXII) and 3-*epi*-deoxynegamycin (XXIII) have been isolated from both *Streptomyces goshikiensis* (120) and a *Micromonospora* species (121). Leucylnegamycin (XXIV) and leucyl-3-*epi*-deoxynegamycin (XXV) are also produced and the former is postulated to be an intermediate in the biosynthesis of negamycin (120,122).

Since the negamycins are low in toxicity, afford protection *in vivo* and inhibit a broad spectrum of bacteria, a variety of analogues and derivatives has been prepared (123, 124). Deoxynegamycin and negamycin O-methyl ether are approximately 50% as active as negamycin but products modified at the β-amino group, in the chain length of the δ-hydroxy-β-lysine component or in the hydrazinoacetic acid component have little or no bioactivity. In a comparison of *in vitro* activity, miscoding generally showed the same requirements as growth inhibition whereas the requirements for inhibition of termination were less rigid. However, inhibition of termination was only about one-tenth as sensitive as miscoding and thus may not be a factor in growth inhibition at low concentrations. Leucylnegamycin was inert, indicating that its activity in whole cells is due to prior hydrolysis.

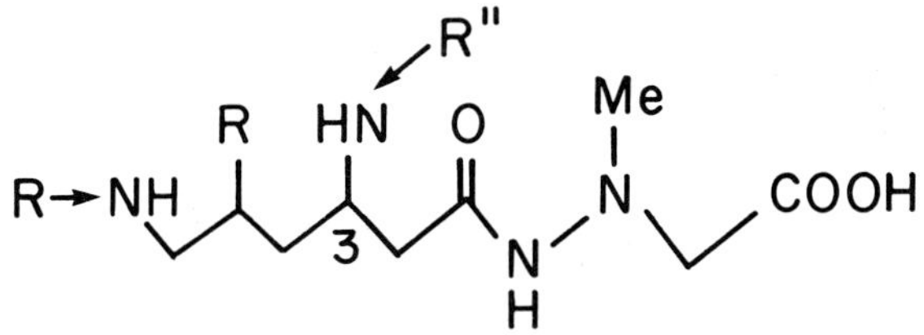

XXII, R = R'' = H, R' = OH; 3-(R)

XXIII, R = R' = R''' = H; 3-(S)

XXIV, R = Leu, R' = R'' = H; 3-(R)

XXV, R = R' = H, R'' = Leu; 3-(S)

C. Virginiamycins

Pairs of modified depsipeptide (type A) and peptidolactone (type B) antibiotics produced by a variety of actinomycetes act synergistically to inhibit protein biosynthesis in prokaryotic organisms. The depsipeptide is known to facilitate binding of the peptidolactone to the ribosome and Cocito (6) suggests that complementary enhancement is due to an ability of the peptidolactone component to lock the depsipeptide into its ribosomal binding site.

The depsipeptide occurs in various modifications of a single structural type. Madumycin I, (antibiotics A2315B and CP36926) has structure (XXVI). Madumycin II, (antibiotics A2315A, A17002F, CP35763 and plauracin) has structure (XXVII). A third structural variant (XXVIII) has been isolated as antibiotic A17002C from *Actinoplanes* species (125). Ostreogrycin G (virginiamycin M_2 and pristinamycin IIB) differs only in having proline in place of the alanine residue of madumycin II, while ostreogrycin A, (virginiamycin M_1, pristinamycin IIA, mikamycin A, streptogramin A, vernamycin A and antibiotic PA 114 A) has 2,3-dehydroproline at this position. The *Streptomyces griseoviridus* metabolite, griseoviridin, has a cysteine residue in place of alanine, with the thiol group cyclized to the adjacent hydroxy acid. Griseoviridin also has different alkyl substituents (126,127).

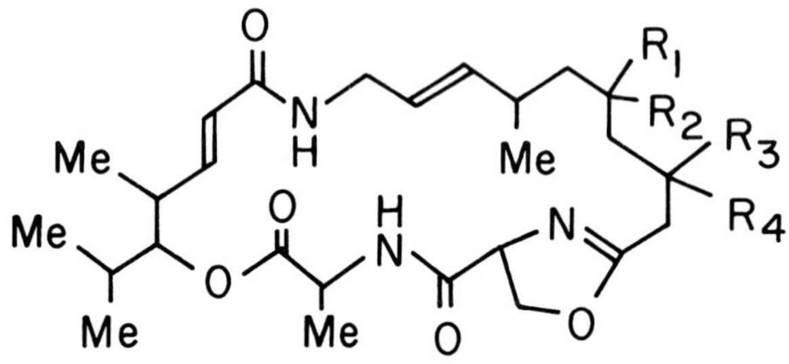

XXVI, $R_1 = H, R_2 = OH, R_3, R_4 = O$

XXVII, $R_1 = R_3 = H, R_2 = R_4 = OH$

XXVIII, $R_1 = R_2 = R_3 = H, R_4 = OH$

The peptidolactone series also consists of a single structural type with the *S. grisoviridus* product, etamycin (viridogrisein), again showing the greatest divergence from the common pattern. All of the 12 remaining compounds can be accomodated within the general structure (XXIX).

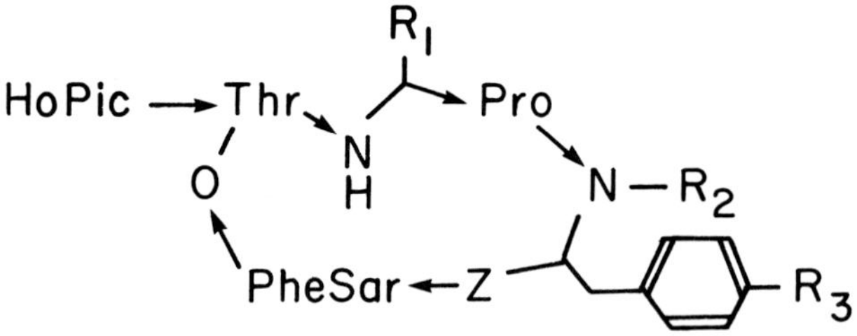

XXIX

Etamycin (XXX), hitherto, the sole representative of its structural type, is now known to undergo the amino acid substitutions usual among peptide antibiotics synthesized by a protein thiotemplate mechanism (128,129). Hydroxyproline is replaced by D-proline to give etamycin B (neoviridogrisein II, XXXI). Under precursor pressure, alanine can be replaced by L-α-amino-n-butyric acid to give neoviridogriseins I (XXXII) and II (XXXIII).

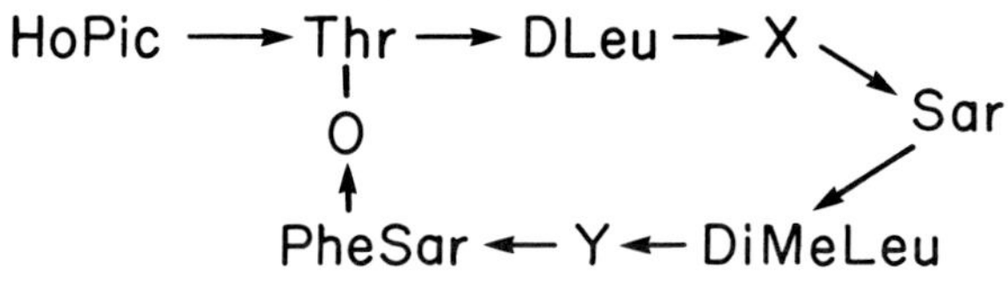

XXX, X = DHoPro, Y = Ala

XXXI, X = DPro, Y = Ala

XXXII, X = DPro, Y = Anb

XXXIII, X = DHoPro, Y = Anb

Streptomyces griseoviridus incorporates L- but no D-leucine into the D-leucine component of etamycin (130). The D-isomer inhibits antibiotic synthesis. *trans*-4-Hydroxy-L-proline, known to be a better precursor than L-proline of the *cis*-4-hydrozy-D-proline, component of etamycin, acquires its budroxyl group in advance of peptide assembly (131). It could be isolated from the free amino acid pool of the organism grown in proline-containing media, even when etamycin was not synthesized. *S. griseoviridus* thus differs from higher plants and animals where hydroxylation of proline follows translation.

IV. INHIBITORS OF DNA SYNTHESIS

Ficellomycin and feldamycin, two metabolites of *Streptomyces ficellus*, inhibit bacteria by interfering with replication of the chromosome. In their presence *E. coli* accumulates 34S DNA molecules that are functionally defective and cannot be incorporated into high molecular weight DNA. Reusser (132) noted a similarity in this respect to the action of nalidixic acid, which inhibits DNA gyrase. Feldamycin (XXXIV) is a dipeptide with feldamycic acid as

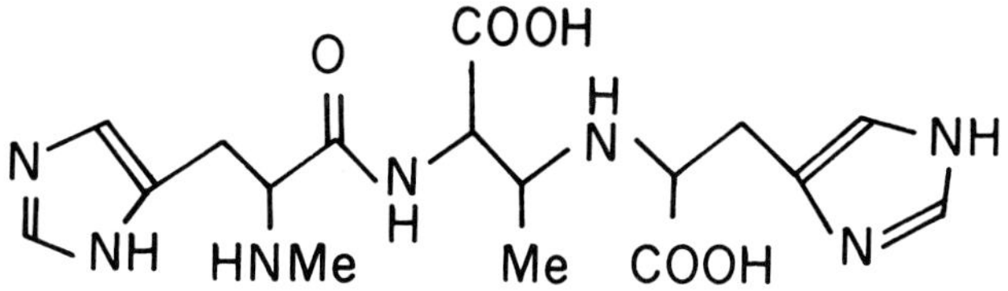

XXXIV

an unusual C-terminal component (133). The structure of ficellomycin is not completely known but the compound appears to be a dipeptide with N-terminal valine (134). Peptides that inhibit DNA synthesis in eukaryotic cells are discussed in the section on antitumour agents.

V. ENZYME INHIBITORS

A series of new peptides has been discovered by screening for inhibitors of proteolytic enzymes. Elastatinal (XXXV), a

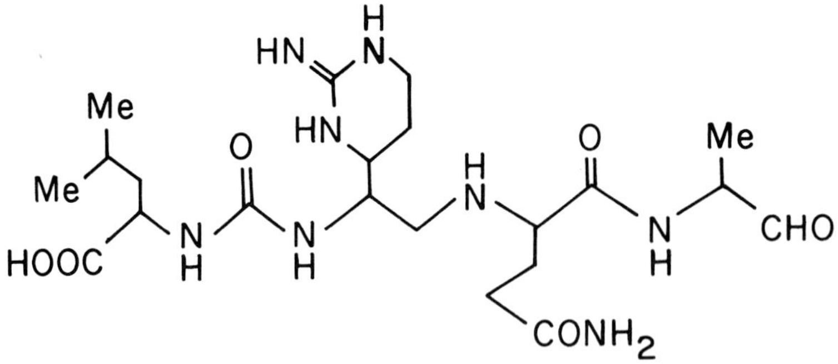

XXXV

specific inhibitor of elastase, is produced by *Streptomyces griseoruber* and other species (135). It resembles antipain, chymostatin and leupeptin in possessing a C-terminal aminoaldehyde matching the substrate specificity of the enzyme inhibited. The corresponding acid is inactive. Antipain (XXXVI), which inhibits papain, trypsin and cathepsins A and B, has been found in cultures of *Planomonospora parontospora* (136) as well as *Streptomyces libani* (137). In the latter it is accompanied by a chymotrypsin-inhibiting analogue, SP chymostatin-B (XXXVII), which has phenylalaninal in place of arginal at its C-terminus. SP chymostatin-B has also been isolated as α-MAPI from *Streptomyces nigrescens* where it occurs with two additional but incompletely characterized inhibitors of alkaline protease activity, β- and γ-MAPI (138,139). The alkaline protease inhibitor AP-1 produced by *Streptomyces pseudogriseolus* appears to be a protein (140).

The structure of an inhibitor designated E-64 isolated from various actinomycetes and specifically active against thiol proteases such as papain and cathepsin B, has been verified as (XXXVIII) by synthesis (141). A relationship between structure and function is also apparent with inhibitors of aminopeptidases. Bestatin (XXXIX) a dipeptide from *Streptomyces olivoreticuli* (142) acts on leucine aminopeptidase and aminopeptidase B whereas amastatin (XL), a tetrapeptide produced by various actinomycetes, inhibits aminopeptidase A which hydrolyses bonds associated with N-terminal aspartic or glutamic acid (143).

Both bestatin and amastatin bind to mammalian cell surfaces where they apparently interfere with the action of exopeptidases. Cells functioning in the immune response are affected since intraperitoneal injection of the inhibitor at the time of immunization markedly increases the reaction to a later challenge. Tests with antitumour agents have shown that bestatin not only provides immunity to a second inoculation of tumour cells but also increases the effectiveness of bleomycin or macromomycin. It has been surmised that destruction of tumour cells by the antitumour agent provides antigen which, in the presence of bestatin, elicits specific antibody synthesis. Because of these effects and its very low toxicity, bestatin shows promise of a useful role in cancer therapy (144).

Antipain and leupeptin have been reported to suppress chemical carcinogenesis, a property which may be related to their ability to prevent mutations (145,146). The likely mechanism suggested by studies with antipain is suppression of error-prone SOS repair of DNA in bacteria due, in turn, to inhibition of proteolytic inactivation of repressor (147). Pepstatin, a potent inhibitor of pepsin, also causes a decrease in ascites formation and may be useful in delaying the buildup of ascites cells in cancer patients (148).

Biosynthetic studies with the leupeptins suggest that these small peptides may be made by a route like that used for glutathione. The acetyl, leucyl and arginal components originate

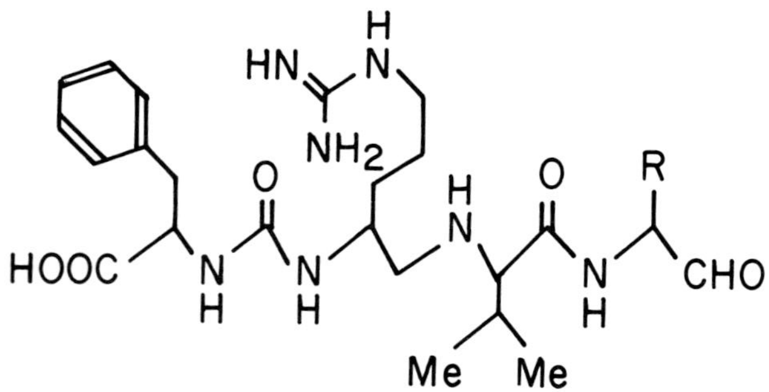

XXXVI , R = guanidinopropyl

XXXVII , R = benzyl

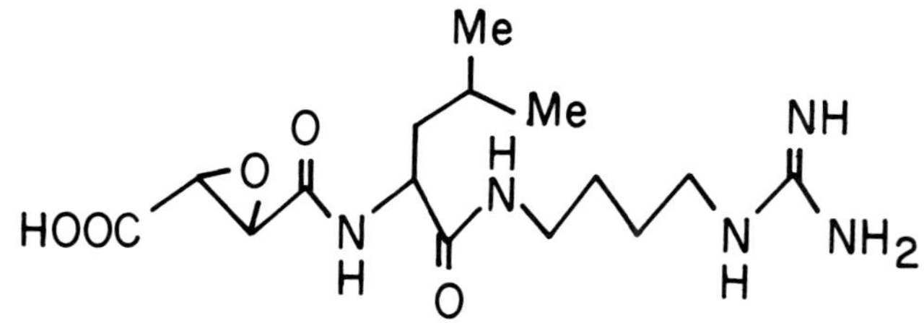

XXXVIII

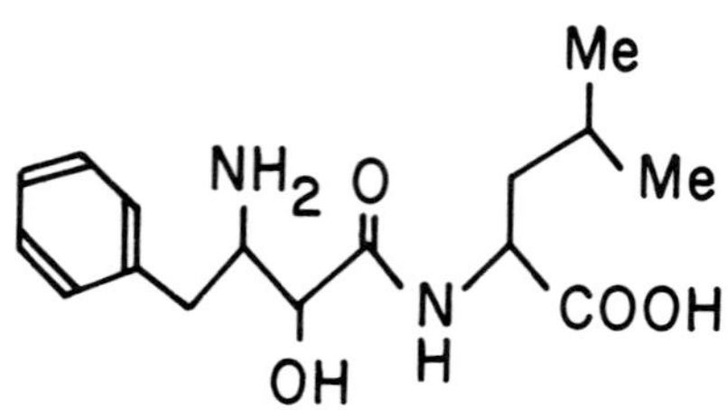

XXXIX

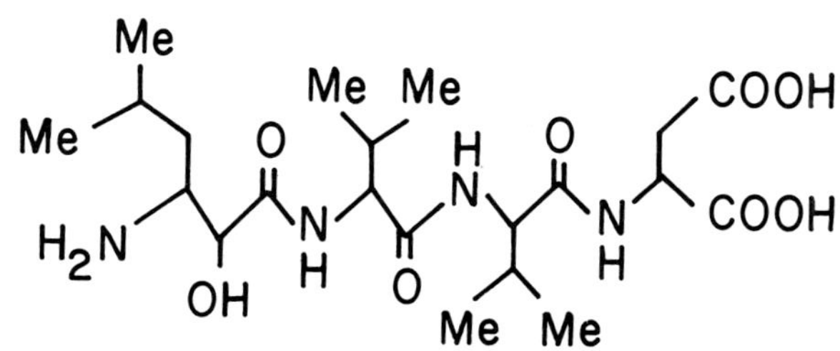

XL

from acetate and the appropriate amino acids. With intact cells of *Streptomyces roseus*, 1-^{14}C)acetyl-L-(G-^{3}H)leucine and (1-^{14}C)acetyl-L-leucyl-L-(G-^{3}H)leucine suffered at least partial hydrolysis during incorporation. However, with cell extracts (^{14}C)arginine incorporation increased progressively in incubation mixtures containing acetate and leucine, N-acetylleucine and leucine, and N-acetylleucylleucine (149). Such evidence of stepwise assembly and equilibration of intermediates with exogenously supplied precursors is incompatible with ribosomal or protein thiotemplate mechanisms of peptide assembly (13). In the *in vitro* experiments the main product was leupeptin acid. Small amounts of leupeptin were detected in fresh extracts, suggesting that an unstable enzyme may reduce the C-terminal carboxyl as the last step in the pathway (149).

The ability to produce leupeptins is found in 17 species of *Streptomyces* and there is evidence that genes coding for this activity are located on a plasmid. Umezawa and coworkers (150) observed increased loss of leupeptin production among strains exposed to acriflavine and also showed with auxotrophically marked strains that these deficient cultures could regain the ability to produce leupeptins when grown in mixed culture with competent strains.

VI. ANTIMETABOLITES

Although many antibiotics might be included in this category, only substances causing a metabolic inhibition which can be reversed by adding an essential primary intermediate are considered here. Most such antimetabolites found in microbial cultures are analogues of amino acids or nucleic acid bases. However, a few are peptides. The plumbemycins (XLI) from *Streptomyces plumbeus* represent interesting new phosphonopeptide antagonists of L-threonine (151). (S)-Alanyl-3-(α(S)-chloro-3-(S)-hydroxy-2-oxo-3-azetidinylmethyl)-(S)-alanine, from an unidentified *Streptomyces* species, interferes with glutamine metabolism (30). In this respect it resembles the tabtoxins (XLII) from *Pseudomonas* species (152) and L-(N-phosphono)methionine-(S)-sulfoximinyl-L-alanyl-L-alanine. Pruess and coworkers (153) noted that L-N-phosphonomethionine alone is a good inhibitor of glutamine synthetase *in vitro* and suggested that the additional alanyl residues merely facilitate uptake, presumably allowing the inhibitory component to enter the cells by an appropriate permease.

A similar explanation has been put forward for the antibiotic action of bacilysin (VIII). The dipeptide, after being transported into the cells, is hydrolysed to anticapsin which strongly inhibits glucosamine synthetase (154). Anticapsin itself is only weakly antibacterial. Diddens and coworkers (155) have demonstrated that bacilysin, (S)-alanyl-3-(α-(S)-chloro-3-(S)-hydroxy-2-oxo-3-asetidinylmethyl)-(S)-alanine and 1-(S)-hydroxy-2-(S,S)-valylamidocyclobutane-1-acetic acid are transported into susceptible bacteria by peptide permeases. The last-named substance is a growth inhibitor produced by *Streptomyces* sp. X1092. Since the inhibition is reversed by cysteine or methionine, it is believed to interfere in the biosynthesis of sulfur-containing amino acids (156).

The tripeptide antibiotics plumbemycin B, L-phosphinothricylalanylalanine and L-(N^5-phosphono)methionine-(S)-sulfoximinylalanylalanine are transported into *E. coli* cells by the action of an oligopeptide permease (157). Dipeptides can also enter by this route should a dipeptide permease be lacking.

VII. TOXINS

A. *Insect Toxins*

Insect death after infection by *Beauveria bassiana* is probably caused by a toxic cyclodepsipeptide produced by the fungus (158). Bassianolide (XLIII), also made by *Verticillium lecani*, elicits atonic symptoms in silkworm larvae at doses of 2 μg per larva. Although the mode of action has not yet been established, it is noteworthy that bassianolide is a ring homologue of enniatin C with three instead of four peptolide units and that the enniatins are ionophones (159). Besides bassianolide, *B. bassiana* produces the nontoxic cyclodepsipeptide beauvericin in which N-methylphenylalanines replace the N-methylleucine residues (160). Three peptidolactones, beauverilide A (161) and beauverolides H and I have been isolated from different strains of the fungus (162). A fourth product, beauvellide, is a mixture of the two beauverolides (163). These peptidolactones are closely related in structure to the isarolides which are produced by an *Isaria* species isolated from an insect larva (164).

Another entomopathogen, *Metarrhizium anisopliae*, which produces the toxic destruxin group of peptidolactones (XLIV and XLV), also produces the non-toxic protodestruxin (XLVI) which lacks both N-methyl substituents. The structures of these metabolites have been confirmed by synthesis (165).

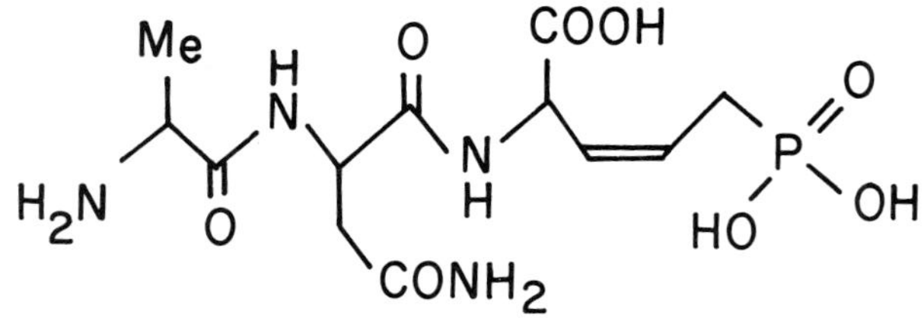

XLI , R = OH or NH_2

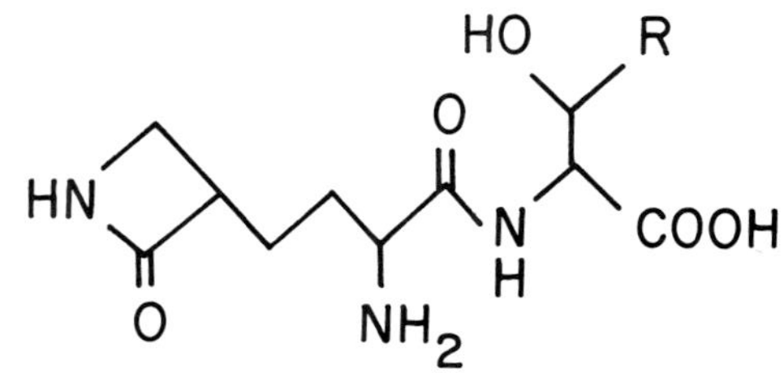

XLII , R = H or Me

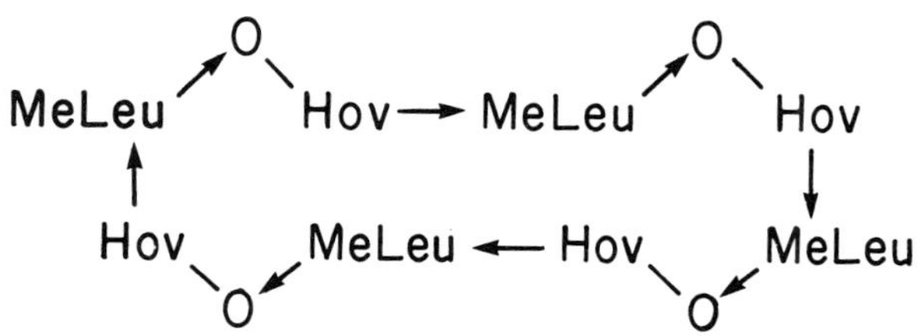

XLIII

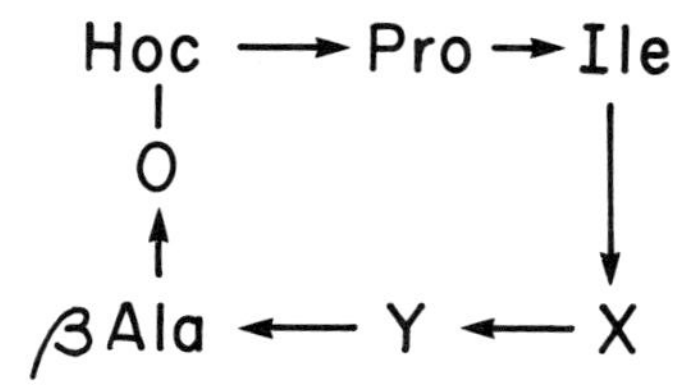

XLIV , X = MeVal , Y = MeAla

XLV , X = Val , Y = MeAla

XLVI , X = Val , Y = Ala

B. Plant Toxins

Alternaria mali produces a group of peptidolactones analogous to the destruxins in containing a small α-hydroxy acid but constituted from fewer amino acids. These AM-toxins, of which AM-1 was independently isolated as alternariolide, are responsible for the brown necrotic spots in infected apples. Several have been synthesized by Lee and coworkers (166).

Toxic peptides are implicated in halo blight of beans and other plant diseases. Infection with pathogenic strains of *Pesudomonas phaseolicola* causes chlorotic regions and an associated accumulation of ornithine. Inhibitors of ornithine transcarbamylase that reproduce the chlorotic response have been isolated from cultures of the bacterium. One is reported to be N-phosphoglutamic acid; a peptide, phaseotoxin, is described as causing chlorosis and inhibition of ornithine transcarbamylase but not accumulation of ornithine *in vivo* (167). However, two recently isolated N-phosphorylated peptides, phaseolotoxin (XLVII) and (2-serine)-phaseolotoxin (XLVIII)

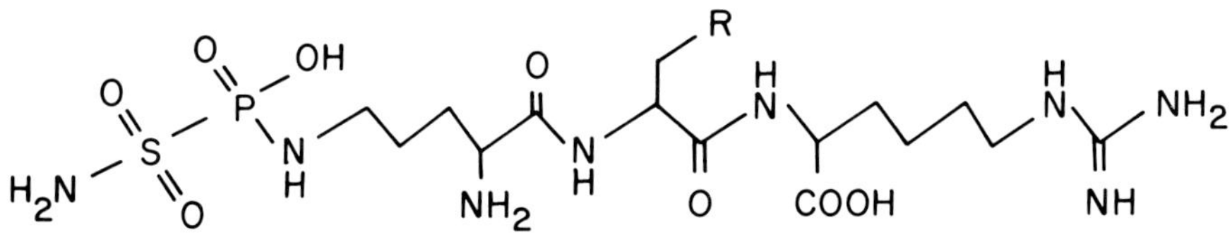

XLVII , R = H

XLVIII , R = OH

reproduce all of the symptoms of *P. phaseolicola* infection (168). These are postulated to act *in vivo* after hydrolysis to N^5-phosphosulfamylornithine since this product is formed rapidly from the toxins in plant tissue (169). Earlier studies on tobacco wildfire toxin from *Pseudomonas tabaci* indicated that tabtoxin and (2-serine)-tabtoxin are similarly involved in the disease symptoms (170). Syringomycin from *Pseudomonas syringae* and the toxins of *Pseudomonas coronofaciens, Pseudomonas glycinea* and *Pseudomonas tomato* also appear to be peptides (171).

C. *Amanita Toxins*

A recent examination of β-amanitin by X-ray crystallography has verified the structure deduced by Wieland and coworkers from chemical studies (172). It also provides important information on the stereo-chemistry of these toxic *Amanita* metabolites. The minor peptides amanullin, proamanullin and amanullic acid have recently been purified from *A. phalloides* and their structures deduced. These substances, though closely related to the toxic and predominant phalloidins and amanitins, are all non-toxic (173).

VIII. ANTITUMOUR AGENTS

A. *Bleomycins*

Tanaka (174) recently reviewed developments in the application of bleomycins as antitumour agents. A large number of analogues has been prepared, either through biosynthetic steering of the fermentation with different side-chain precursors or by chemically introducing new side chains after enzymically removing the natural side chain. A few bleomycins have been obtained by chemical modification of a side-chain amino group after protecting sensitive amino groups elsewhere in the antibiotic. These efforts have yielded products with superior thereapeutic potential. The structures of the bleomycins and phleomycins, based mainly on chemical studies, have been verified by ^{13}C and ^{15}N nmr investigations (175,176). In addition, new structural variants have been discovered from further screening of actinomycetes. The tallysomycins have L-β-lysine and 4-amino-4,6-dideoxy-L-talose as additional components (177), while the victomycins (178), platomycins (179), and antibiotic SS 70 (180), SF 1771 (181 and YA 56 (182) complexes possess other distinguishing features.

A major action of bleomycin *in vivo* is the degradation of DNA by strand scission (183). The same activity *in vitro* is enhanced by ferrous ions and oxygen (184). Recent investigations of the 3-dimensional structure of bleomycin have provided considerable insight into the mechanism by which this occurs. Crystallographic analysis of the copper complex of peptide P-3A, a metabolite isolated from *Streptomyces verticillus* as a putative intermediate in bleomycin biosynthesis, has shown it to have square pyramidal coordination geometry (185). The aprotonated nitrogen of histidine, N-1 of the pyrimidine, N^{π} and the secondary amine nitrogen form the square and the primary amine is at the apex. The same coordination geometry (XLIX) has been deduced for bleomycin itself with

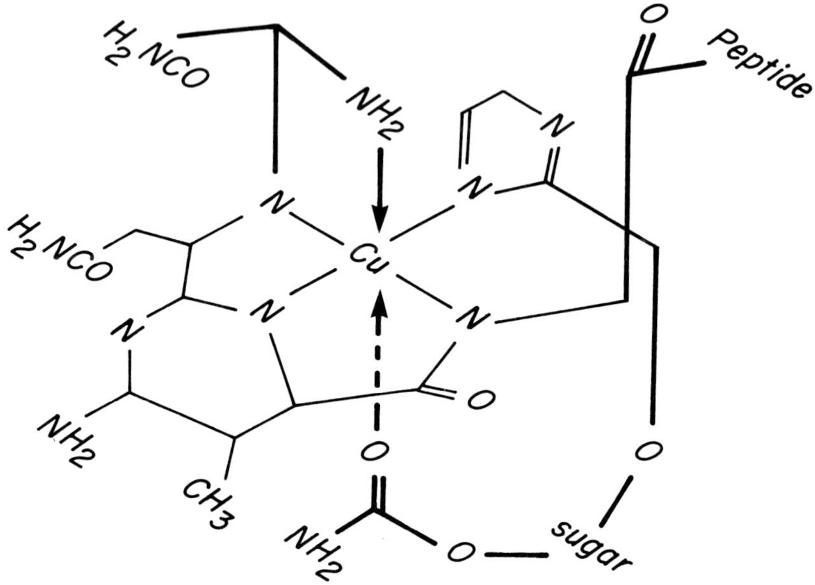

XLIX

a weak bond to the carbamyl oxygen representing the sixth site. Scission of DNA to acid-soluble products by bleomycin occurs only when the complex is formed with ferrous ions. That this complex is oxygen-labile is indicated from ^{13}C nmr evidence of spontaneous decomposition during which the substituent at C-2 of the pyrimidine is eliminated. When DNA is bound to the complex it also becomes a target for the activated oxygen (185,186).

The formation of superoxide and hydroxy radicals in ferrous ion-bleomycin complexes exposed to oxygen has been established by spin-trapping with N-*tert*-butyl-α-phenylnitrone. The ESR examination showed an initial transient and then a stable ferric species which underwent reversible oxidation and reduction by a reducing system to produce $O_2^-\cdot$ and $\cdot OH$ free radicals (187). Although only the ferrous ion complex generates free radicals in the presence of oxygen, similar complexes form between bleomycin and zinc, copper or cobalt. The cobaltous ion-bleomycin complex has been shown by ESR to have pentacoordinated square-pyramidal geometry with oxygen occupying the vacant sixth coordination site in the oxygen adduct (188). Changes in the ESR spectrum when DNA is introduced show that the oxygen molecule shifts orientation relative to the coordination plane of cobalt. This is accounted for by the binding of DNA to the dithiazole and terminal amine moieties, confirming the sequence-specific association indicated by studies on the structural features required for binding and strand scission (186,189,190).

The activity of bleomycin in tumour cells is inversely proportional to the activity of a hydrolase that removes the amide nitrogen at the fifth coordination site (191). This, in turn, affects the coordination and subsequent activation of oxygen at the sixth site, so that free radical production is reduced to about 1%. In agreement with the postulated mode of action of bleomycin the deamido-derivative causes only 1% as much strand scission as its parent. A substantial loss of biological activity also occurs when the methine carbon of the β-lactam ring epimerizes to the R-configuration (192). In cancer chemotherapy with bleomycin one of the adverse drug effects is severe pulmonary fibrosis. A plausible cause is the stimulation of prolyl hydroxylase activity by bleomycin (193). Since the cupric ion complex is inactive and free-radical oxygen is not involved, Takeda and coworkers (194) suggested that the action on prolyl hydroxylase might be due to sequestration of ferrous ions. Comparisons using various semi-synthetic bleomycins and bleomycin hydrolytic fragments indicated that DNA strand scission has the more rigid structural requirements. Stimulation of prolyl hydroxylase needs only those elements of structure that chelate ferrous ions.

B. Macromolecular Peptides

Many of the peptides that continue to be added to the group of microbial metabolites with antitumour activity have the size and characteristics of proteins (195). They invariably affect DNA or DNA synthesis. Neocarzinostatin, with 109 amino acids in a single polypeptide chain cross-linked by two disulfide bonds, has been sequenced (196). It shows considerable homology with actinoxanthin, a polypeptide that contains 107 amino acids and, like neocarzinostatin, inhibits Gram-positive bacteria as well as tumours (197). In addition to the polypeptide moiety neocarzinostatin contains a naphthalenic component that can be removed by mild acid extraction (198). The biological activity resides in this chromophone but is markedly enhanced by the polypeptide component. Neocarzinostatin causes single-strand nicking of DNA both *in vitro* and *in vivo* (199); as with bleomycin, the scission is sequence specific and is probably due to localized generation of free radicals (187). It has also been postulated that neocarzinostatin binds at receptor sites on the cell surface and elicits an effect on microtubules (200). The rigid conformation of neocarzinostatin limits proteolytic attack but inactivation occurs at a measurable rate in blood. It is reduced by succinylation which acylates the -amino function of Lys-20 and evidently masks the main site of proteolyses. Co-administration of proteolytic inhibitors such as pepstatin, antipain or leupeptin also prolongs blood levels (201).

Streptomyces macromyceticus produces a yellow polypeptide, auromomycin, with activity against Gram-positive and Gram-negative bacteria as well as tumours (202). Chromatography on polystyrene resin removes a chromophore, along with the activity against Gram-negative bacteria, to give the more stable and less toxic macromomycin. This, like automomycin, has a molecular weight near 12,500 daltons and lacks methionine and arginine. Macromomycin has common structural features with neocarzinostatin but has the advantage of not losing activity in serum. Both auromomycin and macromomycin inhibit DNA synthesis and cause strand breakage, as does sporamycin from *Streptosporangium pseudovulgare* (203). The latter has a molecular weight of 8500-9000 daltons, but contains only 12 different amino acids. Antibiotic FCRC 53 from *Streptomyces griseus* is a phoshochromopeptide of about 5000 daltons, with some resemblance to prunacetin A (204). Actinocarcin has a molecular weight closer to 11,000 daltons. It contains a full complement of protein amino acids but high proportions of alanine and valine (205). Two other *Streptomyces* antitumour agents, renastacarcin (206) and macracidmycin (207), are larger (approximately 35,000 daltons). Along with actinocarcin they have no antibiotic activity.

IX. SEX FACTORS

Substance I_A (L), a sexual agglutination factor participating in the life cycle of *Saccharomyces cerevisiae,* is produced by the α-type of strain T22 (208). α-Substance I of strain H 15 has been resolved into two factors, I_B (LI) and I_C (209). *Rhodospirillium toruloides* type A cells secrete a substance, rhodotorucine A, which induces mating tube formation in type-a cells of this yeast (210). Rhodotorucine A has been characterized as a linear oligopeptide with C-terminal cysteine attached by thioether linkage to a farnesyl chain (LII). In *Tremella mesenterica,* a jelly fungus of the Heterobasidiomycetidae, formation of conjugation tubes is controlled by sex hormones. One such substance, tremerogen A-10, which induces germ tubes in mating type a is a dodecapeptide (LIII) with a modified C-terminal cysteine, possibly also associated with an isoprenoid group (211).

H–His Trp Leu Glx Leu Lys Pro Gly Glx Pro Met Tyr–OH

I

H–Arg Gly Pro Phe Pro Ile

II

H–Tyr Pro Glu Ile Ser Trp Thr Arg Asn Gly Cys–OH

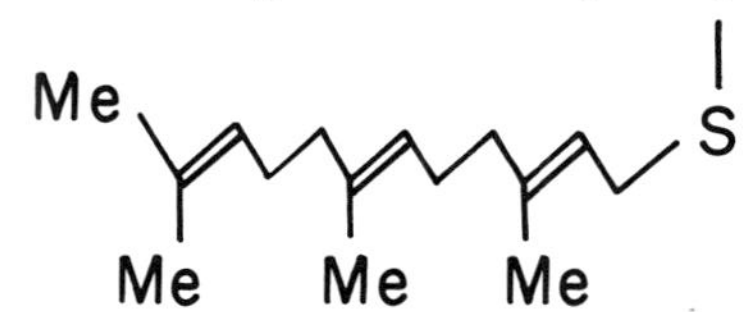

III

H–Glu His Asp Pro Ser Ala Pro Gly Asn Gly Tyr Cys–X
|
R

IIII, R = possible isoprene, X = unknown

X. OTHER PEPTIDES

Along with glutamic acid, *Corynebacterium glutamicum* accumulates a family of γ-glutamyl di- and tri-peptides that reduce the yield of product crystallizing from the fermentation. These are formed by the action of γ-glutamyltransferase acting in reverse of its usual hydrolytic role (212). The more remarkable lentinic acid (LIV) has now been recognized in the

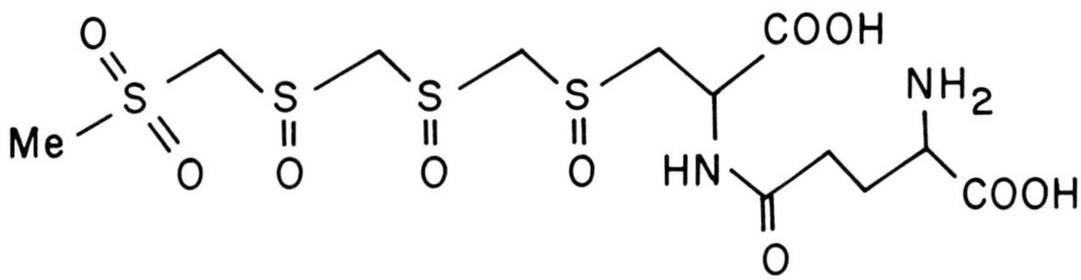

LIV

basidiomycetes *Micromphale perforans, Collybia hariolorum* as well as in *Lentinus edodes* (213). The variety of diketopiperazines has been extended by the addition of cairomycin B (214). Many diketopiperazines found in fermentations are not microbial products but are introduced in the medium. Sakamura and coworkers (215) report that the roasted malts used for black beer fermentations contain bitter prolydiketopiperazines whereas caramel malts used for lager brews do not. Two epidithiadiketopiperazines with antibacterial activity, epicorazines A and B, have been isolated from *Epicoccum nigrum* (216); an antifungal representative, A 30641, is produced by *Aspergillus tamari* (217).

The pheganomycins are a group of tetrapeptide antimycobacterial agents which are extended at the carboxyl end in individual compounds by one or more amino acids (218). An N-terminal phenylguanidinoacetic acid residue carries either a p-methyl or p-hydroxymethyl substituent. The unusual trimethylalanine component in the pheganomycins is also found in bottromycins. A revised structure (LV) based on information from chemical ionization-mass spectrometry is proposed for bottromycin A_2 (219). A key result was the identification of dipeptide and cyclic tetrapeptide fragments from partial acid hydrolysis. The absence of a free C-terminus in the tetrapeptide is accounted for by reaction between the carboxyl group and a ring amidine nitrogen. Bottromycin B_2 has proline

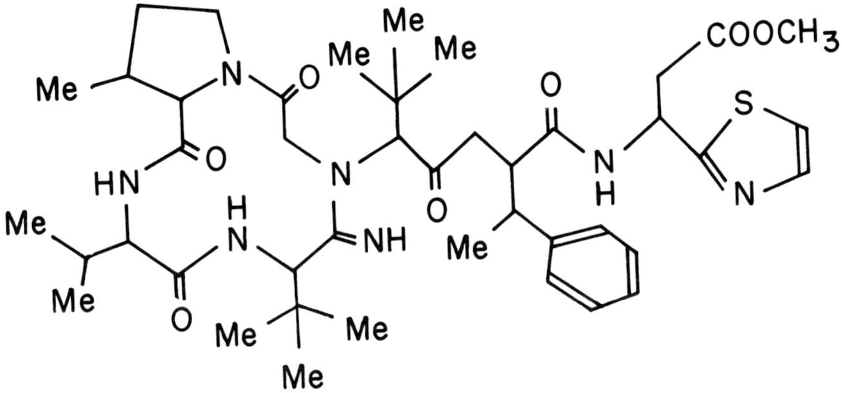

LV

in place of the 3-methylproline residue but the structures of bottromycins A_1,B_1 and C_2 will need to be reexamined. *Trichoderma polysporum* produces a group of related cyclic undecapeptides with antifungal activity. A single amino acid replacement distinguishes cyclosporins B,C,D and E from the main component, cyclosporin A (220).

Polylysine with a chain length of about 25 residues accumulates in cultures of a *Streptomyces* species (221). A number of other peptides with more complex but incompletely known structures have been isolated in screening for antibiotics. These include the galantins from *Bacillus pulvifaciens* (222), antibiotic 339-29 from *Bacillus pumilis* (223) and the 61-26 complex from an unnamed *Bacillus* species (224) as well as antibiotics S-19 (225), KM-8 (226), A-10947 (227), AM-2504 (228), MC 902 I and MC 902 I' (229) from various actinomycetes.

REFERENCES

1. Bricas, E. and Fromageot, C., *Adv. Prot. Chem. 8*, 1 (1953).
2. Studer, R.O., *Prog. Med. Chem. 5*, 1 (1967).
3. Bodanszky, M. and Perleman, D., *Science 163*, 352 (1969).

4. Wright, J.L.C. and Vining, L.C., in "The Filamentous Fungi, Volume II. Biosynthesis and Metabolism" (J.E. Smith and D.R. Berry, Eds.) p. 475, Edward Arnold Ltd., London, (1976).
5. Storm, D.R., Rosenthal, K.S. and Swanson, P.E., *Ann. Rev. Biochem. 46,* 723 (1977).
6. Cocito, C., *Microbiol. Rev. 43,* 145 (1979).
7. Gorman, M. and Huber, F., *Ann. Rep. Fermentation Processes 1,* 327 (1977); *2,* 203 (1978).
8. Abraham, E.P., *J. Antibiot. 30,* S-1 (1977).
9. Nash III, C.H., *Ann. Rep. Fermentation Processes 3,* . . . (1979).
10. Katz, E. and Demain, A.L., *Bacterial. Rev. 41,* 449 (1977).
11. Shoji, J., *Adv. Appl. Microbiol. 24,* 187 (1978).
12. Vining, L.C. and Wright, J.L.C., in "Biosynthesis" (J.D. Bu'Lock, Senior Reporter) Vol. 5 p. 240. Specialist Periodical Reports, The Chemical Society, London, 1977.
13. Zimmer, T.L., Frøyshov, Ø. and Laland, S.C., in "Economic Microbiology, Volume 3. Secondary Products of Metabolism" (A.H. Rose, Ed.) p. 124. Academic Press, London, (1979).
14. Nagarajan, R., Boeck, L.D., Gorman, M., Hamill, R.L., Higgens, C.E., Hoehn, M.M., Stark, W.M. and Whitney, J.G., *J. Am. Chem. Soc. 93,* 2308 (1971).
15. Aoki, H., Sakai, H., Kohsaka, M., Konomi, T., Hosoda, J., Kubochi, Y., Iguchi, E., and Imanaka, H., *J. Antibiot. 29,* 492 (1976).
16. Reading, C. and Cole, M., *Antimicrob. Agents Chemoth. 11,* 852 (1977).
17. Butterworth, D., Cole, M., Hanscomb, G. and Rolinson, G.N., *J. Antibiot. 32,* 287 (1979).
18. Umezawa, H., Mitsuhashi, S., Hamada, M., Iyobe, S., Takahashi, S., Utahara, R., Osata, Y., Yamazaki, S., Ogawara, H. and Maeda, K., *J. Antibiot. 26,* 51 (1973).
19. Kahan, J.S., Kahan, F.M., Goegelman, R., Currie, S.A., Jackson, M., Stapley, E.P., Miller, T.W., Miller, A.K., Hendlin, D., Mochales, S., Hernandes, S., Woodruff, H.B. and Birnbaum, J., *J. Antibiot. 32,* 1 (1979).
20. Albers-Shonberg, G., Arison, B.H., Hensens, O.D., Hirshfield, J., Hoogsteen, K., Kaczka, E.A., Rhodes, R.E., Kahan, J.S., Kahan, F.M., Ratcliffe, R.W., Walton, E., Ruswirkle, R.B., Morin, R.B. and Christensen, B.G., *J. Am. Chem. Soc. 100,* 6491 (1978).
21. Cassidy, P.J., Stapley, E.O., Geogelman, R.T., Miller, T.W., Arison, B.H., Albers-Schonberg, G., Zimmerman, S.P. and Birnbaum, J. Abs. 81, *17th Intersc. Conf. Antimicr. Agents Chemoth.* New York, (1977).
22. Brown, A.G., Corbett, D.F., Eglington, A.J. and Howarth, T.T., *J.C.S. Chem. Comm.* p. 523 (1977).

23. Corbett, D.F., Eglington, A.J. and Howarth, T.T., *J.C.S. Chem. Comm.* p. 953 (1977).
24. Brown, A.G., Corbett, D.F., Eglington, A.J. and Howarth, T.T., *J. Antibiot. 32,* 961 (1979).
25. Okamura, K., Hirata, S., Okamura, Y., Fukagawa, Y., Shimauchi, Y., Kouno, K. and Ishikura, T., *J. Antibiot. 31,* 480 (1978).
26. Maeda, K., Takahaski, S., Sezaki, M., Iinuma, K., Naganawa, H., Kondo, S., Ohno, M. and Umezawa, H., *J. Antibiot. 30,* 770 (1977).
27. Brown, A.G., Butterworth, D., Cole, M., Hanscomb, G., Hood, J.D., Reading, C. and Rolinson, G.M., *J. Antibiot. 29,* 668 (1976).
28. Umezawa, H., *Pure Appl. Chem. 28,* 665 (1971).
29. Leisinger, T. and Margraff, R., *Microbiol. Rev. 43,* 422 (1979).
30. Scannell, J.P., Preuss, D.L., Blount, J.F., Ax, H.A., Kellett, M., Weiss, F., Demney, T.C., Williams, T.H. and Stempel, A., *J. Antibiot. 28,* 1 (1975).
31. Kikuchi, T. and Uyeo, S., *Chem. Pharm. Bull. 15,* 549 (1967).
32. Elson, S.W., and Oliver, R.S., *J. Antibiot. 31,* 586 (1978).
33. Arnstein, M.R.V., Artman, M., Morris, D. and Toms, E.J., *Biochem. J. 76,* 353 (1960).
34. Loder, P.B. and Abraham, E.P., *Biochem. J. 123,* 177 (1977).
35. Elander, R.P., *Rev. Ind. Microbiol. 16,* 355 (1975).
36. Eriquez, L.A. and Pisano, M.A., *Antimicrob. Ag. Chemother. 16,* 392 (1979).
37. Shirafugi, H., Fujisawa, Y., Kida, M., Kanzaki, T. and Yoneda, M., *Agric. Biol. Chem. 43,* 155 (1979).
38. Hayashi, T., Harada, Y. and Ando, K., *J. Antibiot. 28,* 503 (1975).
39. Omura, S., Tanaka, H., Tanaka, Y., Spiri-Nakagawa, P., Oiwa, R., Takahashi, Y., Maysuyama, K. and Iwai, Y., *J. Antibiot. 32,* 985 (1979).
40. Spiri-Nakagawa, P., Tanaka, Y., Oiwa, R., Tanaka, H. and Omura, S., *J. Antibiot. 32,* 995 (1979).
41. Neszmelyi, A., Sztaricskai, F., Liptak, A. and Bognar, R., *J. Antibiot. 31,* 974 (1978).
42. Sztaricskai, F., Harris, C.M. and Harris, T.M., *J. Antibiot. 32,* 446 (1979).
43. Nieto, M. and Perkins, H.R., *Biochem. J. 124,* 845 (1971).
44. Smith, G.A., Smith, K.A. and Williams, D.H., *J.C.S. Perkin I* 2108 (1975).
45. Hlavka, J.J., Betha, P., Boothe, J.H. and Morton, G., *Tetrahedron Lett.* p. 175 (1974).
46. Tamura, A. and Takeda, I., *J. Antibiot. 28,* 395 (1975).

47. Meyers, E., Weisenborn, F.L., Pansy, F.E., Slusarchyk, D.S., von Saltza, M.H., Rathnum, M.L. and Parker, W.L., *J. Antibiot. 23,* 502 (1970).
48. Horii, S. and Kameda, Y., *J. Antibiot. 21,* 665 (1968).
49. Matsuhashi, M., Ohara, I. and Yoshiyama, Y., *Agric. Biol. Chem. 33,* 134 (1969).
50. Bodanszky, M., Sigler, F.G. and Bodanszky, A., *J. Am. Chem. Soc. 95,* 2352 (1973).
51. Tanaka, H., Iwai, Y., Oiwa, R., Shinohara, S., Shimizu, S., Oka, T. and Omura, S., *Biochem. Biophys. Acta. 497,* 633 (1977).
52. Somma, S., Merati, W. and Parenti, F., *Antimicrob. Agents Chemoth. 11,* 396 (1977).
53. Someya, A., Iseki, M. and Tanaka, N., *J. Antibiot. 31,* 712 (1978).
54. Someya, A., Iseki, M. and Tanaka, N., *J. Antibiot. 32,* 402 (1979).
55. Muller, B.W., Zak, O., Kump, W., Tosch, W. and Wacker, O., *J. Antibiot. 32,* 689 (1979).
56. Someya, A., Tanaka, K. and Tanaka, N., *Antimicrob. Ag. Chemother. 16,* 84 (1979).
57. Tanaka, N., Iseki, M., Miyoshi, T., Aoki, H. and Imanaka, H., *J. Antibiot. 29,* 155 (1976).
58. Nakajima, M., Inukai, M., Haneishi, T., Terehara, A., Arai, M., Kinoshita, T. and Tamura, C., *J. Antibiotics 31,* 426 (1978).
59. Omoto, S., Suzuki, H. and Inouye, S., *J. Antibiot. 32,* 83 (1979).
60. Inukai, M., Takeuchi, M., Shimizu, K. and Arai, M., *J. Antibiot. 31,* 1203 (1978).
61. Mizuno, K., Yagi, A., Satoi, S., Takada, M., Hayashi, M., Asano, K. and Matsuda, T., *J. Antibiot. 30,* 297 (1977).
62. Mizoguchi, J., Saito, T., Mizuno, K. and Hayano, K., *J. Antibiot. 30,* 308 (1977).
63. Satoi, S., Yagi, A., Asano, K., Mizuno, K. and Watanabe, T., *J. Antibiot. 30,* 303 (1977).
64. Benz, F., Hnusel, F., Nuesch, J., Treichler, H., Voser, W., Nyfeler, R. and Keller-Schlierlein, W., *Helo. Chem. Acta. 57,* 2459 (1971).
65. Keller-Juslen, C., Kihn, M., Loosli, H.R., Petcher, T.J., Weber, H.P. and von Wartburg, A., *Tetrahedron Lett.* p. 4147 (1976).
66. Shoji, J., Kato, T., Matsumoto, K., Takahashi, Y. and Mayama, M., *J. Antibiot. 29,* 1281 (1976).
67. Kato, T., Hinoo, H. and Shoji, J., *J. Antibiot. 31,* 652 (1978).
68. Atsumi, K., Oiwa, R. and Omura, S., *J. Antibiot. 28,* 77 (1978).

69. Neuss, N., Molloy, B.B., Shah, R. and De LaHiguera, N., *Biochem. J. 118,* 571 (1970).
70. Meyers, E., Parker, W.L. and Brown, W.E., *J. Antibiot. 29,* 1241 (1976).
71. Swanson, P.E. and Storm, D.R., *J. Antibiot. 32,* 511 (1979).
72. Brown, D.A. and Tsang, J.C., *J. Antibiot. 31,* 603 (1978).
73. Kusano, T., Isaki, K. and Takahashi, H., *J. Antibiot. 29,* 674 (1976).
74. Besson, F., Peypoux, F., Michel, G. and Delcambe, L., *J. Antibiot. 31,* 284 (1978).
75. Besson, F., Peypoux, F., Michel, G. and Delcambe, L., *J. Antibiot. 32,* 828 (1979).
76. Sogn, J.A., *J. Med. Chem. 19,* 1228 (1976).
77. Takeuchi, Y., Murai, A., Takahara, Y. and Kainosho, M., *J. Antibiot. 32,* 121 (1979).
78. Takahara, Y., Takeuchi, Y., Komura, I., Hirose, Y. and Murao, S., *J. Antibiot. 32,* 115 (1979).
79. Shoji, J. and Kato, T., *J. Antibiot. 29,* 380 (1976).
80. Shoji, J., Hinoo, H., Wakisaka, Y., Koizumi, K. and Mayama, M., *J. Antibiot. 28,* 126 (1975).
81. Pandey, R.C., Cook, Jr., J.C. and Rinehart, Jr., K.L., *J. Am. Chem. Soc. 99,* 8469 (1977).
82. Jung, G., Konig, W.A., Liebfritz, D., Ooka, T., Janko, K. and Boheim, G., *Biochim. Biophys. Acta. 433,* 164 (1976).
83. Boheim, G., Janko, K., Liebfritz, D., Ooka, T., Konig, W.A. and Jung, G., *Biochim. Biophys. Acta. 433,* 182 (1976).
84. Mueller, P. and Rudin, D.O., *Nature 217,* 713 (1968).
85. Hladky, S.B. and Haydon, D.A., *Nature 225,* 451 (1970).
86. Haynes, D.H., Kowalsky, A. and Pressman, B., *J. Biol. Chem. 244,* 502 (1969).
87. Pandey, R.C., Meng, H., Cook, Jr., J.C. and Rinehart, Jr., K.L., *J. Am. Chem. Soc. 99,* 5203 (1977).
88. Pandey, R.C., Cook, Jr., J.C. and Rinehart, Jr., K.L., *J. Antibiot. 31,* 241 (1978).
89. Argoudelis, A.D. and Johnson, L.E., *J. Antibiot. 27,* 274 (1974).
90. Pandey, R.C., Cook, Jr., J.C. and Rinehart, Jr., K.L., *J. Am. Chem. Soc. 99,* 5025 (1977).
91. Argoudelis, A.D., Dietz, A. and Johnson, L.E., *J. Antibiot. 27,* 321 (1974).
92. Argoudelis, A.D., Misak, S.A. and Baczynskyj, L., *J. Antibiot. 28,* 733 (1975).
93. Cox, R.E., Chexal, K.K. and Holker, J.S.E., *J. Chem. Soc. Perkin I* p. 578 (1976).
94. Inoue, N., Inoue, A., Furukawa, M. and Kanda, N., *J. Antibiot. 29,* 618 (1976).
95. Trejo, W.H., Dean L.D., Pluscec, J., Meyers, E. and Brown, W.E., *J. Antibiot. 30,* 639 (1977).

96. Tori, K., Tokura, K., Okabe, K., Ebata, M., Otsuka, H. and Lukacs, G., *Tetrahedron Lett.* p. 185 (1976).
97. Miyairi, N., Miyoshi, T., Aoki, H., Kohsaka, M., Ikushima, H., Kunugita, K., Sakai, H. and Imanaka, H., *Antimicrob. Agents* and *Chemother. 1,* 192 (1972).
98. Hamill, R.L. and Stark, W.M., *German patent* 2,703,938; *Japanese patent* 77-108,094 (1977).
99. Tori, K., Tokura, K., Yoshimura, Y., Okabe, K., Otsuka, H., Inagaki, F. and Miyazawa, T., *J. Antibiot. 32,* 1072 (1979).
100. Muramatsu, I., Motoki, Y., Aoyama, M. and Suzuki, H., *J. Antibiot. 30,* 383 (1977).
101. Hensens, O.D. and Albers-Schönberg, G., *Tetrahedron Lett.* p. 3649 (1978).
102. Pascard, C., Ducruix, A., Lunel, J. and Prange, T., *J. Am. Chem. Soc. 99,* 6418 (1977).
103. Endo, T. and Yonehara, H., *J. Antibiot. 31,* 623 (1978).
104. Depaire, H., Thomas, J.-P., Brun, A. and Lukacs, G., *Tetrahedron Lett.* p. 1395 (1977).
105. Depaire, H., Thomas, J.-P., Brun, A., Oleskar, A. and Lukacs, G., *Tetrahedron Lett.* p. 1397 (1977).
106. Depaire, H., Thomas, J.-P., Brun, A., Hull, W.E., Oleskar, A. and Lukacs, G., *Tetrahedron Lett.* p. 1401 (1977).
107. Prange, T., Ducruix, A. and Pascard, C., *Nature 265,* 189 (1977).
108. Walker, J., Oleskar, A., Valente, L., Babanal, R. and Lukacs, G., *J.C.S. Chem. Comm.* p. 706 (1977).
109. Abraham, E.P., Heatley, N.G., Brookes, P., Fuller, A.T. and Walker J., *Nature, 178,* 44 (1956).
110. Breiter, J., Metz, H. and Grigo, J., *Arzneim-Forsch. 25,* 1244 (1975).
111. Shoji, J., Hinoo, H., Wakisaka, Y., Koizumi, K., Mayama, M., Matsuura, S. and Matsumoto, K., *J. Antibiot. 29,* 366 (1976).
112. Kirst, H.A., Szymanski, E.F., Dorman, D.E., Occolowitz, J.L., Jones, N.D., Chaney, M.O., Hamill, R.L. and Hoehn, M.M., *J. Antibiot. 28,* 286 (1975).
113. Sakakibara, H., Naganawa, H., Ohno, M., Maeda, K. and Umezawa, H., *J. Antibiot. 27,* 897 (1974).
114. Bycroft, B.W., and Pinchin, R., *J.C.S. Chem. Comm.* p. 120 (1975).
115. Liesch, J.M., McMillan, J.A., Pandey, R.C., Paul, I.C., Rinehart, Jr., K.L. and Reusser, F., *J. Am. Chem. Soc. 98,* 299 (1976).
116. Liesch, J.M., Millington, D.S., Pandey, R.C. and Rinehart, Jr., K.L., *J. Am. Chem. Soc. 98,* 8237 (1976).
117. Abe, H., Ikeda, M., Takaishi, T., Ito, Y. and Okuda, T., *Tetrahedron Lett.* p. 735 (1977).

118. Uehara, Y., Kondo, S., Umezawa, H., Suzukake, K. and Hori, M., *J. Antibiot.* *25,* 685 (1972).
119. Uehara, Y., Hori, M. and Umezawa, H., *Biochim. Biophys. Acta.* *442,* 251 (1976).
120. Kondo, S., Yoshida, K., Ikeda, T., Iinuma, K., Honma, Y., Hamada, M. and Umezawa, H., *J. Antibiot.* *30,* 1137 (1977).
121. Maehr, H., Smallheer, J., Chin, M., Palleroni, N., Weiss, F. and Liu, C., *J. Antibiot.* *32,* 531 (1979).
122. Kondo, S., Yamamoto, H., Maeda, K. and Umezawa, H., *J. Antibiot.* *24,* 732 (1971).
123. Uehara, Y., Hori, M., Kondo, S., Hamada, M. and Umezawa, H., *J. Antibiot.* *29,* 937 (1976).
124. Curran, W.V. and Boothe, J.H., *J. Antibiot.* *31,* 914 (1978).
125. Martinelli, E., Zerilli, L.F., Volpe, G., Pagani, H. and Cavalleri, B., *J. Antibiot.* *32,* 108 (1979).
126. Bycroft, B.W. and King, T.J., *J.C.S. Perkin I* p. 1996 (1976).
127. Bycroft, B.W., *J.C.W. Perkin 1* p. 2464 (1977).
128. Okamura, Y., Takei, T., Sakamoto, M., Ishikura, T. and Fukagawa, Y., *J. Antibiot.* *32,* 1002 (1979).
129. Chopra, C., Hook, D.J., Vining, L.C., Das, B.C., Shimizu, S., Taylor, A. and Wright, J.L.C., *J. Antibiot.* *32,* 392 (1979).
130. Kamal, F. and Katz, E., *J. Antibiot.* *29,* 944 (1976).
131. Katz, E., Kamal, F. and Mason, K., *J. Biol. Chem.* *254,* 6684 (1979).
132. Reusser, F., *Biochemistry* *16,* 3406 (1977).
133. Argoudelis, A.D., Mizsak, S.A., Baczynskyj, L. and Wnuk, R.J., *J. Antibiot.* *29,* 1117 (1976).
134. Argoudelis, A.D., Mizsak, S.A., Baczynskyj, L. and Wnuk, R.J., *J. Antibiot.* *29,* 1119 (1976).
135. Okura, A., Morishima, H., Takita, T., Aoyagi, T., Takeuchi, T. and Umezawa, H., *J. Antibiot.* *28,* 337 (1975).
136. Wingender, W., von Hugo, H., Frommer, W. and Schäfer, D., *J. Antibiot.* *28,* 611 (1975).
137. Oka, S., Suzuki, H., Hirata, Y. and Tanabe, O., *Agric. Biol. Chem.* *43,* 691 (1979).
138. Watanabe, T., Fukuhara, K. and Murao, S., *Tetrahedron Lett.* p. 625 (1979).
139. Watanabe, T. and Murao, S., *Agric. Biol. Chem.* *43,* 243 (1979).
140. Uyeda, M., Yoshida, S., Suzuki, K. and Shibata, M., *Agric. Biol. Chem.* *40,* 1237 (1976).
141. Hanada, K., Tamai, M., Ohmura, S., Sawada, J., Seki, T. and Tanaka, I., *Agric. Biol. Chem.* *42,* 529 (1978).
142. Umezawa, H., Aoyagi, T., Suda, H., Hamada, M. and Takeuchi, T., *J. Antibiot.* *29,* 97 (1976).

143. Aoyagi, T., Tobe, H., Kojima, F., Hamada, M., Takeuchi, T. and Umezawa, H., *J. Antibiot. 31,* 636 (1978).
144. Umezawa, H., *J. Antibiot. 30,* S-138 (1977).
145. Hozumi, M., Ogawa, M., Sugimura, T., Takeuchi, T. and Umezawa, H., *Cancer Res. 32,* 1725 (1972).
146. Umezawa, K., Matsushima, T. and Sugimura, T., *Proc. Japan. Academy 53,* 30 (1977).
147. Meyn, M.S., Rossman, T. and Troll, W., *Proc. Nat. Acad. Sci. U.S.A. 74,* 1152 (1977).
148. Greenbaum, L.M., Grebow, P., Johnston, M., Prakash, A. and Semente, G., *Cancer Res. 35,* 706 (1975).
149. Hori, M., Hemmi, H., Suzukake, K., Hayashi, H., Uehara, Y., Takeuchi, T. and Umezawa, H., *J. Antibiot. 31,* 95 (1978).
150. Umezawa, H., Okami, Y. and Hotta, K., *J. Antibiot. 31,* 99 (1978).
151. Park, B.K., Hirota, A. and Sakai, H., *Agric. Biol. Chem. 41,* 573 (1979).
152. Sinden, S.L. and Durbin, R.D., *Nature 219,* 379 (1968).
153. Pruess, D.L., Scannell, J.P., Ax, H.A., Kellett, M., Weiss, F., Demney, T.C. and Stempel, A., *J. Antibiot. 26,* 261 (1973).
154. Kenig, M., Vandamme, E. and Abraham, E.P., *J. Gen. Microbiol. 94,* 46 (1976).
155. Diddens, H., Zahner, H., Kraas, E., Gohring, W. and Jung, G., *Env. J. Biochem. 66,* 11 (1976).
156. Pruess, D.L., Scannell, J.P., Blount, J.F., Ax, H.A., Kellett, M., Williams, T.H. and Stempel, A., *J. Antibiot. 27,* 754 (1974).
157. Diddens, H., Dorgerloh, M. and Zähner, H., *J. Antibiot. 32,* 87 (1979).
158. Kanaoka, M., Isogai, A. and Suzuki, A., *Agric. Biol. Chem. 43,* 1079 (1979).
159. Dobler, M., Dunitz, J.D. and Karjewski, J., *J. Mol. Biol. 42,* 603 (1969).
160. Champlin, F.R. and Grula, E.A., *Appl. Envir. Microbiol. 37,* 1122 (1979).
161. Isogai, A., Kanaoka, M., Matsuda, H., Hori, Y. and Suzuki, A., *Agric. Biol. Chem. 42,* 1797 (1978).
162. Elsworthy, J.F. and Grove, J.F., *J.C.S. Perkin 1* p. 270 (1977).
163. Frappier, F., Pais, M., Elsworthy, J.F. and Grove J.F., *Phytochemistry 17,* 545 (1978).
164. Briggs, L.H. and Fergus, B.J., *Tetrahedron S-8 (I)* p. 269 (1966).
165. Lee, S., Izumiya, N., Suzuki, A. and Tamura, S., *Tetrahedron Lett.* p. 883 (1975).

166. Lee, S., Aoyagi, H., Shimohigashi, Y., Izumiya, N., Ueno, T. and Fukami, H., *Tetrahedron Lett.* p. 843 (1976).
167. Patil, S.S., Youngblood, P., Christiansen, P. and Moore, R.E., *Biochem. Biophys. Res. Commun. 69,* 1019 (1976).
168. Mitchell, R.E. and Parsons, E.A., *Phytochemistry 16,* 280 (1977).
169. Mitchell, R.E. and Bieleski, R.L., *Plant Physiol. 60,* 723 (1977).
170. Durbin, R.D., Uchytil, T.F., Steele, J.A. and Ribeiro, R.de L.D., *Phytochemistry 17,* 147 (1978).
171. Strobel, G.A., *Ann. Rev. Microbiol. 31,* 205 (1977).
172. Kostanek, E.C., Lipscomb, W.N., Yocum, R.R. and Thiessen, W.E., *J. Am. Chem. Soc. 99,* 1273 (1977).
173. Buku, A. and Wieland, T., *Liebigs Ann. Chemie,* p. 1587 (1974).
174. Tanaka, W., *J. Antibiot. 30,* S-41 (1977).
175. Naganawa, H., Muraoka, Y., Takita, T. and Umezawa, H., *J. Antibiot. 30,* 388 (1977).
176. Naganawa, H., Takita, T., Umezawa, H. and Hull, W.E., *J. Antibiot. 32,* 539 (1979).
177. Kawaguchi, H., Tsukiura, H., Tomita, K., Konishi, M., Saito, K., Kobaru, S., Numata, K., Fujisawa, K., Miyaki, T., Hatori, M. and Kosiyama, H., *J. Antibiot. 30,* 779 (1977).
178. Takasawa, S., Kawamoto, I., Okachi, R., Kohakura, M., Yahashi, R. and Nara, T., *J. Antibiot. 28,* 366 (1977).
179. Takasawa, S., Kawamoto, I., Sato, S., Yahashi, R., Okachi, R., Yamamoto, M., Sato, T. and *J. Antibiot. 28,* 662 (1975).
180. Ohba, K., Kojima, M., Shomura, T., Tsuruoka, T., Ogawa, Y., Watanabe, H., Inouye, S. and Niida, T., *Meiji Seika Kenkyu Nempo 16,* 29 (1976). (*Chem. Abs. 88,* 103302).
181. Ohba, K., Shomura, T., Watanabe, H., Totsukawa, K., Kojima, M., Omoto, S., Tsuruoka, T., Inouye, S., and Niida, T., *Ger. Offen 2,* 649,604 (1977) (*Chem. Abs. 87,* 100672).
182. Ito, Y., Ohashi, Y., Kawabe, S., Sakurazawa, M., Ozawa, T., Egawa, Y. and Okuda, T., *J. Antibiot. 26,* 77 (1973).
183. Umezawa, H. in "Antibiotics III" (J.W. Corcoran and F.E. Hahn, Eds.) pp. 21-33, Springer-Verlag, Berlin, 1975.
184. Sausville, E.A., Peisach, J., and Horwitz, S.B., *Biochem. Biophys. Res. Commun. 73,* 814 (1976).
185. Takita, T., Muraoka, Y., Nakatani, T., Fujii, A., Intaka, Y., and Umezawa, H., *J. Antibiot. 31,* 1073 (1978).

186. D'Andrea, A.D. and Haseltine, W.A., *Proc. Natl. Acad. Sci. U.S.A. 75,* 3608 (1978).
187. Sugiura, Y. and Kikuchi, T., *J. Antibiot. 31,* 1310 (1978).
188. Sugiura, Y., *J. Antibiot. 31,* 1206 (1978).
189. Chien, M., Grollman, A.P. and Horwitz, S.B., *Biochemistry 16,* 3641 (1977).
190. Asakura, H., Umezawa, H. and Hori, M., *J. Antibiot. 31,* 156 (1978).
191. Sugiura, Y., Muraoka, Y., Fujii, A., Takita, T. and Umezawa, H., *J. Antibiot. 32,* 756 (1979).
192. Kunishima, M., Fuji, T., Nakayama, Y., Takita, T., and Umezawa, H., *J. Antibiot. 29,* 853 (1976).
193. Takeda, K., Kawai, S., Kato, F., Tetsuka, T. and Konno, K., *J. Antibiot. 31,* 884 (1978).
194. Takeda, K., Kato, F., Kawai, S. and Konno, K., *J. Antibiot. 32,* 43 (1979).
195. Berdy, J., *Adv. Appl. Microbiol. 18,* 309 (1974).
196. Meienhofer, J., Maeda, H., Glaser, C.B., Czombos, J. and Kuromizu, K., *Science 178,* 875 (1972).
197. Khokhlov, A.S., Reshetov, P.D., Chupova, L.A., Cherches, B.Z., Zhigis, L.S., and Stoyachenko, I.A., *J. Antibiot. 29,* 1026 (1976).
198. Koide, Y., Ishii, F., Hasuda, K., Koyama, Y., Edo, K., Katamine, S., Kitame, F. and Ishida, N., *J. Antibiot. 33,* 342 (1980).
199. Beerman, T.A., and Goldberg, I.H., *Biochim. Biophys. Acta. 475,* 281 (1977).
200. Ishida, N. and Ebina, T., *J. Antibiot. 30,* S-93 (1977).
201. Maeda, H. and Takeshita, J., *J. Antibiot. 29,* 111 (1976).
202. Yamashita, T., Naoi, N., Hidaka, T., Watanabe, K., Kumada, Y., Takeuchi, T. and Umezawa, H., *J. Antibiot. 32,* 330 (1979).
203. Komiyama, K., Sugimoto, K., Takeshima, H. and Umezawa, I., *J. Antibiot. 30,* 202 (1977).
204. Chan, J.A., Wei, T.T., Kalita, C.C., Warnick, D.J., Garretson, A.L. and Aszalos, A.A., *J. Antibiot. 30,* 1140 (1977).
205. Kihara, T., Suzuki, S. and Yonehara, H., *J. Antibiot. 29,* 428 (1976).
206. Sasaki, T. and Otake, N., *J. Antibiot. 28,* 553 (1975).
207. Oki, T., Yoshimoto, A., Matsuzawa, Y., Hori, S., Tone, H., Takamatsu, A., Takeuchi, T., Ishizuka, M., Hamada, M., and Umezawa, H., *J. Antibiot. 28, 479 (1975).*
208. Sakurai, A., Tamura, S., Yamagishima, N. and Shimoda, C., Agric. Biol. Chem. 41, *395 (1977).*
209. Sakurai, A., Sakata, K., Tamura, S., Aizawa, K., Yanagishima, N. and Shimoda, C., Agric. Biol. Chem. 40, *1451 (1976).*

210. Kamiya, Y., Sakurai, A., Tamura, S., Takahashi, N., Tshuchiya, E., Abe, K and Fukui, S., *Agric. Biol. Chem. 43,* 363 (1979).
211. Sakagami, Y., Isogai, A., Suzuki, A., Tamura, S., Tsuchiya, E. and Fukui, S., *Agr. Biol. Chem. 42,* 1301 (1978).
212. Hasegawa, M. and Matsubara, I., *Agric. Biol. Chem. 42,* 383 (1978).
213. Hofle, G., Gmelin, R., Luxa, H.H., and Hatanaka, S.I., *Tetrahedron Lett.* p. 3129 (1976).
214. Shimi, I.R., Abedallah, N. and Fathy, S., *Antibiot. Ag. Chemother. 11,* 373 (1977).
215. Sakamura, S., Furukawa, K. and Kasai, T., *Agric. Biol. Chem. 42,* 607 (1978).
216. Baute, M.-A., Deffieux, F., Baute, R., and Neveu, A., *J. Antibiot. 31,* 1099 (1978).
217. Berg, D.H., Massing, R.P., Hoehn, M.M., Boeck, L.D. and Hamill, R.L., *J. Antibiot. 29,* 394 (1976).
218. Takita, T., Shimada, N., Yagisawa, N., Kato, K., Naganawa, H., Maeda, K. and Umezawa, H., in "Peptide Chemistry, 1977" Proceedings of the 15th Symposium on Peptide Chemistry (T. Shiba, Ed.) pp. 121-126. Protein Research Foundation, Osaka, 1978.
219. Takahashi, Y., Naganawa, H., Takita, T., Umezawa, H. and Nakamura, S., *J. Antibiot. 29,* 1120 (1976).
220. Traber, R., Kuhn, M., Loosli, H.-R., Pache, W. and von Wartburg, A., *Helv. Chim. Acta. 60,* 1568 (1977).
221. Shima, S. and Sakae, H., *Agric. Biol. Chem. 41,* 1807 (1977).
222. Shoji, J., Sakazaki, R., Wakisaka, Y., Koizumi, K., Mayama, M. and Matsuura, S., *J. Antibiot. 28,* 122 (1975).
223. Shoji, J., Hinoo, H., Wakisaki, Y., Koizumi, K., Mayama, M. and Matsuura, S., *J. Antibiot. 29,* 809 (1976).
224. Shoji, J., Sakazaki, R., Wakisaka, Y., Koizumi, K. and Mayama, M., *J. Antibiot. 28,* 129 (1975).
225. Suzuki, S., Kibara, T., Nakamura, G., Kusakabe, H., Mizuno, T. and Yahagi, T., Japan. Kokai 77-12,993 (*Chem. Abs. 87,* 4025).
226. Oiwa, R., Katagiri, M., Tanaka, N., Takahashi, Y., Sato, K., Masuma, R. and Omura, S., *J. Antibiot. 28,* 819 (1975).
227. Yaginuma, S., Muto, N. and Otani, M., *J. Antibiot. 32,* 969 (1979).
228. Omura, S., Iwai, Y., Hirano, A., Awaya, J., Suzuki, Y. and Matsumoto, K., *Agric. Biol. Chem. 41,* 1827 (1977).
229. Umezawa, H., Hamada, M., Kondoh, S., Matsuoka, M. and Satoh, K., Japan. Kokai 77-102,201 (1977). (*Chem. Abs. 88,* 20528).

CHAPTER 6

DIRECT MICROBIOLOGICAL CONVERSION OF CELLULOSICS TO ETHANOL

George C. Avgerinos
Daniel I. C. Wang

Department of Nutrition and Food Science
Massachusetts Institute of Technology
Cambridge, Massachusetts

I. INTRODUCTION

The production of ethanol as a liquid transportation fuel has received a great deal of attention in recent years. Presently, the programs in ethanol production have focused mainly on the use of grain as the feedstock. In long range planning, however, one cannot neglect alternate renewable resources as feedstocks for alcohol production. In particular, cellulosic biomasses such as agricultural, forestry, and municipal residues represent a vast quantity of carbohydrate potentially convertible to liquid fuels. This resource has not yet been commercially exploited to any significant extent for alcohol production.

Research and development activities in the use of these cellulosic residues for ethanol production have traditionally focused in the areas of pretreatment, saccharification, and fermentation. The biomass is first pretreated through physical or chemical processes. These involve size reduction through a variety of mechanical means. Selective solvent pretreatment processes have also been proposed. In general, these pretreatment methods are needed to render the biomass more susceptible to subsequent hydrolysis of the polymeric species such as cellulose and hemicellulose. The degradation of the biomass

ISBN 0-12-040304-8

components is then accomplished through either enzymatic or chemical techniques. The soluble sugars are subsequently fermented, traditionally through the action of yeasts, to produce the desired product ethanol.

A simpler approach for the utilization of cellulosic biomass can also be envisioned. The keys to such a scheme are based on the existence of anaerobic microorganisms which possess cellulolytic enzymes able to degrade biomass producing soluble sugars. More importantly, these anaerobic microorganisms are known to possess catabolic pathways leading to the production of alcohols and organic acids. In theory, one can therefore envision a possible single step fermentation where biomass is directly degraded and simultaneously fermented for the production of ethanol.

It is the intent of this review to assess some of the published literature dealing with this concept for the direct microbiological conversion of cellulosic biomass to ethanol. However, it is beyond our scope to consider every microbial system which is theoretically capable of achieving such a conversion process. Instead, we have focused on a model but realistic microbial system which might be potentially useful for the direct production of ethanol from cellulosic residues. The example for this "case study" being the anaerobic thermophiles *Clostridium thermocellum* and *Clostridium thermosaccharolyticum*.

II. ANAEROBIC DEGRADATION OF CELLULOSE

A. Historical Perspectives

The first observation of the thermophilic anaerobic fermentation of cellulose is generally attributed to MacFayden and Bloxell in 1899. A number of investigators subsequently isolated organisms with these properties from ruminant fecal samples (1,2). The first vigorous promotion of thermophilic cellulolytic fermentation for the production of useful chemicals resulted from the work of Langwell and Lynn in the 1920's (3). These investigators isolated numerous thermophilic mixed cultures from nature capable of degrading waste cellulosics to acetic acid, ethanol, and fuel gas. However, they could not obtain a practical cellulose fermentation with "purified" cultures. Their use of unknown mixed isolates, nevertheless, demonstrated ethanol yields of 27% from sulfite pulp, 8% from filter paper, 15% with xylose from rice straw, and 89% from corn cob during pilot plant trials. However, reproducible fermentations were difficult to achieve with these ill-defined cultures, and further industrial interest in this approach waned. Meanwhile, attempts to isolate pure cultures were continued by other investigators in the 1920's and 1930's. In 1926, Wiligoen, Fred, and Petersen isolated a thermophilic, cellulolytic, anaerobe and designated it *Clostridium thermocellum* (4). Subsequent attempts by Tetrault, Snieszko, and Imsenecki, to verify this isolate as a pure culture or to obtain one themselves were all unsuccessful (5,6). Improvements in anaerobic techniques brought about by Hungate in 1944 (7) subsequently enabled Enebo and McBee to independently isolate and characterize an organism with similar properties identified as *Clostridium thermocellulaseum* and *Clostridium thermocellum*, respectively (8,9).

Meanwhile, the fermentative production of ethanol from cellulosics was commercialized in an alternate manner. Chemical hydrolysis of woodchips with dilute H_2SO_4 followed by yeast fermentation was patented in 1913. A 5,000 gallon 95% ethanol/day plant was built in the United States. Unfortunately, the sawdust feedstock the plant was originally designed to use soon was exhausted, and transportation cost proved to be too high forcing abandonment of the process by the mid 1920's.

Widespread raw material shortages during World War II revived interest in this technology in both Germany and the United States. Numerous process improvements were added by Schoeller. But, with the war coming to an end, fermentatively derived ethanol rapidly lost ground to ethanol obtained from ethylene hydration. Today the major fraction of industrial alcohol is produced from ethylene. However, it is evident that the supply picture for petroleum based feedstocks is again changing, making it increasingly important to examine alternate technology for ethanol production with the aid of today's better understanding of these biological systems.

B. Properties of Clostridium thermocellum

Although *Clostridium thermocellum* was first named in 1926, the definitive characterization of this obligately anaerobic, thermophilic, and cellulolytic bacterium was not accomplished until the 1950's by McBee, Enebo, and subsequent workers. *C. thermocellum* (ATCC 27405) is described as a gram negative rod 0.6 x 4 microns in size forming terminally spherical spores on solid but not in liquid medium. The optimum temperature for growth has been reported between 58 and 64°C and optimum pH between 6.4 and 7.4. The organism has been reported to grow on cellulose, cellobiose, glucose, fructose, and mannitol by Patni and Alexander (10). Glucokinase activity was shown to be inducible by glucose but absent during growth on cellobiose or cellulose and was believed to be responsible for the accumulation of this sugar. Fructose and mannitol utilization were shown to occur by PEP: phosphofructotransferase and PEP: phosphomannotranserase/dehydrogenase activity also inducible in the presence of these substrates. Two recent literature reports, however, indicated that *C. thermocellum* was only capable of growth on cellulose and cellobiose (11,12). Other investigators have also reported growth on xylose, but this capability usually coincides with the simultaneous ability to use a variety of carbohydrates including starch which is uncharacteristic of this organism (13). Although these reports may be due to actual strain variations, it should also be pointed out that they may also arise from two unusual features of this organism. These include the relative ease in which single colony isolates can sporadically contain morphologically similar non-cellulolytic contaminants (such as *Clostridium thermosaccharolyticum*) and the very long lag time in the presence of at least 0.4% yeast extract required to demonstrate growth on glucose (14). A defined media has been developed for *C. thermocellum* incorporating thiamine, riboflavin, pantothenate, pyridoxine, biotin, folic acid, para aminobenzoic acid, and the amino acids phenylalanine, tryosine, tryptophan, cystine, cysteine, and methionine (15). However, more recent studies have shown that these requirements can be reduced to only biotin and pyridoxine in minimal salts media (16).

The major fermentation products of the wild type of organism (Figure 1) are acetate, ethanol, lactate, CO_2 and H_2, along with smaller amounts of butyric and formic acids (17). However, through a program of adaptation by serial transfer in media containing increasing concentrations of ethanol, and mutation and selection for ethanol tolerance and reduced lactate production, a strain has been developed which produces ethanol as the major product (18). Dissimilation of fermentable carbohydrate occurs primarily through the Embden-Myerhoff pathway to pyruvate; although the enzymes of the hexose monophosphate path have also been demonstrated in lesser amounts (19). Pyruvate can be further metabolized to lactate with NAD linked lactate dehydrogenase or decarboxylated to acetyl CoA with ferredoxin oxidoreductase. This enzyme has been shown in *C. acidi urici* to contain 1 mole of thiamine pyrophosphate, iron, and sulfur and catalyzes the decarboxylation of pyruvate with a concomitant two electron transfer to oxidized ferredoxin (20). The reduced ferredoxin may be reoxidized with a hydrogenase liberating hydrogen or by reducing NAD^+ or $NADP^+$ (21). Acetyl CoA may be subsequently converted to acetyl-PO_4 and acetic acid with the production of ATP or twice reduced with NADH to form ethanol.

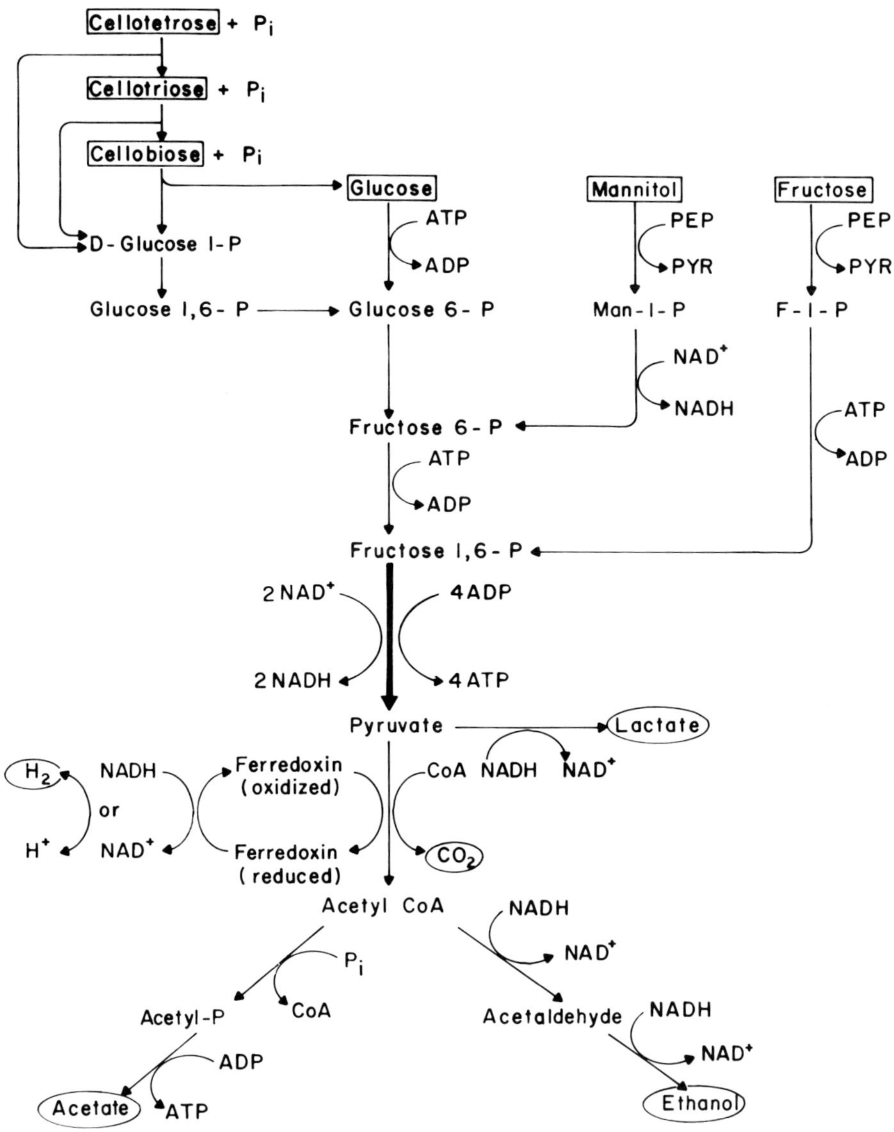
Cellotetrose + P_i
Cellotriose + P_i
Cellobiose + P_i
Glucose
Mannitol
Fructose
ATP
ADP
PEP
PYR
PEP
PYR
D-Glucose 1-P
Glucose 1,6-P
Glucose 6-P
Man-1-P
F-1-P
NAD^+
NADH
ATP
ADP
Fructose 6-P
ATP
ADP
Fructose 1,6-P
2 NAD^+
4 ADP
2 NADH
4 ATP
Pyruvate
Lactate
H_2
NADH
or
H^+
NAD^+
Ferredoxin (oxidized)
Ferredoxin (reduced)
CoA
NADH
NAD^+
CO_2
Acetyl CoA
NADH
NAD^+
P_i
CoA
Acetyl-P
Acetaldehyde
NADH
NAD^+
ADP
Acetate
ATP
Ethanol

FIGURE 1.

In summary, the overall balanced fermentation from 1/2 mole of glucose can be theoretically made through any of the reactions shown in Table 1.

C. Cellulases of Clostridium thermocellum

C. thermocellum produces a large quantity of extracellular cellulase recently purified by Ait et al. (22). These investigators demonstrated that after 13 hours of growth on purified cellulose (MN 300) only 10% of the total cellulase activity could be recovered in the supernatant. After 60 hours, approximately 30% could be freed after removal of the broth through centrifucation from the substrate and cells. Although strong adsorption of cellulase to the substrate was evident, it could not be discerned whether this was the result of saturation of cellulose with a single enzyme or the production of different enzymes during the course of the fermentation.

Gel filtration of the protein remaining in solution at the end of the fermentation on Bio Gel P-200 (1 x 80 cm) revealed a single protein peak of 125,000 $\pm$ 10,000 MW. Subsequent sodium dodecyl sulfate gel electrophoresis of this peak yielded at least five discrete bands (some with cellulase activity) from this unusually large complex. Reducing sugar analysis of the H_2SO_4 hydrolyzed protein revealed the carbohydrate content of the complex to be 9.9% (W/W). Whether this indicates the presence of glycoproteins stabilizing this thermophilic cellulase, or represents cellodextrins strongly bound in the active sites of a cellulase complex of homogeneous molecular weight was not concluded by these authors. The cellulase from this organism is also unique as compared to fungal cellulases in that it is constitutively produced and uninhibited by glucose (23).

TABLE 1. Net Stoichiometry of Glucose Fermentation

	ΔG per ½ mole Glucose	Net ATP per ½ mole Glucose
½ glucose →		
1 Ethanol + 1 CO_2	- 23.7	1
1 Acetate + 1 CO_2 + $2H_2$	- 27.0	2
1 Lactate	- 24.7	1

Although the entire mechanism of sugar hydrolysis and transport in *C. thermocellum* is not completely understood, isolation of a number of intermediate enzyme activities have been obtained. Sheth and Alexander partially purified a cellodextrin phosphorylose activity from cell sonicates with the following activity (24):

$$\text{Pi} + \text{Cellohexose} \rightleftarrows \text{Cellopentose} + \text{Glucose - 1 - Phosphate}$$

This enzyme also displayed analogous activity with cellopentose, cellotetrose, and cellotriose, but was unable to cleave cellobiose. However, a separate enzyme, cellobiose phosphorylase, was also isolated from sonicated cells and catalyzed the phosphorolytic cleavage of cellobiose (25):

$$\text{Cellobiose} + \text{Pi} \rightleftarrows \text{Glucose} + \text{Glucose - 1 - Phosphate}$$

The Km's for cellobiose and phosphate during phosphorolysis were 7.3 mM and 29 mM, respectively, and a K_I of 1.2 mM was reported for glucose. This enzyme was also capable of using a variety of glucosyl-PO_4 acceptors in the reverse reaction including D-xylose. From this evidence, it seems reasonable to speculate that cellodextrins up to 6 glucose units in length can be transported into the cell where they are phosphorolytically cleaved conserving the ATP equivalent of each glycosidic bond. Studies with *Cellovibrio gilvus* growing on cellodextrins show an increased cellular yield as the size of the cellodextrin is increased further supports this hypothesis (26). The final cleavage of each cellodextrin by cellobiose phosphorylase results in a glucose-1-phosphate and a free non-phosphorylated D-glucose. In addition to these enzymes, a number of investigators have detected the presence of a β-glucosidase in the periplasmic space of *C. thermocellum* (27,28). Due to the uninduced state of glucokinase and/or the glucose transport system under these conditions, D-glucose produced from this reaction and through the action of β-glucosidase is seen to accumulate and may further inhibit the action of cellobiose phosphorylase thus allowing cellobiose to accumulate as well. D-glucose, and to a lesser extent cellobiose, are free to diffuse from the cell and are indeed found in the media during growth on cellulose (29). Since these reactions are reversible *in vivo*, it is likely that the accumulation of polysaccharides as demonstrated during growth of *C. thermocellum* on glucose by Wang can result as well (18).

D. Properties of Clostridium thermosaccharolyticum

C. thermosaccharolyticum was first isolated by McClung in 1935. This obligately anaerobic thermophile (ATCC 7956, NCIB 9385) displays a temperature optimum between 58 to 64°C and a pH optimum between 5.8 and 7.0. This gram negative motile rod 0.4 x 7 microns in size and forms terminal spores under conditions of restricted growth (30). *C. thermosaccharolyticum* has been shown to ferment a variety of carbohydrates including glucose, cellobiose, fructose, xylose, arabinose, maltose, mannose, and lactose, as well as being able to grow on starch and xylan. The major products of fermentation are acetate, ethanol, and lactate, with occasional smaller amounts of butyrate (31). Mutation and selection of this organism for increased tolerance to ethanol has resulted in a strain HG-4 in which ethanol is the predominant product of fermentation (32). The fermentation of hexoses is believed to occur primarily through the Embden Meyerhoff pathway in which all of the enzymes have been demonstrated (31). The metabolism of pentoses by some Clostridia has been thoroughly investigated by Cynkin and Giggs (33). The fermentation of C^{14} labeled D-xylose and D-ribose by *C. perfringens, C. beijerinckii,* and *C. botulinum* was demonstrated to result from the action of transaldolase and transketolase in the pentose phosphate cycle followed by dissimilation of the fructose -6- phosphate through the Embden-Myerhoff path. In this respect, 6 xylose units are effectively equivalent to 5 glucose units resulting in the net stoichiometry shown in Table 2. The thermodynamics of this conversion are similar to that seen from glucose as are the yields of ATP per mole of end product thus produced.

TABLE 2. Net Stoichiometry of Xylose Fermentation

	ΔG per mole Alcohol or Acid Produced	*Net ATP per mole Alcohol or Acid Produced*
6 Xylose →		
10 Ethanol + 10 CO_2	*- 27*	*1*
10 Acetate + 10 CO_2 *+ 20* H_2	*- 24.7*	*2*
10 Lactate	*- 23.7*	*1*

III. MICROBIOLOGICAL AND ENZYMATIC ASPECTS OF CELLULOSE UTILIZATION

Model systems for the utilization of renewable biomass, producing fermentable sugars and chemicals, are often studied with pure defined substrates. However, real cellulosics represent complex ill-defined systems in which a variety of secondary effects may be observed. A large number of agricultural and forestry residues have been considered as potential sources of fermentable carbohydrates. These materials are mostly composed of cellulose, hemicellulose, and lignin. Current knowledge concerning the composition and structure of these components has been amply reviewed by Tsao et al. and Flickinger and Tsao in Volume 2 of this series (34,35). It is an inescapable conclusion that physical factors such as cellulose crystallinity and association of lignacious components in the real biomass present significant barriers to the rate and extent of its utilization. However, in addition to these major factors, the balance or real biomasses contains a diverse fraction which includes tannins and other low molecular weight phenolics related to lignin, coloring compounds, oils, sugars, starches, pectins, gums, nitrogen compounds, free organic acids and minerals. These components possess a high degree of variability between species and seasons. For this reason, only a limited amount of information is available on the specific content of these compounds in particular species of interest. However, the significance of these materials in the context of microbial degradation of plants has been reported on numerous occasions and will be elaborated upon in the following sections.

A. Factors Affecting the Rate of Degradation of Cellulosics

It is widely recognized that actual physical contact between cellulases and cellulose is required for enzymatic hydrolysis. Therefore, it is not surprising that resistance of natural cellulosics to enzymatic hydrolysis is derived in part from the close physical association of lignin with cellulose (36). The multiple anisotropic layers of cellulose give the plant strength and rigidity, while the hydrophobic lignin residues offer chemical protection primarily against water. Hemicellulose is believed to further stabilize this structure by hydrogen bonding to both lignin and cellulose. In a number of surveys, the lignin content of the substrate has been shown to correlate with the resistance of a variety of agricultural materials to the action of cellulolytic organisms, cell free enzymes, and the yield of free sugar obtained (37,38).

The influence of cellulose crystallinity has been demonstrated as another limiting factor to cellulose action (39). X-ray diffraction measurements show that crystallinity increases as the substrate is enzymatically cleaved. The rate of hydrolysis of amorphous regions was demonstrated to occur much faster than that of crystaline ones (40). Cellulose has been shown to exist in four basic crystaline forms: Cellulose (I) native cellulose, (II) regenerated cellulose, (III) anhydrous ethanolamine treated, and (IV) high temperature treated. Rautela has obtained interesting results by culturing *Trichoderma viride* on all four of these substrates (41). He subsequently determined the activation energies for hydrolysis of each substrate. The results demonstrated that the organism was capable of producing an adapted cellulase requiring a minimal energy of activation for hydrolysis of each crystaline form.

The moisture content of cellulosic materials also affects their susceptability to enzymatic degradation. Moisture swells the fibers by hydrating the cellulose molecules causing their fine structure to open up making the surface more accessible to celluloses and other reagents (42). The fiber saturation point for hydration is between 24 and 32% of the oven dry weight of the substrate. The total surface area exposed in water saturated cellulosics comprises gross capillaries greater than 200 Å in diameter as well as substantially finer cell wall capillaries formed from spaces between microfibrils and cellulose molecules. The surface area of cellulose including the gross capillaries is estimated by Cowling to be approximately 2×10^3 cm^2 per gram of wood or cotton while the area including the fine capillaries ranges up to 3×10^6 cm^2 per gram (43). Determination of the pore size distribution before and after the pulping of sprucewood showed a shift from a narrow distribution centered around 10 Å after pulping. An estimation of cellulase size compiled from various sources by Whitaker et al. reveals an equivalent average size of 59 Å in diameter for many cellulases (44).

B. Effects of Natural Cellulosics on Microorganisms and Cellulase

It has been repeatedly shown that physical characteristics of natural cellulosics are to a large extent responsible for limiting the rate of cellulolytic degradation and the subsequent microbial growth. However, there is also evidence that tannins, coloring compounds, and lower molecular weight phenolics related to lignin may inhibit cellulase action as well. In addition, many of these compounds are believed to be toxic or inhibitory to the growth of many microbial species. While few generalizations of these events as yet appear warranted, many specific reports have been made. Although no work to date on this subject has been reported with *C. thermocellum* or *C. thermosaccharolyticum*, evidence has been accumulated for other mesophillic, cellulolytic species.

1. *Inhibition of Cellulase Activity.* Isolation of powerful natural inhibitors of β-amylase and pectinase activity have been reported in wheat, cereals, and grape leaves (45,46,47). These inhibitory compounds have been shown to be simple phenolics such as trans-cinnamic or salicyclic acid, leucoanthocyanins, or unidentified complex tannins. The β-amylase inhibitor of grape leaves has also been determined to inhibit cellulase as well (48). This non-protein inhibitor was non-dialyzable, resistant to autoclaving at pH 5 or 9 and was strongly adsorbed by activated charcoal. Mandels and Reese examined over 500 species of plants for the presence of cellulase inhibitors and found them in widespread occurrence (49). Plants were extracted with methanol and water (10 ml/gm) and assayed for inhibition of *Trichoderma viride* cellulose activity. Of the plant species tested, 17% of the extracts were capable of inhibiting 20 international units of cellulase by at least 50%. (1 International Unit - 1 μ mole glucose/min). In particular, the inhibition by extracts of bayberry leaves and berries was over 200x this level and was further characterized. The activity was shown to be non-dialyzable, alkali labile over pH 8, but relatively stable in acid. The inhibitor complexed strongly with gelatin, polyamides, charcoal, sephadex, and various forms of cellulose. From U.V. spectra and chemical tests, it was shown to be polyphenolic in character, probably a leuco-anthocyanin. However, its complete structure was not elucidated. Inhibition kinetics suggest that it acts in a non-competitive manner, reversible forming a complex with the enzyme.

Inhibition of *T. viride* cellulase was also examined with two types of polymeric leuco-anthocyanins isolated from pears and persimmon. Both of these inhibitors were as potent as the isolate from bayberry and possessed similar chemical and physical properties. Inhibition studies with these extracts on cellulases from the various other fungal sources demonstrated that *T. viride* possessed the most sensitive enzyme studied. In addition, the relative potency of each inhibitor tested remained proportional with respect to each different cellulase tested. Inhibition of cellulases, β-amylases, and pectinases, by leuco-anthocyanins and oxidized catechins, was also reported by Williams (50). While many reports of cellulase inhibition *in vitro* have been made, little evidence regarding the importance of these effects on the overall rate of enzymatic decomposition of the substrate has been presented. Indeed, inhibition of cellulase by higher molecular weight phenolic polymers may not be due to active site binding or even allosteric binding affecting the active site. Instead, the binding of the phenolic to the cellulase may simply result in a sterically hindered complex unable to effectively orient on the substrate surface. In terms of the final activity, however, the difference between this inhibition and "physical" inhibition by sterically hindered cellulase may only be one of point of view.

2. *Microbial Inhibition by Natural Cellulosics.* In addition to cellulase inhibition, the inhibition of microbial growth by natural products of real biomass has also been reported for a variety of fungal and bacterial species (51). As with cellulase inhibition, each separate case is usually demonstrated with one microbial strain and particular plant extract *in vitro*. These results often cannot be justifiably generalized to explain the resistance of plants to pathological infection in general as an extract toxic to one strain has sometimes proven to be stimulatory to another. In addition, very seldom has a mechanism of action been offered. A well-documented case, however, demonstrated that two simple phenols, protocatechuic acid and catechol, found in the outer scales of harvested, pigmented onions inhibited the germination of the fungus *Colletotrichum circinans* on the onions and in extracts at 2×10^{-3}M (52). Alternatively, white onions not containing these phenols were susceptible to infection. The same effect was also demonstrated with *Diplodia natalensis* and *Botrytis allii* (53) (54). On the other hand, these compounds or crude extracts had little effect on *Aspergillis niger*, a pathogen in both of these strains of onions.

Extracts of tannins from a variety of plants have been shown *in vitro* to inhibit species of Fusarium, Gloeosporium, Colletotrichum, Cladosporium, and Penicillium and provide some degree of protection from infection *in vitro* (55). However, many other plant species containing large amounts of tannins are readily attacked by these fungi. Rust fungus resistance of wheat species has also been shown to correlate with the content of phenolic compounds (mainly flavones, protocatechuic acid, and catechols) while in susceptible strains pyrogallol has been shown to predominate (56). Other phenolics, chlorogenic, and isochlorogenic acid and their oxidation products, D and L, catechin were isolated from a variety of seed plants and subsequently shown to strongly inhibit nitrogen fixing strains of Azotabacter, Rhizobium, Nitrosomonas, and Nitrobacter (59). Gallotannin and gallic acid were also shown to inhibit these strains. These compounds were also implicated in the differential resistance of various types of apples to rotting infection by *Sclerotinia fructogena,* however, in this case it was believed that microbial inhibition was due to inactivation of extracellular fungal pectinases rather than any direct toxicity to the organism. In addition, these compounds could not be detected in whole apples but were shown to be produced by oxidation of polyphenols with host enzymes following infection of the apple (58). Schaal and Johnson correlated inhibition of *Streptomyces scaries* in potato tubers with the intensity of green color produced by reaction of extracts with 5% $FeCl_3$ (indicator of phenols) (59). However, *in vitro* tests with chlorogenic acid, caffeic acid, catechol, and p-hydroxybenzoic acid, the predominant phenolics detected, all failed to retard growth at concentrations less than 10^{-2} M.

Perhaps the most relevant example of microbial inhibition by plant extracts was recently shown in the case of *Erwina carotovora,* a causative agent of stalk rot of corn. Aqueous extracts of hydroxyamic acids from hardy corn strains were shown to strongly inhibit the growth of two species of Erwinia *in vitro* (60). A number of related hydroxyamic acids, principally a glycoside of 2,4-dihydroxy-7-methoxy-2H-1, 4-benzoxazin-3-(4H)-one can make up over 1% of the dry weight of hardy corn varieties (61). Extracts of hybrid strains of corn unable to make DIMBOA and susceptible to stalk rot are not able to inhibit *E. carotovora.* Inhibition of these strains was demonstrated with less than 0.1 gms/liter of DIMBOA extracted from plant material and resulted in an increased lag time of up to 46 hours over that of the control. The final growth rate, however, was unaffected by this compound. Subsequent investigators pointed out that DIMBOA was unstable in solution, decomposing to MBOA with a half-life of five hours at 25°C (62). Growth of Erwinia species in the presence of MBOA showed no lag but showed a strong effect on the growth rate decreasing the maximum specific growth rate by 50% at 0.5 g/l MBOA. Further examination of the eventual growth of Erwinia in the presence of DIMBOA could not be mechanically accounted for by its decomposition, however (63). The effect of benzoxazolinone and its 6-methoxy derivative on a number of microbial strains was also tested on glucose agar. Complete inhibition of *Fusarium nivale, Selerotinia trifolarium, Aspergillus niger, Penicillium roquefortii,* and *Pseudomonas fluorescens* was observed with 0.5 to 1.0 mg/ml (64,65).

From these results we see that microbial inhibition by plant polyphenolics is of widespread occurrence (66). What effect these materials have on a mechanistic level is at best poorly understood in a few instances. Whether or not this is a major mechanism of resistance of plants to pathogenic infection is also unclear. However, any effort to use these materials as substrates for fermentation would be well advised to investigate the possibility of these effects as well.

IV. PRETREATMENT OF CELLULOSICS

In order to overcome the major obstacles limiting the rapid enzymatic degradation and yield of soluble sugar from natural cellulosics, a pretreatment procedure comprising several steps is often employed (67). Mechanical size reduction is required in order to ensure short diffusion paths for pretreatment chemicals. Lignin must be decomposed into smaller units better able to elute from this structure. The second major limiting factor is the crystallinity of the cellulose itself. Thus, reagents able to cause simultaneous swelling of the carbohydrate structure and destruction of the crystallinity are extremely effective in creating pores for chemical attack, lignin removal, and subsequent cellulose hydrolysis (68,69). It must be remembered, however, that these reagents only alter the cellulose to a new thermodynamically stable state and recrystallization may rapidly occur to cellulose II upon their removal.

It is not surprising that a standard inexpensive pretreatment of biomass prior to enzymatic saccharification has been accomplished under alkaline conditions causing both lignin and hemicellulose removal as well as cellulose swelling. The soaking of waste straws in dilute (< 5%) alkali for 2 weeks was the basis for the Beckman process (1919) for increasing the digestability of these materials as animal feed. This strategy also forms the basis for the commercial soda or kraft pulping industry. Experiments with isolated lignin show that aryl-alkyl ether cleavage of lignin occurs at the para position of the phenyl ring at temperatures between 120 and 160°C. The addition of sulfite to this reaction allows the cleavage to occur at lower temperatures. Hydrogen sulfide facilitates the lignin disolution by reacting with the cleaved ethers and preventing possible recondensation reactions which would otherwise occur to hinder the diffusion of the lignin from the structure. Of course, alkali cooking also causes rapid solubilization and degradation of the hemicelluloses as well as lignin present.

Treatment of sugar cane with 0.1 N NaOH at 120°C for 15 minutes resulted in greatly increased susceptability (10 fold) of the substrate to the action of cellulase and hemicellulase from *Penicillium variable* (70). Extended treatment, however, resulted in reduced sugar yields. The kinetics of enzymatic sugar hydrolysis demonstrated more rapid xylose production than release of glucose or cellobiose residues by this treatment. Increased rates of degradation and yield were also shown by Ghose et al. on studies with bagasse pretreated with 1% NaOH (1:7 w/w) at 80°C for 3 hours followed by washing (71). Under these conditions, the bulk of the hemicellulose was reported to be removed (72).

Pretreatment of wheat straw with a variety of delignifying and swelling agents was shown to increase the yield of glucose achieved by saccharification with cellulases (38). As seen in Table 3, there is a strong correlation between the amount of lignin removed and cellulose hydrolyzed. Under these conditions, however, large losses of potentially utilizable hemicelluloses are correlated with lignin removal. The effect of temperature during treatment on the rate of hydrolysis of this material was also examined. From these results, we see the positive effect of increasing temperature during NH_3 and NH_4OH treatment, but little effect during ethylene-diamine treatment. Similar results were obtained with rye grass for 5% NH_3 and 4% NaOH (w/w) treated at room temperature for 4 weeks and 1 hour respectively (73).

Numerous additional methods have been reported for biomass pretreatment prior to enzymatic saccharification. Acid treatment extensively used for hydrolysis of cellobiose has been employed at lower concentrations with mechanical disruption of biomass in short contact time high temperature extrusion reactors (10 sec. at 200°C) (74). Subsequent enzymatic hydrolysis demonstrated higher rates and yields as compared to

TABLE 3. Effect of Pretreatment of Wheat Straw on Composition and Enzymatic Hydrolysis (38)

Treatment at 30 C 4 Hours	*% Loss of Cellulose*	*Hemicellulose*	*Lignin*	*% Enzymatic Conversion of Residual Cellulose*
None (H_2O)	-	-	-	*10%*
24% NH_3 (w/v)	*2*	*46*	*20*	*40*
28% ethylene diamine	*4*	*51*	*60*	*80*
2% NaOH	*13*	*32*	*7*	*55*
5% H_2SO_4	*2*	*15*	*6*	*40*
5% HCl	*6*	*15*	*10*	*44*

non-treated biomass. Unfortunately high temperature acid treatment often produces toxic or inhibitory compounds such as furfurals, formic acid, and other carbohydrate degradation products from secondary reactions of the rapidly hydrolyzed hemicelluloses.

A simple two step pretreatment procedure of wheat straw by autohydrolysis at 160 to 200°C with water 2:1 (w/w) followed by extraction of lignin with 50% aqueous ethanol at 120 to 160°C has also recently been reported (75). The authors claim that these conditions first depolymerize lignin and solubilize hemicellulose and secondly extract the depolymerized lignin. Subsequent analysis of enzymatic hydrolysis kinetics revealed twice the initial yield of glucose after the first treatment and four times the yield of glucose following lignin extraction.

Another method of substrate pretreatment has been reported which utilizes 2 MeV electron irradiation of cellulosics. Rice straw exposed to greater than 5 x 10^8R shows an increase in reducing sugar yield from 2 to 12% after cellulose digestion (76). Although novel, this process is not believed to be economically attractive at present.

Lastly, the use of cadoxen, a solution of 5% cadmium oxide in 28% aqueous ethanolamine has been reported by Ladisch and Tsao as an extremely effective agent for the complete solubilization of cellulose. Cellulose dissolved in this manner shows extremely high effective rates of saccharification by cellulases from *T. viride*. Complete conversion to glucose has been demonstrated in this manner (77).

V. ECONOMICS OF THE ETHANOL FERMENTATION

Traditional yeast fermentations are capable of converting glucose and many other hexoses, but not pentoses, to ethanol at over 95% of the theoretical yield of 0.5 gms of ethanol per gram of glucose. At this yield, approximately 13 lbs. of sugar are required per gallon of ethanol produced. With industrial ethanol currently selling at approximately $1.60 per gallon and sugar priced between 9 and 15 cents per pound (with government price supports), the raw material costs alone represent between 70 and 120% of the selling price. Due to the substrate intensive nature of this fermentation, molasses and various sources of starch have been under investigation as alternate substrates for the production of ethanol. The starches require hydrolysis, usually with amylases prior to fermentation. A comparison of the manufacturing cost of ethanol obtained from these alternative substrates along with credits obtained from the sale of byrpdoucts is presented in Table 4 (78). The relatively small quantities of these

TABLE 4. Raw Material Costs and By-Product Credits for Various Feedstocks (78)

Raw Material	Salable By-Product	By-product Credit (per gal. EtOH)	Net Manufacturing Cost (per gal. EtOH)
Sugar	--	--	$ 1.16
Molasses	Molasses stillage	$ 0.12	0.85
Corn	Corn distillers dry solubles	0.36	0.60
Wheat	Wheat distillers dry solubles	0.44	0.85
Potatoes	Potato distillers by-product	0.03	0.66
Sugar Beets	Sugar beet distillers by-product	0.60	0.68

substrates, their volatile price behavior (especially if diverted for industrial ethanol production), as well as social concerns regarding the use of foodstuffs for industrial ethanol production all serve to limit the potential of these carbohydrate sources. It has long been recognized, however, that wood or waste cellulosics form a plentiful, inexpensive, and unutilized source of carbohydrate. A comparison of the availability of these materials is given in Table 5. Utilization of this biomass has fostered research into a variety of chemical and enzymatic means of hydrolysis. Although considerable research into these routes of saccharification has been conducted, these processes have not resulted in the demonstration of commercial feasibility. Consistently, one of the major

TABLE 5. U.S. Agricultural Production of Potential Feedstocks[a]

Feedstocks	*Annual Production*	*Equivalent Ethanol Production*[b] *(million gallons)*
Cane Sugar	5.2×10^{9} *lbs.*	*416*
Molasses	1.64×10^{8} *gallons*	*65.6*
Corn	6.357×10^{9} *bushels*	*24,450*
Wheat	2.026×10^{9} *bushels*	*7,790*
Potato	3.52×10^{10} *lbs.*	*507*
Sugar Beets	25.1×10^{6} *tons*	*510*
Bagasse	2×10^{10} *lbs.*	*800*
Corn Stover	7×10^{9} *bushels*[c]	*14,000*
Wheat Straw	2.0×10^{9} *bushels*[c]	*4,000*
Industrial Ethanol Production (million gallons)		*300*
Gasoline Consumption (U.S. million gallons)		*(150,000)*

[a]*Statistical Abstract of the United States (1978) (79) U.S. Department of Commerce.*

[b]*At theoretical yield utilizing all carbohydrate available.*

[c]*Estimates.*

drawbacks has been the low yield of fermentable sugar produced per pound of cellulosic material processed. This has been due in part to the inability of the yeast cells to ferment the pentose fraction of the biomass (up to 35% of the substrate weight) as well as the low yields of sugar and resultant concentration required during pretreatment and hydrolysis. Thus, the final cost per pound of utilizable glucose is often quite high.

Corn stover, for example, contains 40% cellulose, 30% hemicellulose, and is quoted to be available at 1.5 cents per pound. If only the cellulose portion of this carbohydrate is utilized, it represents sugar at 3.75 cents per pound plus pretreatment and hydrolysis costs.

High temperature short contact time acid hydrolysis in an extrusion reactor represents one of the recent technologies for chemical hydrolysis. An economic evaluation of recent data with 55% conversion of wood to fermentable sugars has yielded raw material cost of $1.25 to $1.40 per gallon of ethanol produced (see Table 6) (80). With the addition of fermentation costs of 30 cents per gallon, this method represents a manufacturing cost between $1.55 and $1.70 per gallon. Alternatively, these investigators have proposed to hydrolyze the biomass under milder conditions subsequently use enzymatic

TABLE 6. Economics of Ethanol Production for Alternate Methods of Hydrolysis of 41% Glucan Biomass[a] (80)

	Acid Hydrolysis	*Dilute Acid Pretreatment Followed by Enzymatic Hydrolysis*
Conversion	*55%*	*90%*
Cost per gallon of ethanol	*--*	*--*
Substrate	*88.4 ¢*	*57¢*
Dilute Acid Pretreatment	*--*	*28¢*
Hydrolysis Cost	*36-51¢*	*40¢*
Sugar Concentration	*--*	*1.5¢*
Raw Material Cost	*1.25-1.40*	*1.26*
Fermentation	*30¢*	*30¢*
Total Manufacturing Cost	*1.55-1.70*	*1.56*

[a]*No credits have been taken for sale of byproducts.*

hydrolysis to free the remaining sugars. Economically, this yields a raw material cost of \$1.26 and final cost of \$1.56 per gallon of ethanol produced (Table 6). Thus, in the best case reported, the raw material cost represents 79% of the selling price of ethanol at \$1.60 per gallon. It is clear, however, that an intrinsic process advantage could be gained if both the cellulose and hemicellulose could be utilized for ethanol production. Raw materials under these conditions could represent as little as 27 cents per gallon of ethanol plus pretreatment, hydrolysis, and fermentation costs. Thus, by combining the cellulolytic and hexose fermenting capability of *C. thermocellum* with the ability of *C. thermosaccharolyticum* to ferment pentoses, this process strives to take advantage of two major factors. These are the potential for complete utilization of the substrate, as well as obviating the need for separate acid or enzymatic hydrolysis prior to the fermentation.

An economic evaluation of this process design, recently made by Battelle Laboratories for a 1,500 ton per day plant, gives a manufacturing cost of \$1.05 per gallon of ethanol (Table 7) (81). In order to achieve this, a continuous culture system was designed based on an ethanol productivity of 1.37 g/l hr operating at a dilution rate of 0.03 hr^{-1}. Under these conditions, the mixed cultures were assumed to produce 45 g/l ethanol, 22 g/l acetic acid, 30 g/l lactic acid, and 7.7 gms/l cells (Table 8). An alternate design was also costed in which the microorganisms were assumed to be capable of fermenting the carbohydrates solely to ethanol. This design yielded a cost of 79 cents per gallon of ethanol produced (Table 9). The authors of this economic evaluation pointed out that no provision was included for the cost of storage or sterilization of the stover, although surplus steam was generated from combustion of the lignin produced.

IV. CONCLUSIONS

In conclusion, there exists an alternate technique for the utilization of cellulosic materials for the production of useful chemicals. Direct microbiological degradation of biomass by cellulolytic microorganisms represent the first step for such a process. In this manner, direct fermentative production of alcohols or acids is possible. Through the use of mixed but defined anaerobic cultures, a more directed approach towards the production of the desired product can be further designed. This approach offers the advantage of effectively utilizing all of the fermentable carbohydrates in cellulosic biomass. It should be mentioned, however, that although ethanol was used

TABLE 7. Manufacturing Economics, MIT Ethanol Process Case A, 1979 % Plant Capacity 1500 Ton Stover/Day, 83,400 Gal Ethanol/Day (Operation - 24 Hrs/Day, 330 Days/Year) (Assumes a Yield of 0.18 GM EtOH/GM Corn Stover) (81)

	Feed Preparation	Fermentation	Distillation	Anaerobic Digestion	Offsites	Total
Fixed Investment, Million $	0.91	8.22	4.01	6.18	14.99	34.31
Dollars/Gallon Ethanol						
Fixed Capital Related	0.007	0.069	.034	0.051	0.126	0.29
Stover, $30/Dry Ton	--	0.541	--	--	--	0.54
Soda Ash, $80/Ton	--	0.044	--	--	--	0.04
Corn Steep Liquor, $101/Ton	--	0.149	--	--	--	0.15
Other Materials	--	0.060	--	--	--	0.06
Steam, $0.96/1000#[a]	--	0.004	0.023	--	(0.03)	--
Power, 3.2¢/kwhr	0.002	0.002	0.002	--	(0.007)	--
Power (Surplus) 1.6¢/kwhr	--	--	--	--	(0.05)	(0.05)
Direct Labor Related	0.009	0.018	0.009	0.004	(0.018)	0.06
Maintenance	0.002	0.018	0.008	0.004	(0.011)	0.04
Fuel Gas Credit, $2/MM Btu	--	--	--	(0.10)	--	(1.00)
TOTAL	0.02	0.91	0.08	0.04	0.07	1.03

[a]Low steam cost results from the use of low cost bagasse boilers and no fuel charges for combustible residue.

TABLE 8. Basis for Battelle Process Design (81)

Case	Productivity, g EtOH/l-hr	Dilution hr^{-1}	Initial Stover g/l	Product Concentrations, g/l					Yield[a] (gms EtOH/ gm C.S. Added)
				EtOH	Acetic Acid	Lactic Acid	Cells	Residue	
A	1.37	0.03	245	45	22	30	7.7	98	.18
B	1.37	0.03	165	45	0	0	7.9	66	.27

[a] The maximum theoretical yield based on a corn stover's composition of 77% holocellulose is 0.38 gm EtOH/gm Corn Stovers.

TABLE 9. *Manufacturing Economics, MIT Ethanol Process Case B, 1979 % Plant Capacity 1500 Ton Stover/Day, 123,800 Gal Ethanol/Day (Assumes a Yield of 0.27 gms EtOH/gm Corn Stover) (81)*

	Feed Preparation	Fermentation	Distillation	Offsites	Total
Fixed Investment, Million $	0.91	9.63	5.21	15.45	31.20
Dollars/Gallon Ethanol					
Fixed Capital Related	0.005	0.054	0.029	0.087	0.176
Stover, $30/Dry Ton	--	0.364	--	--	0.364
Soda Ash, $80/Ton	--	--	--	--	--
Corn Steep Liquor, $101/Ton	--	0.149	--	--	0.149
Other Materials	--	0.059	--	--	0.060
Steam, $0.96/1000#	--	0.004	0.023	(0.027)	--
Power, 3.2¢/kwhr	0.001	0.002	0.002	(0.005)	--
Power (Surplus), 1.6¢/kwhr	--	--	--	(0.030)	--
Direct Labor Related	0.006	0.015	0.006	0.012	0.039
Maintenance	0.002	0.014	0.008	0.008	0.031
TOTAL	0.013	0.665	0.067	0.044	0.79

in this review as a case study, there is no reason to believe that other chemicals cannot be produced using this concept. On the other hand, there still remain many technical and economic unknowns to be resolved. These include increasing product concentration, conversion efficiency, and rate of production. However, the potential advantages shown by these direct process concepts will continue to foster investigation to overcome these probelms.

REFERENCES

1. Woodman, J., *J. Agr. Science 17,* 333 (1927).
2. Khouvline, Y., *Ann. Inst. Pasteur 37,* 711 (1923).
3. Lynn, A.H. and Langwell, H., *J. Soc. Chem. Ind. 42,* 279 (1923).
4. Viljoen, J.A., Fred, E.B., and Peterson, W.H., *J. Agric. Science 16,* 1-17 (1926).
5. Tetrault, P.A., *Zenti. Bakt. Parasitenk III 81,* 28 (1930).
6. Snieszko, S., *Zent. Bakt. Parasitenk II 88,* 403.
7. Hungate, R.E., *J. Bact. 48,* 499 (1944).
8. Enebo, L., *Suensk Kem. Tod. 60,* 176 (1948).
9. McBee, R.H., "Studies on Thermophilic Cellulose Decomposing Bacteria", Ph.D. Thesis, State College of Washington (1948).
10. Patni, N.J. and Alexander, J.K., *Journal Bact. 105,* 226-231 (1971).
11. Ng, T.K., Weimer, P.J., and Zeikus, J.G., *Arch. Microbiology 114,* 1-7 (1977).
12. Shinmyo, A., Garcia-Martinez, D.V., Demain, A.L., *J. Appl. Biol. 1,* 202-209 (1979).
13. Lee, B.H. and Blackburn, T.H., *App. Micro. 30,* 346-353 (1975).
14. Gomez, R.F., Hernandez, P., Man. Submitted to Proceedings of the VIth Int. Ferm. Symp., London, Ontario, July 1980.
15. Fleming, R.W. and Quinn, L.Y., *Applied Microbiology 21,* 967 (1971).
16. Wang, D.I.C., Cooney, C.L., Demain, A.L., Gomez, R.F., and Sinskey, A.J. Degradation of Cellulosic Biomass and Its Subsequent Utilization for the Production of Chemical Feedstocks. Report No. COO-4198-9, Distribution Category UC-61, February (1979).
17. Weimer, P.J. and Zeikus, J.C., in "Anaerobic Degradation of Cellulose by Thermophilic Bacteria" (1975).
18. Wang, S.D., in "Production of Ethanol from Cellulose by *Clostridium thermocellum*" (M.S. Thesis M.I.T.) (1978).

19. Patni, N.J. and Alexander, J.K., *J. Bact. 105,* 220-225 (1971).
20. Uyeda, K. and Rabinowitz, J.C., *J.B.C. 246,* 3120-3125 (1971).
21. Erbes, D.I., Burris, R.H., and Orme-Johnson, W.H., *Pivas 72,* 4795-4799 (1975).
22. Ait, N., Creuzet, N., and Forget, P., *J. Gen. Microbiology 113,* 399 (1979).
23. Wang, D.I.C., Cooney, C.L., Demain, A.L., Gomez, R.F., and Sinskey, A.J., Degradation of Cellulosic Biomass and Its Subsequent Utilization for the Production of Chemical Feedstocks, Report No. COO-4198-9, Distribution Category UC-61, August (1977).
24. Sheth, K. and Alexander, J.K., *B.B.A. 148,* 808 (1967).
25. Sheth, K. and Alexander, J.K., *J.B.C. 244,* 457-464 (1969).
26. Schaefer, M.C. and King, K.W., *J. Bac. 89,* 113 (1963).
27. Gordon, J., Cooney, C.L., Man. Submitted to Proceedings of the VIth Int. Ferm. Symp., July (1980).
28. Ait, N., Creuzet, N. and Cattaneo, J., *B.B.R.C. 90,* 537 (1979).
29. Gordon, J.G., Jiminez, M., Cooney, C.L., and Wang, D.I.C., in "Biochemical Engineering: Renewable Sources of Energy and Chemical Feedstocks" (J.M. Nystrom and S.M. Barnett, eds.), Vol. 74 (1978).
30. Pheh, C.G. and Ordal, Z.J., *Applied Microbiology 15,* 893 (1967).
31. Lee, C.K. and Ordal, Z.J., *J. Bact. 94,* 530 (1967).
32. Fang, H.Y., "Ethanol Production Using *Clostridium thermosaccharolyticum*", M.S. Thesis, M.I.T. (1980).
33. Cynkin, M.A. and Gibbs, M., *J. Bact. 75,* 335 (1957).
34. Tsao, G.T., Ladisch, M., Ladisch, C., Hsu, T.A., Dale, B., Chou, T. Fermentation Substrates from Cellulosic Materials: Production of Fermentable Sugars from Cellulosic Materials (D. Perlman, ed.) in Annual Reports on Fermentation Processes, Vol. 2, Academic Press (1978).
35. Flickinger, M.C. and Tsao, G.T. Fermentation Substrates from Cellulosic Materials: Fermentation Products from Cellulosic Materials (D. Perlman, ed.) in Annual Reports on Fermentation Processes., Vol. 2, Academic Press (1978).
36. Shurz, I., Proceedings of the Bioconversion Symposium IIT Delhi 37, 58 (1978).
37. Siu, R.G.H., in "Microbial Decomposition of Cellulose, with Special References to Cotton Textiles" Rheinhold, New York (1951).
38. Detroy, R.W., Lindenfelser, L.A., Julian, G. St., and Orton, W.L. Presented at the Second Symposium on Biotechnology in Energy Production and Conservation, Gatlinberg, Tennessee, October (1979).

39. Fan, L.T., Lee, Y.H., and Beardmore, D.H., *Biotech. and Bioengineering 23,* 177 (1980).
40. Norkrans, B., *Physid. Plant 3,* 75 (1950).
41. Rautela, G.S., "The Importance of Crystal Lattic Structure of Cellulose in the Production and Action of Cellulose", Ph.D. Thesis, V.P.I. Blacksburg, Virginia (1967).
42. Etheridge, D.E., *Can. J. Bot. 35,* 615 (1957).
43. Cowling, E.B., in "Cellulose as a Chemical and Energy Resource" (C.R. Wilke, ed.) Biotechnology and Bioengineering Symposium #5, Wiley & Sons, New York (1974).
44. Whitaker, D.R., Colvin, J.R., and Cook, W.H., *Arch. Bioch. & Biophy. 49,* 257 (1954).
45. Militzer, W., Ikeda, C., and Kneed, E., *Arch. Biochem. 9,* 321 (1946).
46. Kneen, E. and Sandstadt, R.T., *Arch. Biochem. 9,* 235 (1946).
47. Porter, W., *J. Food Sci. 26,* 600-605 (1961).
48. Bell, T.R., Aurand, L., Etchells, J., *Bot. Gazz. 122,* 143-148 (1960).
49. Mandels, M. and Reese, E.T., in "Advances in Enzymatic Hydrolysis of Cellulose and Related Materials" (E.T. Reese, ed.) McMillan Company, New York (1963).
50. Williams, A.H., in "Methods in Polyphenol Chemistry" Symposium, Oxford, McMillan, New York (1963).
51. Stoessl, A. "Antifungal Compounds Produced by Higher Plants" (C. Steelink and V.C. Runeckles, eds.) Recent Advances in Phytochemistry, Vol. 3, Appleton Century Crofts (1970).
52. Link, K.P. and Walker, J.C., *J. Biol. Chem. 100,* 379 (1933).
53. Ramsey, G.B., Hieberg, B.C. and Wiant, J.S., *Phytopathology 36,* 245.
54. Owen, J.H., Sniker, J.C., and Stahmann, M.A., *Phytopathology 40,* 292 (1950).
55. Cruickshank, I.A.M. and Perrin, D.R., in "Biochemistry of Phenolic Compounds" (I.B. Harbourne, ed.) Academic Press, New York (1964).
56. Kiraly, Z. and Farkas, G.L., *Phytopathology 52,* 657-664 (1962).
57. Rice, E.L., *Physiol. Plant 18,* 255 (1965).
58. Byrde, R.J.W., Fielding, A.H., and Williams, A.H., in "Phenolics in Plants in Health and Disease" (J.B. Pridham, ed.) Pergamon Press, New York (1959).
59. Schaal, L.A. and Johnson, G., *Phytopathology 45,* 626 (1955).
60. Corcuera, L.J., Woodward, M.D., Helgeson, J.P., Kelman, A., and Upper, C.D., *Plant Physiology 61,* 791-795 (1978).

61. Woodward, M.D., Coruera, L.J., Helgeson, J.P., Kelman, A., and Upper, C., *Plant Physiol. 63,* 14-19 (1979).
62. Woodward, M.D., Corcuera, L.J., Helgeson, J.P., and Upper, C.D., *Plant Physiol. 61,* 796-802 (1978).
63. Woodward, M.D., Corcuera, L.J., Helgeson, J.P., Kelman, A., and Upper, C.D., *Plant Physiol. 61,* 803-805 (1978).
64. Virtanen, A.I., Hietala, P.K., and Wahlrous, O., *Arch. Bio. and Biophysics 68,* 486 (1959).
65. Wahlroos, O. and Virtanen, A.I., *Actachem. Scand. 13,* 1906 (1959).
66. Malekzadeh, *Mycopathology Myco. Appl. 54,* 73 (1974).
67. Lipinsky, E.S., in "Hydrolysis of Cellulose" (R. Brown and Jurasek, L., eds.) Advances in Chemistry Series - 181 ACS (1979).
68. Schurz, J. and John, K., *Cellulose Chem. and Technology 9,* 493 (1975).
69. Bartunek, R., *Koll. Z. 146,* 35 (1956).
70. Martinez, D.V.G., Ogawa, T., Shinmyo, A., and Enatsu, T., *J. Ferm. Tech. 52,* 378-387 (1974).
71. Ghose, T.K. and Bisaria, V.S., *Biotech. Bioengineering 21,* 131-146 (1979).
72. Toyama, N. and Ogawa, K., Proc. IV IFS, Ferm. Tech. Today (G. Tervi, ed.) Soc. Ferm. Tech., Kyoto, Japan (1972).
73. Han, Y.W., Tu, P.L. and Smith, S.K., *Biotechnology and Bioengineering 20,* 1015-1026 (1978).
74. Converse, A.O. and Grethlein, H.E., "Acid Hydrolysis of Cellulosic Biomass" Solar Energy Research Institute, 3rd Annual Biomass Energy Systems Conference, Golden, Colorado, 5-7 June 1979.
75. Linden, J.C., Murphy, V.G., Moreira, A.R., and Henk, L.L., "Combined Autohydrolysis and Organosolv Treatment of Wheat Straw" ACS Meeting - Microbial & Biotechnology Division, Washington, D.C., September 1979.
76. Kumakura, M. and Kaetsu, I., *B.B. 20,* 1309-1315 (1978).
77. Ladisch, M.R., Ladisch, C.M., and Tsao, G.T., *Science 201,* 743 (1978).
78. King, S.R., "Gasohol: Ethanol from Plant Matter as Motor Fuel" Industry Report, F. Eberstadt and Company, October 1969.
79. Statistical Abstract of the United States, U.S. Department of Commerce, Bureau of the Census (1978).
80. Knappert, D.R., Grethlein, H.E., and Converse, A.O., "Partial Acid Hydrolysis of Cellulosic Materials as a Pre-treatment for Enzymatic Hydrolysis" (Submitted to Bio-technology and Bioengineering).
81. Jenkins, D.M. and Reddy, T.S., "Economic Evaluation of the M.I.T. Process" Battle Columbus Laboratories (1979).

CHAPTER 7

THE NEW BIOLOGY: OPPORTUNITIES FOR THE FERMENTATION INDUSTRY[1]

Arnold L. Demain

Laboratory of Fermentation Microbiology
Department of Nutrition and Food Science
Massachusetts Institute of Technology
Cambridge, MA

I. INTRODUCTION

Being chosen as the first David Perlman Memorial Lecturer is the greatest honor ever bestowed upon me. I was a devoted observer and admirer of David Perlman throughout his scientific life. My work on β-lactam antibiotics, streptomycin, vitamin B_{12} and riboflavin was strongly influenced by his earlier research in these areas. David Perlman was a giant in his field, a fine researcher, teacher and historian. All of us in the fermentation field are indebted to him for his willingness to periodically recall in print the past, evaluate the present, and predict the future. I feel that Dave would have liked me to continue in this spirit today. Therfore, in memory of David Perlman, I have chosen to look at the past, present and future of genetics, fermentation, and the interaction between these two fields.

[1]*The first David Perlman Memorial Lecture of the Microbial and Biochemical Technology Division of the American Chemical Society. Presented in Las Vegas, Nevada on August 26, 1980.*

ISBN 0-12-040304-8

II. PHASES OF GROWTH OF THE INDUSTRY

The field of industrial fermentations can be divided on a historical basis into 3 phases. Phase I, which began in biblical times and ended in the 1930's brought us beer, wine, cheese, pickles, vaccines, alcohol and organic solvents. Phase II began with penicillin in the early 1940's and ended in the early 1970's. The second phase of growth was rich in microbial discoveries, processes and products, yielding antibiotics, organic acids, amino acids, vitamins, enzymes, insecticides, nucleotides, coccidiostats, gums, alkaloids, plant growth factors, animal growth promotants, single cell protein and transformed steroids. Of great importance in phase II was one aspect of microbial genetics, i.e. mutation, which shared center stage with major developments in microbial screening techniques and biochemical engineering.

Phase III began in the early 1970's and is being propelled almost exclusively by revolutionary discoveries and applications of molecular genetics. In 1970, I presented a plenary lecture in Prague optimistically titled "The Marriage of Genetics and Industrial Microbiology: After a Long Engagement, a Bright Future" (9). I am pleased to report to you that ten years later, there is no divorce in sight! We see today more and more conferences featuring papers by both technologists and basic geneticists. Since the establishment of Cetus Corporation in Berkeley in 1971, many new companies have appeared combining the talents of business people, geneticists, microbiologists and biochemical engineers--indeed, a genetic engineering industry has been born.

III. DEVELOPMENTS IN MICROBIAL GENETICS

To understand this new emphasis on genetics, we have to go back to Gregor Mendel in the mid-19th century. This Austrian monk began it all by his studies on green peas, discovering that their characteristics could be passed on from one generation to the next without change. His work led to the concept of genes as separately functioning units that dictate hereditary characteristics. Although many advances were made in the genetics of higher plants and animals in the late 1800's and early 1900's, we can jump to 1944 when Avery and his coworkers (1) at the Rockefeller Institute in New York discovered in bacteria that deoxyribonucleic acid (DNA) is the hereditary material. Today we know that DNA is made up of segments (genes) each of which codes for a particular polypeptide.

This was soon followed by the 1946 discovery of Lederberg and Tatum (31) at Yale that bacteria have sex. Up until that time, it was thought that bacteria propagate only asexually. The Lederberg-Tatum discovery made it possible to cross different parental types of the bacterium, *Escherichia coli*. In 1953, Watson and Crick (55) in Cambridge, England, discovered that DNA exists as two strands twisted about each other in the form of a double helix, giving us for the first time insight into the relationship between structure and function of DNA. Watson and Crick correctly proposed that four repeating compounds which appear throughout the DNA strands make up a four-letter alphabet that stores the genetic information in cells. In the early sixties, Nirenberg (36) at NIH and others cracked the genetic code, showing how this four-letter DNA alphabet is transcribed into the 4 letter alphabet and 3 letter words of messenger ribonucleic acid (mRNA) and how these three letter words of RNA are translated into the 20 amino acids of protein. During the 1960's, Monod and Jacob (28) in Paris elucidated the means by which cells control their metabolism, i.e. how the transcription and translation of the DNA code is turned on and off so that only at the proper time and under the proper conditions, does the cell convert the DNA-coded information into instructions that lead to the synthesis of enzymes and other proteins. Around 1970, the first natural gene was isolated by Shapiro and Beckwith (47) at Harvard, and Khorana (29) created the first artificial gene at Wisconsin.

IV. APPLICATION OF GENETICS TO THE FERMENTATION INDUSTRY

Before I mention the next discovery, let me take a moment to remind you that microorganisms can generate new genetic characters ("genotypes") by two means: mutation and sexual recombination. In mutation, a gene is modified either unintentionally ("spontaneous mutation") or intentionally ("induced mutation"). Although the change is usually detrimental and eliminated by selection (i.e. Darwin's survival of the fittest), some mutations are beneficial to the microorganism. Even if not beneficial to the organism, but beneficial to us, we can detect the mutation by screening and preserve it indefinitely. This is indeed what the fermentation microbiologists did in the strain development programs which led to the great expansion of the fermentation industry in phase II (10). Not only was mutation of fermentation organisms extremely valuable in strain improvement, it also was useful in eliminating undesirable fermentation products and in elucidating the biosynthetic pathways of natural products. The second way to mix genes, i.e. by genetic recombination between parents of different sexes, didn't work out well during the fermentation industry's phase II due to the extremely low frequency of genetic recombination in industrial microorganisms. For example, the frequency in streptomycetes is usually 10^{-6} or even less (26).

In phase III, there has been a heightened interest in the application of genetics to the production of important microbial products such as antibiotics. An extended use of mutation is the technique of mutational biosynthesis applied for production of new antibiotic derivatives. Here antibiotic-negative mutants ("idiotrophs") are fed precursor analogues and produce antibiotic molecules never before formed in nature. One such antibiotic, hydroxygentamicin, is a less toxic gentamicin in animals and is about to undergo clinical testing (4).

Fermentation organisms such as the actinomycetes, which were ignored for years in basic genetic studies, have come under recent genetic scrutiny. Much work has been done on the new technique of polyethylene glycol-mediated protoplast fusion which increases recombination frequency [to 10^{-2} - 10^{-1} in the streptomycetes (26)] and expands the breadth of recombination, thus encouraging industry to devote more time to this second means of increasing genetic diversity. Protoplast fusion was first used with animal cells and plant cells, later with fungi (12), unicellular bacteria (14;45) and finally with actinomycetes (26). Although one cannot get viable and stable recombinants by fusing completely unrelated species, successful interspecific protoplast fusion and recombination has been accomplished between *Penicillium chrysogenum* and *Penicillium cyaneo-fulvum*, *Aspergillus nidulans* and *Aspergillus rugolosus* (37), various streptomycetes (25) and even between *Candida* and *Endomycopsis* (39). This broadening of the recombination spectrum can probably be increased even further by the recent finding that ultraviolet irradiation of *Streptomyces* protoplasts before fusion selectively favors the regeneration of recombinants and can increase recombination frequencies after fusion by ten-fold (24).

Recombination is especially useful when combined with conventional mutation programs for two reasons. The first deals with usual production of "sickly" organisms as a result of accumulated genetic damage over a series of mutated generations. By making a cross between a high-producing sick strain and a low-producing vigorously growing strain, it is possible to make a healthy high producer. For example, a cross via protoplast fusion has been carried out with strains of *Cephalosporium acremonium* from the Glaxo strain improvement program (21). A low-titer, rapidly-growing, spore-forming strain which requires methionine to optimally produce caphalosporin C was crossed with a high-titer, slow-growing, asporogenous strain which could use the less expensive inorganic sulfate. The progeny included a recombinant which grew rapidly, sporulated, produced cephalosporin C from sulfate and made 40% more antibiotic than the high-titer parent. The second application is the recombination of improved producers from a single mutagenesis treatment. In the usual strain improvement program, after considerable testing of survivors of a mutational step, the best producing mutant is retained for further mutagenesis and the rest of the improved producers are discarded. By recombination, one is able to combine the yield-increase mutations and obtain a superior producer before carrying out further mutagenesis. This has recently been demonstrated (56) with *Nodardia lactamdurans*; two improved cephamycin-C producing strains from the Merck strain development program were fused and among the recombinants were two cultures which produced 10-15% more antibiotic than the best parent.

Another useful application of genetic recombination is the discovery of new antibiotics by fusing two producers of different or even the same antibiotics. For example, the fusion of two different species of *Streptomyces* led to a new anthracycline antibiotic; only one of the parents could make an anthracycline (13). In another example, Mazieres *et al.* (32) fused *Streptomyces rimosus* f. *paromomycinus* (paromomycin producer) with *Streptomyces kanamyceticus* (kanamycin producer) and obtained a recombinant producing neomycin. Schupp and Nüesch (46) obtained three new rifamycins by fusing divergent lines of *Nocardia mediterranei* non-producing mutants from the rifamycin strain development program at Ciba-Geigy. According to Hopwood (25), these examples may reflect more the different expression of genes from parent A in the cytoplasm of parent B than the formation of hybrid antibiotics.

Plasmids are another important part of the new biology. Plasmids are bits of extrachromosomal DNA, carrying as little as two and as many as 250 genes, which can exist autonomously in the cytoplasm of a cell or can integrate into the chromosome. When they are present in the autonomous state, they usually reproduce at the same rate or at a somewhat higher rate than chromosomes do. Therefore, they exist at 1 to 30 copies per cell. However under certain conditions or in certain cells, they can be forced into reproducing much faster than chromosomal DNA yielding as many as 3000 copies of a plasmid gene per cell. This is what is meant by gene amplification, and it has been widely exploited in bacteria such as *E. coli*. Indeed it is possible to transfer any chromosomal gene (or cluster or genes) in *E. coli* to a plasmid and to amplify the gene, thus tremendously increasing gene dosage and enzyme formation to very high levels. *Bacillus* plasmids can be transferred from one cell to another by transformation in the presence of polyethylene glycol (6). Certain plasmids from *Pseudomonas* are termed "promiscuous" because they can be transferred to other Gram-negative genera such as *Escherichia, Salmonella, Klebsiella, Rhizobium, Agrobacterium, Acinetobacter* and *Proteus* (5,8). Indeed, they can put sex into bacteria that never had a generalized recombination system, they can mobilize chromosomal genes, and are very useful in genetic mapping (24,49). Plasmids from *Staphylococcus aureus* have been transformed into *Bacillus subtilis* cells where they replicate and express themselves (11). Some yeast cells contain as many as fifty plasmid molecules per cell (23).

Plasmid DNA has been detected in virtually all antibiotic-producing species and found to contain either structural genes or genes regulating the expression of the structural genes of antibiotic biosynthesis. In *Streptomyces*, a number of plasmids have been found (25): SCP1 and SCP2 in *Streptomyces coelicolor*, SRP1 in *S. rimosus* and SLP1 in *Streptomyces lividans*. The first 3 are sex plasmids and constitute an essential part of the sexual recombination process. Efforts are going on to try to exploit these discoveries, e.g. studies on plasmid gene amplification in streptomycetes aim at increasing the number of genes coding for antibiotic production. It should be noted that bacterial viruses can also be used for gene transfer and gene amplification. Success with plasmids or phage could markedly reduce the cost of antibiotics as well as that of development of new antibiotics.

V. RECOMBINANT DNA AND MONOCLONAL ANTIBODIES

Let me now return to my historical account of genetic discovery. In nature, genetic recombination occurs only between organisms of the same species or closely-related species. Indeed, even in the laboratory, protoplast fusion is restricted to genetically-related species. All organisms have enzymes known as restriction endonucleases which recognize foreign DNA and destroy it so that "illegitimate recombination" does not occur. The revolutionary discovery in 1973 by Cohen and Boyer (7) at Stanford did away with all that. These investigators discovered how to use restriction enzymes to cut DNA molecules, how to use another enzyme, called DNA ligase, to join DNA molecules of different organisms, and how to reintroduce the recombinant DNA via a vehicle such as a plasmid or a phase into *E. coli*. In this way, they defied nature and carried out recombination across species barriers. To repeat, they removed a gene from one species and in the test tube they enzymatically stitched ("spliced") it into plasmid DNA of *E. coli*, and introduced the hybrid DNA into *E. coli* where it was replicated ("cloned") and expressed so as to produce a new protein which *E. coli* had never produced before. Within three years (40,51), it was shown that eukaryotic (yeast)DNA not only could replicate in a prokaryote but could express itself in the bacterium. By 1977, genetically-engineered bacteria were making a mammalian gene for rat insulin (54) and within two years, human insulin itself (19). Somatostatin, a human brain hormone (27), and human growth hormone (18) were soon produced in genetically-modified *E. coli* strains. In 1978, bacterial DNA was successfully inserted into yeast chromosomes (22). In 1979, it was shown that yeast protoplasts could be transformed by a hybrid *E. coli* plasmid carrying a yeast gene (23). Earlier this year, the production of human interferon by *E. coli* was accomplished (34).

I should also backtrack to 1975 to mention a development of importance in the field of immunology. In that year, Kohler and Milstein fused a mouse skin cancer cell ("myeloma") with an antibody-producing white cell (30). The result was a hybrid cell ("hybridoma") which produces a pure specific antibody. Never before could such pure ("monoclonal") antibodies be produced, i.e. we had to rely on impure mixtures of antibodies to provide immunological protection against disease.

VI. OPPORTUNITIES FOR THE FERMENTATION INDUSTRY

The tools of the recombinant DNA practitioner are DNA sequencing, chemical DNA synthesis, DNA enzymology and cloning. Thus the DNA to be introduced can be isolated from a chromosome, chemically synthesized, or enzymatically synthesized from messenger RNA by use of the enzyme reverse transcriptase. The coming of the "new biology" offers many opportunities to the fermentation industry. Some of the more evident possibilities are described in the following paragraphs.

A. Mammalian Polypeptides.

The development of recombinant DNA technology will have its major impact on the production of rare polypeptides which are either enzymes, hormones, antibodies or other agents. Recombinant DNA technology should yield purer proteins than other technologies. For example, it avoids the problem of contamination with unwanted polypeptide hormones, serum albumin and other serum proteins, and viruses associated with tissue culture or tissue and blood extraction techniques. It should be more economical than organic synthesis. However, the products still have to be separated from microbial antigens, polypeptides and endotoxins. The most well-known example is human interferon which costs today about 22 billion dollars per pound but which offers hope against viral disease and cancer. It will be manufactured much more cheaply by the use of recombinant bacteria in the near future. Genetically-engineered interferon, despite the fact that it lacks the carbohydrate of the natural material, is biologically active in combatting viral infection in monkeys (50). I must point out that the success or failure of interferon in the clinic will be no reflection on genetic engineering. Indeed, genetic engineering is probably the cheapest way to determine whether interferon has a future in therapy.

Another molecule of importance is human insulin. Diabetics are normally administered pig or cattle insulin which is a 140 million dollar annual market in the U.S.A. Not only is the deamnd exceeding the supply, but one of every 20 diabetics is allergic to animal insulin.. Human insulin via genetic engineering will certainly be of medical importance. Eli Lilly & Co. has committed 40 million dollars for plant construction to produce genetically-engineered insulin by 1982.

Other products of immediate medical, if not economic, importance are somatostatin and human growth hormone. The latter is used to treat pituitary dwarfism in children. Human growth hormone is being produced in *E. coli* cells at the level of 100,000 molecules per cell or 20 mg/liter of broth; this amounts to 1% of *E. coli* soluble protein (42). The product has been reported to be comparable in structure, purity and activity with the pituitary material (50). Mouse-beta endorphin, a pituitary painkiller, can be made in *E. coli* (48). Although it can be made chemically, the cost is 50 million dollars per pound. Monoclonal antibodies are already on the market. Vaccine production will certainly yield to the new technology. These protein antigens can be made by cloning and expressing genes coding for viral, bacterial and parasite surface proteins. Already in progress is the development of hepatitis B and influenza vaccines and in the agricultural area, vaccines for foot and mouth disease and swine fever will yield increases in animal productivity. Silk, urokinase, casein, calf rennin, serum albumin, enzymes involved in blook coagulation and the complement system, and proteins (or genes) to cure genetic diseases including hemophelia are possible products of future engineered microbes.

B. Primary Metabolites

New processes for production of amino acids and vitamins could result either from recombinant DNA or from gene amplification technology. Many of the enzymes coding for structural genes of amino acid and vitamin biosynthesis are clustered on the chromosome of *E. coli* and other bacteria. In the fungus *Neurospora*, five consecutive enzymes of the early common pathway of aromatic amino acid synthesis exist as a single polypeptide coded by a single gene (15). Transfer of these operons to plasmid DNA or to phage followed by amplification could yield effective new processes. One example is the use of a lambda transducing phage into which was incorporated the *trp* operon of *E. coli* which normally occurs on the chromosome. Amplification of the phage resulted in the overproduction of *trp* enzymes to the remarkable degree that they constituted 50% of the cell's soluble protein (33). Other examples of amplification of *trp* operon genes have been carried out via plasmid amplification. Indeed if these enzymes can be made to act at their maximum potential, an economic tryptophan process could be at hand. Efforts are also underway to develop processes for the production of proline, biotin and riboflavin, among others. Improvement in the yield of single cell protein from methanol has been achieved by introduction of the gene for glutamate dehydrogenase into *Methylophilus methylotrophus* (16). This enzyme, in contrast to glutamine synthetase, does not waste any ATP to assimilate ammonia.

C. Microbial Enzymes

The area of enzyme production will certainly feel the impact of the new biology. Enzymes such as streptomycete glucose isomerase, whose development allowed the corn wet milling industry to convert corn starch to high fructose corn syrup and to take away 30% of the sweetener business from the sugar industry, could be made more efficiently by recombination, plasmid amplification or recombinant DNA technology. Amplification of the gene coding for penicillin acylase, an *E. coli* product, could result in a much more economical process for this important enzyme. The power of recombination was recently demonstrated by Yoneda (57) who recombined individual mutations (*amy*R3, *amy*S, *pap*S, *tmr* and *pap*M118), each of which increase α-amylase production by 2-7 fold in *B. subtilis*. A strain constructed by genetic transformation which contains all 5 mutations produced 250-fold more α-amylase. The transfer of genes coding for the amylases of *Aspergillus*, the glucoamylase of *Aspergillus* or *Rhizopus*, the glucose isomerase of *Streptomyces*, or the renin of *Mucor* into rapidly-growing bacteria could result in more economical enzyme processes. A recent example of the use of plasmid amplification was done using yeast plasmids. The *ura*3 gene was attached to the 2μm yeast plasmid and the complex was incorporated into a bacterial plasmid. When yeast protoplasts were transformed with this hybrid plasmid, the specific activity of orotidine-5'-monophosphate decarboxylase was increased up to 30-fold (17). Other work underway at various institutions includes the transfer of α-amylase, glucoamylase, and glucose isomerase genes into a single fructose-producing organism, the introduction of amylase and glucoamylase genes into *Saccharomyces cerevisiae* to allow alcohol production from starch, and the introduction into bacilli of genes forming commercial enzymes which are stable to both heat and extremes of pH. As pollution and energy become more and more serious problems, there will be a greater need for replacement of chemical processes by enzymatic ones. Such examples include the enzymatic production of alkene oxides and fructose (35). There are also opportunities for development of new enzyme processes for degradation of cellulose, hemicellulose and lignin to sugars and chemicals. Although there has been considerable work done on cellulases and hemicellulases, there is need for a real breakthrough in the lignin degradation area. It is conceivable that production of lignin-degrading enzymes via recombinant DNA techniques could solve the problem and convert this widespread and troublesome polymer into a source of aromatic chemicals.

D. New Processes for Liquid Fuel and Chemicals

Some of the most useful microorganisms for the future production of liquid fuels and chemicals from plant materials are the clostridia, which as a group, are capable of producing ethanol, lactic acid, acetic acids, acetone and butanol. Since most of these strains cannot grow on hemicellulose or cellulose and are mesophilic, it would be useful to be able to transfer genes coding for cellulase, hemicellulase, and thermophily from *Clostridium thermocellum*, for example, to other clostridia. Since hemicellulose is made up of pentose units, it would also be of advantage to be able to transfer the pentose-assimilation character between the clostridia. The high temperature optima (65-75°C) of some of the thermophilic clostridia present an opportunity for lessening the cost of distillation of ethanol and other solvents, helping to make these processes economic. Unfortunately, nothing is known about the genetics of the genus *Clostridium*. Attempts to correct this situation have begun at MIT (20).

E. Antibiotics and Other Secondary Metabolites

I have already mentioned the importance of crossing different improved mutants in maximizing the efficiency of strain development programs. Several other possibilities exist which could lead to new production processes for known antibiotics. It might be useful to transfer antibiotic-producing operons from slow-growing streptomycetes to rapidly-growing eubacteria (such as *E. coli* or *B. subtilis*) to achieve rapid growth and more reproducible antibiotic production. Other advantages could be more rapid nutrient uptake due to a greater surface/volume ratio, better oxygen transfer since filamentous organisms produce viscous non-Newtonian broths, better mixing and thus more reliable control of pO_2, pCO_2 and pH, and a better organism for mutagenesis. Another possibility is the transfer of such operons from one streptomycete to another in the hope that the structural genes might be better able to express themselves in another species. For example, a newly discovered aminoglycoside may be produced at very low levels, e.g. 10μg/ml, and a traditional strain-improvement program might take years to raise the titer to an economically feasible one. Transfer of the structural genes to a high kanamycin producer, which already possesses resistance mechanisms to aminoglycoside antibiotics, might yield a major increase in antibiotic titer. These possibilities would be unrealistic if antibiotic synthetase genes were scattered around the actinomycete genetic map, but fortunately they do not appear to be so. Recent studies have revealed clustering of the biosynthetic structural genes of actinorhodin (43) and the red prodiginine antibiotic of *S. coelicolor* (44), of oxytetracycline in *S. rimosus* (38,41), and of rifamycin in *N. mediterranei* (46). Thus it should be possible to incorporate chromosomal biosynthetic operons from actinomycetes into plasmids or phage and transfer them to *E. coli*, ot to other actinomycetes. The latter will require actinomycete cells. With regard to plasmids, these are being constructed from plasmid SLP 1.2 of *S. lividans* and plasmid SCP2* from *S. coelicolor* (2,3,53). Uptake of plasmid DNA into *Streptomyces* is best accomplished by the use of protoplasts in the presence of polyethyleneglycol (25). A phage vector is also being developed. Temperate *Streptomyces* phage C31 has been converted to a deletion mutant and an *E. coli* plasmid (pBR322) has been introduced into it. This bifunctional replicon reproduces as a phage in *Streptomyces* and as a plasmid in *E. coli*. Additional deletions have been selected which allow the insertion of DNA segments into the hybrid plasmid (52).

With respect to the production of new antibiotics, the techniques of mutational biosynthesis and protoplast fusion have already been mentioned. Of course, plasmid transfer and recombinant DNA techniques can also be used to introduce genes coding for antibiotic synthetases into producers of other antibiotics or into non-producing strains. Additional possibilities include introducing genes coding for enzymes which catalyze addition or elimination of particular chemical functions and those which catalyze formation and/or attachment of new sidechains or of new moieties (e.g. sugars) (Hopwood, personal communication).

F. Agricultural Products

The major agricultural advance which could come from the "new genetics" is the replacement of synthetic fertilizers by better nitrogen fixation. The hope of incorporating nitrogen fixing (*nif*) genes into nonleguminous plants is not meeting with great success but other possibilities exist. One is the transfer of *nif* genes into microorganisms growing symbiotically with crops. Other possibilities include the pelleting of plant seeds with genetically-engineered bacteria containing amplified *nif* genes.

The above items merely scratch the surface with respect to possible applications of the new biology. Certainly the future will also include exploitation of cell culture of both plant and animal cells.

In his writings and lectures, David Perlman was a great optimist concerning the future of fermentation. I predict that his vision will be realized and confirmed many times as we enter growth phase III of the industry and combine the new biology with the ancient art of fermentation.

VII. SUMMARY

The field of industrial fermentations is in a new phase of growth, propelled by revolutionary discoveries and applications of molecular genetics. Never before has there been such a close working relationship between basic and applied scientists and engineers. Although mutation had been successfully applied to industrial processes since the 1940's, recombination had not. This new growth phase of the industry will feature the application of genetic recombination in many ways. Recombination is being carried out via protoplast fusion, plasmid and viral transfer, and recombinant DNA technology. Success has already been obtained in antibiotic strain-improvement and in the discovery of new antibiotic derivatives. Amplification of genes is another valuable technique which will be used by itself or together with recombination. Recombinant DNA methodology has already succeeded in designing bacteria which make insulin, somatostatin, human growth hormone, interferon, albumin and endorphin. New opportunities include the microbial production of mammalian polypeptides, antibodies, vaccines, and new processes for production of primary metabolites, microbial enzymes, industrial chemicals, liquid fuel and antibiotics. Applications in the agricultural area are also anticipated. The vision of David Perlman and his faith in the future of fermentation will be realized in this new growth phase of the industry.

ACKNOWLEDGMENTS

I thank Ronald E. Cape and David A. Hopwood for suggestions and James Stramondo for preprints.

REFERENCES

1. Avery, O.T., Macleod, C.M. and McCarty, M. *J. Exp. Med.*, *79* 137. (1944).
2. Bibb, M.J., Schottel, J.L. and Cohen, S.N. *Nature, 284,* 526 (1980).
3. Bibb, M.J., Ward, J.M. and Hopwood, D.A. *Devel. Ind. Microbiol. 21,* 55 (1980).
4. Came, P.E., O'Connor, J.R., Dobson, R.A., Wagner, R.B. and Fabian, R.J. *Antimicrob. Agents Chemother., 16,* 813 (1979).
5. Chakrabarty, A.M. *In:* Microbiology-1976 (Schlessinger, ed.) pp. 579-582. Am. Soc. Microbiol., Washington, D.C. (1976).
6. Chang, S. and Cohen, S.N. *Molec. Gen. Genet., 168,* 111 (1979).
7. Cohen, S.N., Chang, A.C.Y., Boyer, H.W. and Helling, R.B. *Proc. Natl. Acad. Sci.* USA, *70,* 32 (1973).
8. Datta, N., Hedges, R.W., Shaw, E.J., Sykes, R.P. and Richmond, M.H. *J. Bacteriol., 108,* 1244 (1971).
9. Demain, A.L. *In:* Genetics of Industrial Microorganisms: Bacteria (Vanek, Z., Hostalek, Z. and Cudlin, J., eds.) pp. 19-32, Academia, Prague. (1973).
10. Demain, A.L. *Adv. Appl. Microbiol., 16,* 177 (1973).

11. Erhlich, S.D. *Proc. Natl. Acad. Sci.* USA, *74*, 1680 (1977).
12. Ferenczy, L., Kevei, F. and Zsolt, J. *Nature, 248*, 793 (1974).
13. Fleck, W.F. *In*: Genetics of Industrial Microorganisms (Sebek, O.K. and Laskin, A.I., eds.) pp. 117-122, Am. Soc. Microbiol., Washington, D.C. (1979).
14. Fodor, K. and Alfoldi, L. *Proc. Natl. Acad. Sci.* USA, *73*, 2147 (1976).
15. Gaertner, F.H. and Cole, K.W. *Biochem. Biophys. Res. Commun., 75*, 259 (1977).
16. Felfand, D.H. Abstract No. MICR-60. 2nd Chem. Cong. N. Amer. Continent, Las Vegas, Nevada. (1980).
17. Gerbaud, C., Fournier, P., Blanc, H., Aigle, M., Heslot, H. and Guerineau, M. *Gene, 5*, 233 (1979).
18. Goeddel, D.V., Heyneker, H.L., Hozumi, T., Arentzen, R., Itakura, K., Yansura, D.G., Ross, M.J., Miozzari, G., Crea, R. and Seeburg, P. *Nature, 281*, 544 (1979).
19. Goeddel, D.V., Kleid, D.G., Bolivar, F., Heyneker, H.L., Yansura, D.G., Crea, R., Hirose, T., Kraszewski, A., Itakura, K. and Riggs, A.D. *Proc. Natl. Acad. Sci.* USA, *76*, 106 (1979).
20. Gomez, R.F., Snedecor, B. and Mendez, B. Abstract No. 7, Soc. Ind. Microbiol. 37th General Meeting, Flagstaff, Arizona. (1980).
21. Hamlyn, P.F. and Ball, C. *In*: Genetics of Industrial Microorganisms (Sebek, O.K. and Laskin, A.I., eds.) pp. 185-191, Am. Soc. Microbiol., Washington, D.C. (1979).
22. Hinnen, A., Hicks, J.B. and Fink, G.R. *Proc. Natl. Acad. Sci.* USA, *75*, 1929 (1978).
23. Hinnen, A., Hicks, J.B., Ilgen, C. and Fink, G.R. *In*: Genetics of Industrial Microorganisms. (Sebek, O.K. and Laskin, A.I., eds.) pp. 36-43. Am. Soc. Microbiol., Washington, D.C. (1979).
24. Hopwood, D.A. *In*: Genetics of Industrial Microorganisms (Sebek, O.K. and Laskin, A.I., ed.) pp. 1-9, Am. Soc. Microbiol., Washington, D.C. (1979).
25. Hopwood, D.A. Symp. Soc. Gen. Microbiol., *31*, in the press. (1981).
26. Hopwood, D.A., Wright, H.M., Bibb, M.J. and Cohen, S.N. *Nature, 268*, 171 (1977).
27. Itakura, K., Hirose, T., Crea, R. and Riggs, A.D. *Science, 198*, 1056 (1977).
28. Jacob, F. and Monod, J. *J. Mol. Biol., 3*, 318 (1961).
29. Khorana, H.G., Agarwal, K.L., Buchi, H., Caruthers, M.H., Gupta, N.K., Kleppe, K., Kumar, A., Ohtsuka, E., Rajbhandary, V.L., VandeSande, J.H., Sgarmella, V., Terao, T., Weber, H. and Yamada, H. *J. Mol. Biol., 72*, 209 (1972).

30. Kohler, G. and Milstein, C. *Nature, 256,* 495 (1975).
31. Lederberg, J. and Tatum, D.L. *Nature, 158,* 588 (1946).
32. Mazieres, N., Peyre, M. and Penasse, L. Abstract No. F-4.5.1(P), VIth Internatl. Ferm. Symp., London, Ontario. (1980).
33. Moir, A. and Brammar, W.J. *Molec. Gen. Genet, 149,* 87 (1979).
34. Nagata, S., Taira, H., Hall, A., Johnsrud, L., Streuli, M., Ecsodi, J., Boll, W., Cantell, K. and Weissman, C. *Nature, 284,* 316 (1980).
35. Neidleman, S.L., Giegert, J., Amon, W.F., Liu, E. and Wolf, B. Abstract No. F-21.4(L) VI Internatl. Ferm. Symp., London, Ontario. (1980).
36. Nirenberg, M.W. and Matthaei, J.H. *Proc. Natl. Acad. Sci.* (U.S.) *47,* 1588. (1961).
37. Peberdy, J.F. *In*: Genetics of Industrial Microorganisms (Sebek, O.K. and Laskin, A.I., eds.) pp. 192-196, Am. Soc. Microbiol., Washington, D.C. (1979).
38. Pigac, J. and Alacevic, M. *Period. Biol., 81,* 575 (1979).
39. Provost, A., Bourguignon, C., Fournier, P., Ribet, A.M. and Heslot, H., *FEMS Microbiol. Lett., 3,* 309 (1978).
40. Ratskin, B. and Carbon, J. *Proc. Natl. Acad. Sci.* USA *74,* 487 (1977).
41. Rhodes, P.M., Winskill, N., Friend, E.J. and Warren, M. (1981) cited in reference 25.
42. Ross, M.J. Paper presented at FDA Meeting (Production of medically important polypeptides using recombinant DNA Technoloty) June 3-4. (1980).
43. Rudd, B.A.M. and Hopwood, D.A. *J. Gen. Microbiol., 114,* 35 (1979).
44. Rudd, B.A.M. and Hopwood, D.A. *J. Gen. Microbiol.,* in the press. (1980).
45. Schaeffer, P., Cami, B. and Hotchkiss, R.D. *Proc. Natl. Acad. Sci.* USA, *73,* 2151 (1976).
46. Schupp, T. and Nüesch, J. 7th Course Internat. School Gen. Genet.: Microbial Breeding II, Erice, Sicily. (1980).
47. Shapiro, J., MacHattie, L., Eron, L., Ihler, G., Ippen, K., Beckwith, J., Arditti, R., Reznikoff, W. and MacGillivray, R. *Nature, 224,* 768 (1969).
48. Shine, J., Fettes, I., Lan, N.C.Y., Roberts, J.L. and Baxter, J.D. *Nature, 285,* 456 (1980).
49. Stanisich, V.A. and Holloway, B.W. *Genet. Res., 17,* 169 (1971).
50. Stebbing, N., Olson, K., Lin, N., Harkins, R.N., Snider, C., Ross, M.J., Fields, F., May L., Fenns, J., Fodge, D. and Prender, G. Paper presented at FDA Meeting (State-of-the-art: insulin and growth hormone) June 3-4. (1980).
51. Struhl, K., Cameron, J.R. and Davies, R.W. *Proc. Natl. Acad. Sci.* USA, *73,* 1471 (1976).

52. Suarez, J.E. and Chater, K.F. *Nature, 286,* 527 (1980).
53. Thompson, C.J., Ward, J.M. and Hopwood, D.A. *Nature, 286,* 525 (1980).
54. Ullrich, A., Shine, J., Chirgwin, J., Rictet, R., Tischer, E., Rutter, W.J. and Goodman, H.M. *Science, 196,* 1313 (1977).
55. Watson, J.D. and Crick, F.H.C. *Nature, 171,* 737 (1953).
56. Wesseling, A.C. and Lago, B.D. Abstract No. 72, Soc. Ind. Microbiol. 37th General Meeting, Flagstaff, Arizona. (1980).
57. Yoneda, Y. *Appl. Environ. Microbiol., 39,* 274 (1980).

ANNUAL REPORTS ON FERMENTATION PROCESSES, VOL. 4

CHAPTER 8

LARGE-SCALE PRODUCTION OF HUMAN INTERFERONS

R. J. White
F. Klein
J. A. Chan
and
R. M. Stroshane

Chemotherapy Fermentation Program
Frederick Cancer Research Center
Frederick, Maryland

I. INTRODUCTION

Interferons are a family of low molecular weight glycoproteins produced by vertebrate cells in response to a variety of stimuli. Initially they were discovered as virus induced factors resulting in resistance to further virus challenge (1). Subsequent work has shown that, in addition to viruses, a wide range of microbes, microbial products, nucleic acids (natural and synthetic), and low molecular weight compounds can induce the synthesis of interferon *in vitro* and *in vivo* (2). Furthermore, it is now apparent that the antiviral activity is but one aspect of a wide range of effects that interferons are capable of exerting on cells.

A. Interferons as Antiviral Agents

A number of important features of interferon's antiviral activity have been established:

(1) Interferon induced by one type of virus renders cells resistant to infection by other types of virus, i.e. activity is virus non-specific.

ISBN 0-12-040304-8

(2) Interferon is only effective in protecting a restricted range of cells from virus infection. Initially interferon's antiviral activity was described as species specific (only active on cells of the species that produced it), but examples now exist in which similar and even greater antiviral activity has been demonstrated on cells from different species (3, page 136). For each interferon there is a defined host range on which it is capable of exerting antiviral activity.

(3) Interferon, *per se*, is not directly responsible for antiviral activity but induces the synthesis of specific proteins in host cells which are responsible for the inhibitory effect (4).

The important role that interferons play in spontaneous recovery from viral infections (3, page 266), coupled with their broad spectrum of activity led to early optimism about their therapeutic potential. In particular, interest has centered on their use against those viral infections intransigent to vaccine treatment, e.g. certain respiratory and eye infections, and rare life-threatening diseases such as Lassa fever and rabies (5).

B. *Interferons as Anticancer Agents*

Much excitement has been generated recently by preliminary reports of interferon's activity in the treatment of certain human cancers. The first studies on cancer stemmed from the antiviral activity of interferon and were limited to animal tumors of known or suspected viral etiology (6). Encouraging results were obtained but in the majority of cases the anticancer activity demonstrated was prophylactic rather than therapeutic (7). However, the realization that the antiviral effect was only one of a number of biological activities that interferons possess encouraged further experiments with tumors that were of non viral etiology. Interferon was first tested against human malignant disease in Sweden by Strander (8). Osteogenic sarcoma was the cancer chosen for study and interferon was given as an adjuvant following surgery. The objective was to evaluate whether interferon delayed the onset of metastases and/or improved survival. Although this study is still in progress, the results reported so far indicate a definite anticancer activity. However, this clinical trial has been criticized because of the control used. Nonetheless, this pioneering study has had a great psychological impact on the field of interferon research and has been a potent stimulus to the clinical use of exogenous interferon in the treatment of cancer (5). One of the difficulties in ascribing other activities to interferon continues to be the very low purity of the material used. The variety of biological effects observed include inhibition of cell proliferation, alteration of cell surfaces, and modification of the immune response (7). It seems likely that any antitumor activity of interferon is the result of a composite of its effects on the immune system and as an inhibitor of cell growth. Of particular interest is the ability of interferon to stimulate the ability of certain lymphocytes, termed natural killer cells, to specifically lyse tumor cells *in vitro* (9).

C. Characterization and Classification of Interferons

By definition, interferons are relatively small glycoproteins which exert broad spectrum antiviral activity through cellular metabolic processes involving synthesis of both RNA and protein (4). Deglycosylated interferons have been prepared by enzymatic or chemical cleavage, or by preventing glycosylation through the use of glycosylation inhibitors *in vivo* (5). These "interferoids" remain biologically active *in vitro* and studies have revealed that the carbohydrate moieties are not involved in hydrophobicity or antigenicity (3, page 156).

Until very recently, little could be said about the structure of interferons. Some general properties of interferons were used to classify the various types, including: pH stability ("Type I" interferons are stable at pH 2, "Type II" interferons are not); apparent molecular weight (18,500-daltons for "leukocyte-type" interferon, 38,000-daltons for "fibroblast-type"); producer cells (fibroblast, leukocyte, lymphoblastoid); antigenic properties (three antigenically distinct groups are known: α, β and γ interferons). These characteristics can overlap somewhat as in the case of lymphoblastoid cells, where both "fibroblast" and "leukocyte" interferons are produced simultaneously (10). Because of these numerous, nonspecific classifications, an interferon nomenclature committee was organized in an attempt to standardize the interferon classification system (11). Their recommended abbreviations for the known interferons are given in Table 1. The emphasis in this system is on antigenic determinants, with modifiers to denote the animal species and producer cells. Thus, the 18,500-dalton interferon produced by human lymphoblastoid cells is abbreviated HuIFN-α (Ly).

After years of painstaking purification, picomole quantities of homogeneous human interferons were isolated and analyzed using modern approaches to classical amino acid

TABLE 1.

New Nomenclature	*Old Nomenclature*
IFN-α	*Le(leukocyte), type I, pH 2 stable, foreign cell-induced*
IFN-β	*F(fibroblast), Fi, type I, pH 2 stable*
IFN-γ	*IIF(immune), type II, T, pH 2 labile, antigen-induced, mitogen-induced*

sequencing. The results of such approaches have elucidated the partial sequences of lymphoblastoid (12,13) and fibroblast (14-16) interferon proteins.

With the advent of recombinant DNA methods (17), the amino acid sequences of leukocyte (18) and fibroblast (19,20) interferons have been totally determined based upon the DNA sequences. This methodology provides the advantage of an internal confirmation of the described sequence by analyzing the complimentary nucleotide sequence in the double stranded DNA. The leukocyte and fibroblast interferons are both comprised of 165 amino acids. There is a 45% homology between the nucleotide sequence of the two genes and a 29% homology between the amino acid sequence of the two proteins (21). There are four regions of high homology that could be responsible for the common functions of the two proteins, such as the antiviral activity (22). To date, only the sequence of feron has been reported (13). Comparison of this sequence with those of fibroblast and leukocyte interferons (21) show only 10% homology with the fibroblast interferon sequence and 75% homology with the leukocyte interferon sequence. Thus, although the two 18,500-dalton proteins are similar antigenically and structurally (at least in the first 20 amino acids) they are not identical.

D. Need for Large-Scale Production

Progress in evaluating the effect of exogenous interferon as a broad spectrum antiviral and anticancer agent has been severely hampered by the limited supply of material available. Interferon is produced in exceptionally small quantities and has proved difficult to purify. Species specificity (or more correctly, limited host range) has precluded the use of animal-derived interferon in the treatment of human diseases. The encouraging results obtained in the treatment of osteosarcoma coupled with the American Cancer Society's and the National Cancer Institute's decisions to purchase large quantities of interferon for clinical trials, have spawned a budding interferon industry. Currently methodology has been described for the large scale production of leukocyte, lymphoblastoid and fibroblast interferons and this report will confine itself to these materials. This is in no way meant to reflect the relative clinical potential of other types such as immune interferon, but simply represents the status of published production technology at the time of writing. It is these processes that will be described and compared. Interferon for clinical use and the exciting area of genetic engineering for eventual production of interferon by procaryotes will be mentioned.

II. LARGE-SCALE PRODUCTION OF CRUDE INTERFERON

A. *Human Leukocyte Interferon*

An outline of the procedure used to produce leukocyte interferon is shown in Figure 1.

1. *Origin of Cells.* Most of the interferon used to date in clinical trials has been produced from human leukocytes by methods developed by Cantell and coworkers (23). The leukocytes, obtained as a by-product of blood donations for transfusion therapy, are diploid cells and can be used only once for the production of human leukocyte interferon. The production process involves centrifugation of the whole blood in the plastic bags used for collection, and removal of the leukocyte layer ("buffy coat") from each bag. Since only 10-30 ml of "buffy coat" are recovered per unit of blood (400 ml), the "buffy coats" from many bags are pooled and stored overnight at 2-4°C. On the average, one unit of blood contains 3×10^9 leukocytes and approximately one-third (0.95×10^9) of these are recovered in the process.

2. *Preparation of Leukocytes for Production.* The leukocytes in the pooled "buffy coat" layers are contaminated with erthyrocytes. Since these erythrocytes compete with leukocytes for the inducer virus, they must be removed from the pool. This is accomplished by lysing the erythrocytes by addition of 0.83% ammonium chloride at 4°C for ten minutes, followed by centrifugation. The leukocytes are resuspended in phosphate buffered saline and the process is repeated. These

Whole blood from individual donors separated into three layers by centrifugation (plasma, "buffy coat", erythrocyte). "Buffy coat" layers from numerous donations, collected and pooled.

↓

Contaminating erythrocytes removed from pooled "buffy coats" by successive treatments with ammonium chloride.

↓

Leukocytes primed in fresh medium by addition of low concentrations of crude exogenous interferon.

↓

Leukocytes induced with Sendai virus.

↓

Leukocytes removed by centrifugation, interferon-containing supernatant-harvested.

FIGURE 1. Outline of procedure for preparation of crude leukocyte interferon.

steps result in the loss of approximately one-half of the remaining leukocytes, giving a final leukocyte yield of 0.47 x 10^9 cells per "buffy coat".

The leukocytes are suspended at a concentration of 10^7 cells/ml in round bottom 2-liter flasks in Eagle's minimal essential medium (containing no phosphate) to a total volume of 900 ml. The medium is supplemented with 3 mg/ml tricine buffer, 25 μg/ml neomycin, and 4% human "agamma" serum (human serum treated with 35% saturated ammonium sulfate to precipitate γ-globulins, then desalted). The flasks are loosely covered with aluminum foil and stirred at 37.5°C.

Burke and Isaacs (24) reported in 1958 that treatment of cells with interferon prior to virus induction enhances interferon production. This process, known as priming, takes place 1-2 hours before induction and requires 100 units/ml of interferon.

3. *Induction.* Cantell and co-workers have found Sendai virus to be the most effective inducer in their system (25). Production of inducer virus is a tedious and time-consuming operation. Sendai virus is propagated in 11-day-old embryonated chicken eggs and the allantoic fluid is harvested after 65 hours incubation at 37°C. The virus preparations are filtered through glass wool and stored at 4°C for periods of weeks. Alternatively, the virus preparation can be supplemented with 4% human "agamma" serum and stored at -70°C. The virus is added as infectious allantoic fluid 1-2 hours after priming, to a final concentration of 100-150 hemagluttination units per ml culture. The culture remains in the 37.5°C water bath with stirring for 16-20 hours. The cells and cell debris are removed by low-speed centrifugation and the final yield of interferon is approximately 10^6 units/"buffy coat" of 2.1 x 10^3 units/10^6 cells. This crude interferon-containing supernatant is the starting point for purification.

B. *Human Lymphoblastoid Interferon*

An outline of the procedure used to make human lymphoblastoid interferon is shown in Figure 2.

1. *Origin of Cells.* Cantell and co-workers, who have made such important contributions to the production of leukocyte interferon, realized that large-scale production of the leukocyte-derived material was always going to be limited by the availability of human blood (26). To circumvent this limitation, human lymphoblastoid cells have been screened as a potential source of human interferon. These cells have two

Biopsy from Burkitt's lymphoma

↓

Cell line established in vitro

↓

Master stock of lymphoblastoid cells maintained in liquid N_2.

↓

Scale-up of lymphoblastoid culture volume by successive transfer to vessels of increasing capacity.

↓

Growth of lymphoblastoid cells in production-scale vessel.

↙	↘
Induction with paramyxovirus	Cells harvested and primed in serum-free medium with low concentration of crude exogenous interferon.
↓	↓
Dilution in serum free medium	Induction with paramyxovirus
↘	↙

Synthesis of interferon

↓

Lymphoblastoid cells removed by centrifugation, crude interferon-containing supernatant harvested.

FIGURE 2. Outline of procedure for preparation of crude lymphoblastoid interferon.

particular advantages: they are cell lines having an indefinite life span; they grow well in suspension culture. Thus, they are especially suitable candidates for large-scale production. A total of 21 especially suitable candidates for large-scale production. A total of 21 lymphoblastoid lines were examined and the best interferon producer was found to be the Namalva (Namalwa) cell, which came from a patient with Burkitt's lymphoma (27). Almost all lymphoblastoid cell lines contain Epstein-Barr Virus (EBV) DNA, some of them are able to produce intact virus particles when suitably induced, others are non-producers only containing a portion of the viral genome. The Namalva line is a non-producer of EBV and cannot be superinfected with EBV. Several different groups have developed processes for the large-scale production of lymphoblastoid interferon using the Namalva line (28-31). Although the majority of the interferon produced by Namalva is antigenically indistinguishable from the leukocyte-derived material, preliminary data indicates that the amino acid sequences of the proteins are in fact different (18).

2. *Cultivation of Cells.* The Namalva cell line grows well in suspension culture using RPMI 1640 medium supplemented with fetal calf, calf, or human serum. Doubling times at 37°C are dependent on environmental conditions and range from 20-72 hr. The exact concentration and type of serum used has varied for the different groups involved in producing this material. Obviously, availability and cost of the serum component of the growth medium is of critical importance to the scale-up of these processes; fetal calf and human sera are both expensive and in short supply. Fortunately, it has proved possible to find acceptable substitutes and the use of cheaper, more readily available calf serum has been described (31), as also has the use of animal protein extracts (30) and even formulation of a serum-free medium (32). According to the literature Namalva cells have been cultivated in volumes up to 800 L in fermentation vessels (33) and the indications are that growth could be carried out in much larger vessels should the need arise. Obtaining a large quantity of cells to produce interferon basically consists of building up the volume of cell culture through a series of transfers in vessels of increasing size until enough inoculum is generated for the production-scale vessel. Generally inocula for the successive steps are in the 20-25% (v/v) range with initial cell densities *ca.* 5×10^5 viable cells/ml. Final cell densities achieved after 2-4 days growth are in the range $2\text{-}5 \times 10^6$ viable cells/ml.

3. *Induction*. As in the case of leukocyte interferon, synthesis is induced with a paramyxovirus. The two most commonly used inducers are Sendai and Newcastle disease virus (strain B-1), both produced in the allantoic fluid of embryonated eggs. No superinduction procedure similar to that devised for fibroblast interferon (see next section) has been shown to work with leukocytes or Namalva cells. Although the methods used to produce large quantities of Namalva cells are rather similar, there are significant differences between the published procedures as far as the induction process is concerned. The processes have the common objective of reducing the serum concentration present when the interferon is synthesized. The presence of large quantities of serum proteins can cause problems during the subsequent purification.

In one approach (29-31) Namalva cells are recovered from the serum-containing growth medium by centrifugation (or other recovery methods) and resuspended in serum-free medium. The presence of serum is not required for induction and synthesis of interferon by Namalva cells (31). Generally cells are resuspended in serum-free medium at a higher density (*ca.* 1 x 10^7 cells/ml) than was present in the growth medium (*ca.* 2-5 x 10^6 cells/ml).

In another approach cells in growth medium are diluted directly in serum-free medium. Published data has shown that this increased interferon yields on a per cell basis, but that the procedure became somewhat unreliable when the cell concentration dripped below a certain minimum value. Dilution of cultures to give a final cell count of *ca.* 9 x 10^5 cells/ml has been shown to give a reproducible system for interferon production (34). As cell densities in growth medium prior to dilution are in the range of 2-4 x 10^6 cells/ml dilution factors are usually in the range of 2.0 - 4.4 fold. Assuming a concentration of 10% serum in the growth medium, this corresponds to a final serum concentration in the 2.3% - 5.0% range subsequent to induction with virus.

There are advantages and disadvantages associated with both approaches. The recovery and resuspension of cells at a high density in serum-free medium gives rise to induced cultures containing a high concentration of interferon at a higher specific activity (there is less total protein present). However the large-scale recovery of viable cells represents a formidable technical obstacle. On the other hand, direct dilution of cells in growth medium with fresh serum-free medium avoids the problems associated with the large-scale recovery of cells for induction. The price that is paid for this processing convenience is that the interferon subsequently produced is less concentrated and has a lower specific activity; both these factors affect the subsequent purification procedure.

Typically, lymphoblastoid cells are induced with 10-70 hemagglutination units of paramyxovirus per 10^6 cells. In the process which resuspends cells in fresh, serum-free medium a small "priming" quantity of exogenous interferon may be necessary (generally *ca.* 100 units/ml) prior to treatment with the inducing virus. No primer is necessary in the case of cultures that are diluted directly, presumably a result of the presence spontaneously produced, low levels of endogenous interferon. Once induction has occurred, the synthesis and release of interferon is usually complete within 20-33 hours. Published data indicates that interferon titers produced by Namalva cultures range from 2 x 10^3 to 5 x 10^4 units/ml. However, it should be remembered that the concentration of cells present in the different processes used varies between 9 x 10^5 and 1 x 10^7 per ml. If the interferon titers are expressed on a per cell basis the range of titers produced is not as great and yields are typically 2-5 x 10^3 units/10^6 cells.

C. *Human Fibroblast Interferon*

An outline of the procedures used to make human fibroblast interferon is shown in Figure 3.

1. *Origin of Cells.* Fibroblasts for the production of human interferon have been derived from a variety of sources, most commonly from infant foreskins or embryonal tissues such as muscle and lung. They are diploid cells that can be propagated by anchorage-dependent culture. Like other primary diploid cells, they have a finite life span and can only be cultivated *in vitro* for *ca.* 60 population doublings. In practice the productivity of cells declines before this, and frequently cultures are only used up to *ca.* 35 population doublings (35).

2. *Cultivation of Cells.* A major limitation to the large-scale cultivation of anchorage-dependent cells, such as fibroblasts, is the provision of adequate surface area for growth. Conventional methods of growing small quantities of anchorage-dependent cells usually employ roller bottles, but this technique has a seriously limited scale-up potential. A variety of approaches have been made to solving this problem. These include the construction of special multisurface stacked plate propagators (36,37), use of spirals of plastic fiber (38), and recently the use of microcarriers (39,40). In the initial work with microcarriers Van Wezel used particles of a commercial DEAE dextran (Sephadex) to provide increased surface area for the attached growth of cells (39). Problems associated

Biopsy from normal tissue e.g. Infant foreskin.

↓

Diploid fibroblasts cultivated *in vitro*

↓

Master stock of characterized fibroblast cells maintained in liquid N_2.

↓

Scale-up of fibroblast culture volume by successive transfers to vessels of increasing capacity.

↓

Growth of fibroblasts in production-scale vessel.

↓

Priming with low concentration of exogenous interferon.

↓

Superinduction of "primed" fibroblasts:

i) induction with poly(I)-poly(C) in presence of cycloheximide.

ii) addition of actinomycin D

↓

Synthesis of interferon in fresh medium, lacking antibiotics and inducer.

↓

Removal of fibroblasts by centrifugation, interferon-containing supernatant harvested.

FIGURE 3. Outline of procedure for preparation of crude fibroblast interferon.

with the toxicity of these microcarriers have now been overcome by using a diethylaminoethyl substituted dextran microspheres with a significantly diminished exchange capacity (41). Growth on dextran microcarriers in suspension can give cell yields in the 1 x 10^6 cells/ml range with 5 mg/ml of microcarrier (40). Although rapid progress has been made with microcarriers, some important questions remain to be answered: Can the process be scaled-up beyond 3.0 liters (the largest culture volume published to date)? Can microcarriers be used to generate inocula for subsequent steps (the inoculum densities required by microcarriers are high)? Can cells be stripped from dextran and the microcarriers reused (a key economic factor)? Irrespective of the culture method employed, the growth medium normally used for fibroblasts is Eagle's minimal essential medium supplemented with 5-10% fetal calf serum.

3. *Induction.* Although fibroblasts, like leukocytes and lymphoblastoid cells, can be induced to make interferon by exposure to paramyxoviruses such as Sendai, they can also be induced with synthetic polynucleotides. The most commonly used polynucleotide is polyinosinic acid-polyctidylic acid [poly(I)·poly(C)]. Use of a chemically defined inducer instead of an infectious virus is advantageous from a safety and economic standpoint. Attempts to use poly(I)·poly(C) as an inducer with leukocytes and lymphoblastoid cells have not proved successful.

As a result of work done primarily in Vilcek's laboratory (42) a procedure for the superinduction of interferon synthesis by fibroblasts has been devised. This results in a dramatic increase in the interferon titers produced and improvements of up to 100-fold have been realized (42). Prior to carrying out superinduction, it is necessary to prime fibroblasts by incubation with a low concentration of exogenous interferon (typically 100 units/ml).

Superinduction consists of the sequential treatment of interferon-primed cells with poly(I)·poly(C) and the antibiotics cycloheximide and actinomycin D. Although the concentrations of inducer and antibiotics, and other details of the method vary from group to group a typical superinduction is carried out as follows: cells are grown to confluency and primed by preincubation with 100 units/ml of interferon for 16 hours; primed cells are then incubated with 500 μg/ml of the inducer poly(I)·poly(C) in serum-free medium containing 10 μg/ml of cycloheximide for 4 hours at 37°; at this stage actinomycin D is added to give a concentration of 1 μg/ml and the incubation continued for an additional hour; the medium containing inducer and antibiotics is then discarded, and the washed cells finally incubated in serum-containing medium for 24 hours at 37°C, during which time the interferon is synthesized and released. The presence of serum or a serum protein component is necessary during this latter phase of the process. The mechanism by which sequential treatment with cycloheximide and actinomycin D causes superinduction has been shown to involve increased transcription of interferon mRNA and increased translation of interferon mRNA into protein (43). Unfortunately, this superinduction procedure has not proved applicable to leukocytes or lymphoblastoid cells. After superinduction, fibroblast cultures at *ca.* 10^6 cells/ml will contain interferon titers in the range of 2-3 x 10^4 units/ml. Fibroblast containing supernatant is harvested and represents the starting point for purification.

III. LARGE-SCALE PURIFICATION

Although human leukocyte (44,45), fibroblast (15,46), and lymphoblastoid (47) interferons have been purified to homogeneity [based on sodium dodecyl sulfatepolyacrylamide gel electrophoresis (SDS-PAGE)], no large-scale process have been published which produce homogeneous interferons.

Material of *ca.* 10^6 units/mg protein is presently accepted for use in human clinical trials, but a great deal of research effort is being expended to deliver material of purity approaching homogeneity in the range of 10^8 - 10^9 units/mg protein.

A. *Human Leukocyte Interferon*

The most attractive method for the large-scale purification of leukocyte interferon is that of Cantell and Hirvonen (48) because this process does not involve column chromatography. Briefly, the procedure involves potassium thiocyanate precipitation of an interferon preparation with a specific activity, 2.0 x 10^4 units/mg protein. The precipitate is extracted with acidic ethanol and interferon selectively precipitated at pH 8.0 to yield a product with a specific activity of *ca.* 1 x 10^6 units/mg protein and an overall recovery of *ca.* 50%. This is currently the method of choice for producing leukocyte interferon for clinical trials. Another method yielding material of similar quality and which has a scale-up potential is the scheme reported by Paucker and Erickson (49). The procedure involves a modification of the acidic ethanol extraction of Cantell and Hirvonen followed by Blue Sepharose column chromatography. The extraction step removes sufficient protein contaminants to permit the use of a relatively small volume of Blue Sepharose. They report that their method gave a more consistent recovery of interferon with yields of *ca.* 40%.

B. Human Lymphoblastoid Interferon

Zoon and co-workers (33,47) have purified a large-scale lymphoblastoid interferon preparation (*ca.* 1 x 10^3 units/mg protein) using a four-step procedure. The process involves trichloroacetic acid (TCA) precipitation of the crude interferon from the cell-free culture supernatant, resuspension of the crude precipitate in water by slowly raising the pH to 6.8, Sephadex G-25 column chromatography, anti-lymphoblastoid affinity column chromatography and Sephadex G-150 column chromatography. The procedure yields interferon with a specific activity of *ca.* 3 x 10^6 units/mg protein and an overall recovery of *ca.* 25%. Johnston and co-workers (31) have also used an immobilized anti-interferon antibody column chromatography step in their purification scheme of material with a starting specific activity of 1.5 x 10^4 units/mg protein. In their procedure the interferon was also precipitated by TCA but was extracted with ethanol before application to the column. The final product with a specific activity of $>10^7$ units/mg protein has been used in preliminary clinical trials (50). Mizrahi and co-workers (30) have developed a two-step procedure for purifying interferon produced in a serum-free medium and having an initial specific activity of 2 x 10^5 units/mg protein. This process again requires initial TCA precipitation and is then followed by Ultrogel AcA 54 column chromotography and gives a final product with a specific activity of *ca.* 3 x 10^6 units/mg protein and an overall recovery of 80%. The method of Paucker and Erickson (49) was also applied in purifying lymphoblastoid interferon with an overall yield of about 35% and 1,000-fold purification.

C. Human Fibroblast Interferon

Billiau and co-workers (51) were able to purify fibroblast interferon with a starting specific activity of *ca.* 6 x 10^4 units/mg protein by a two-step procedure involving batchwise acid desorption from controlled-pore glass beads and dialysis against polyethylene glycol with an overall yield of 50% and a final product with a specific activity of *ca.* 1 x 10^6 units/mg protein. Knight and co-workers (15) applied crude interferon with a specific activity of *ca.* 8 x 10^4 units/mg protein from serum-free medium directly onto a Blue Sepharose column and recovered the product with a specific activity of *ca*. 5 x 10^7 units/mg protein and yields ranging from 50-100%. Both of these methods are applicable to large-scale purification.

D. Methods for Purification to Homogeneity

Many of the procedures used to purify the interferon preparations to homogeneity are not amenable to large-scale operations or result in very low recovery yields. Zoon and co-workers (47) used SDS-PAGE as the last step in the purification of deglycosylated lymphoblastoid interferon to obtain material with a specific activity of *ca*. 10^8 units/mg protein, as did Knight and co-workers (15) for the purification of fibroblast interferon. This step gives specific activity increases of 10-20 fold, but is neither scalable nor high-yielding. High performance liquid chromatography (HPLC) is rapidly emerging as the method of choice for purifying interferons to homogeneity. Rubinstein and co-workers (45) were able to purify leukocyte interferon to homogeneity using HPLC and this technique for purifying proteins is gaining widespread support (52-55). The method demonstrates excellent resolving power, good loading capacity, efficient recovery, and has the potential for scale-up to preparative liquid chromatography systems.

Antibody affinity chromatography has the potential to effect single step purification of proteins to homogeneity. Although anti-lymphoblastoid interferon affinity columns have been used effectively in large-scale purifications, the technique requires purified antigen in quantities large enough to immunize animals. The purification effected by use of these immobilized antibodies is directly dependent upon the purity of the antigen. Obviously, if there are impurities present in the interferon antigen preparation, antibodies of these impurities will also be produced and the immobilized antibody preparation will be specific for the impurities as well as the desired interferon. The dual problem of obtaining sufficient quantities of antibodies and having antibodies that are pure and specific for interferon has now been circumvented by the use of hybridomas (56). Secher and Burke (57,58) have prepared anti-lymphoblastoid interferon affinity columns with monoclonal antibody. They reported a 5,000-fold purification giving a specific activity of *ca*. 2 x 10^8 units/mg protein in a single step. The monoclonal antibody technique has the advantage of being able to produce specific antibody even when an impure antigen such as crude interferon is used and has the capability of producing unlimited amounts of antibody, and therefore has the potential for use in large-scale purification. This method can also be applied to other types of interferon.

IV. ACCEPTABILITY CRITERIA FOR HUMAN INTERFERONS

Interferon preparations intended for investigative human clinical use in the United States must meet the requirements established by the Bureau of Biologics of the Food and Drug Administration. The Bureau's concern includes the purity, potency and safety of these investigational products (58,59).

A. Purity

The purification procedure used in interferon production should not only increase the interferon concentration relative to other proteins (i.e. increase the specific activity) but also remove undesirable materials from the final product. The process used must withstand a challenge test: known contaminants are added to challenge the purification process and the final product analyzed to demonstrate that the undesirable contaminants have been removed. The types of agents and/or reagents tested for in this way will, of course, vary depending upon the type of interferon produced, the inducer used, and the purification process employed. As examples: if Sendai or Newcastle disease virus are used for interferon induction, the final product must be demonstrated free of infectious virus; if potassium thiocyanate precipitation is used, the residual amount of this compound in the final product must be determined; for lymphoblastoid interferon it should be demonstrated that viral and cellular DNA are effectively removed in the purification scheme.

The choice of antibiotic in the growth medium is important. The antibiotic chosen must be cleared with the Bureau and should be effectively removed in the purification process. Because of widespread allergic reactions to penicillins, these antibiotics should not be used.

At the present time, clinical interferon preparations should be of a specific activity of at least 10^6 units/mg protein. Since such a preparation is less than 1% pure, it is important to know what kind of protein makes up the other 99%. It would not be surprising to see the Bureau's policy change to one demanding a higher specific activity or indeed homogeneous interferon with defined protein stabilizers, as the technology becomes available for obtaining good recoveries of the pure material.

In addition to the above test, the activity of interferon preparations destined for the clinic should have the following characteristics: solubility (non-sedimentable at 100,00 x g for 2 hours); protein dependence (loss of activity after treatment with protease); pH stability (stable at pH 2 for 24 hours at 4°C), this requirement would not apply to the case of immune interferon.

B. Potency

To date, no standard method of determining interferon potency exists. Since interferons exhibit a number of biological actions, including viral inhibition, inhibition of antibody production, inhibition of delayed-type hypersensitivity, etc. it is difficult to determine exactly which activity to score in a biological assay.

The reproducibility, accuracy, sensitivity, precision, and efficiency of such biological assays are difficult to maximize (60-62). Determination of assay results can be estimated in a number of ways: dye uptake (63,64); visual microscopic scoring for cytopathic effects; foci [e.g. uptake of fluorescamine (65)]; enzyme yield (66); microplaque reduction (67,68); RNA synthesis (60,70); quantitative hemadsorption (71); infectivity yield (72); or hemagluttination yield (73).

As an illustrative example, a typical antiviral assay involves planting human fibroblast cells in microtiter plates and incubating them for 24-72 hours until the cells have become confluent. The growth medium is then removed and appropriate dilutions of the interferon-containing samples and reference standard are then added to the cells and left in contact with the cells for a time sufficient to give a response (4-24 hours). The medium is again removed and a solution containing a virus is added to the culture to challenge the cells. The cells are incubated in the presence of the virus for a suitable period (16-24 hours) and the plates are scored for cytopathic effects, either microscopically or by using a vital stain. If the interferon concentration is sufficient to protect the cells from viral attack, viable cells are present in the culture wells. If the concentration of interferon is insufficient to protect the cells, lysis occurs and cells die. The reciprocal of that concentration of interferon which give 50% protection from cell lysis (ED_{50}) is the interferon titer, generally expressed in $\log_{10}$ units.

Assay systems can vary in cell type, virus strain, cell passage, pH and nutritional elements. Some of these variables can be difficult to control precisely and this gives rise to rather large standard deviations in the assay results.

A more reliable method of determining interferon concentration would be radioimmunoassay (RIA). To date this concept has not been realized because of the scarcity of the pure interferon. This problem might be circumvented by the use of hybridomas (56) for the production of pure interferon antibodies. Application of this new technique has been reported recently (57). With the production of monoclonal antibody should come the long-awaited RIA assay for interferons. This step will go a long way towards improving interferon production, recovery and purification techniques which have been seriously hampered by the lack of a truly reliable assay method.

C. *Safety*

Safety tests on the material destined for clinical trials includes sterility and pyrogenicity testing. This latter test may be somewhat misleading since it is possible that pyrogenicity is an inherent property of interferons.

In addition, for the production of fibroblast and lymphoblastoid interferons, the identity of the cells in that passage used for production should be established and should be confirmed to be free of adventitious agents such as mycoplasma.

V. ADVANTAGES AND DISADVANTAGES ASSOCIATED WITH PRODUCTION OF DIFFERENT INTERFERONS

Ignoring for the moment any differences that exist in the biological properties of leukocyte, lymphoblastoid, and fibroblast interferons, it is of interest to compare limitations associated with their large-scale production. The most important aspects of the different processes in this respect are summarized in Table 2. Clearly the major limitation associated with the large-scale production of leukocyte interferon is the availability of human blood. Even with the best possible collaboration between blood banks and interferon producers, it is doubtful that this approach can do more than supply material for limited clinical trials; the blood from up to 50 separate donations can be required to treat a single cancer patient. Lymphoblastoid cells, such as Namalva, can be propagated indefinitely in suspension culture. The technology is already available within the pharmaceutical industry to cultivate cells on a large scale. Limitations associated with such large-scale cell production will be the design of suitable media with readily available and reasonably priced constituents, and the provision of adequate quantities of the appropriate inducing virus. In the past the use of transformed cell lines for the production of biological agents, such as vaccines, has been prohibited. In the case of lymphoblastoid interferon it

TABLE 2. Important Factors Governing Practicability of Scale-Up of Interferon Production by Human Cells

Interferon Type	Origin and Type of Cells	Life Span of Cells	Culture Method	Inducer
Leukocyte	Blood donations, diploid cells	Used only once	Suspension	Paramyxovirus
Lymphoblastoid	Biophys from cancer patient, transformed cell line	Can be propagated indefinitely	Suspension	Paramyxovirus
Fibroblast	Biopsy from infant tissue, diploid cells	Can be propagated for a limited number of passages	Anchorage-dependent	Synthetic polynucleotide

appears that an exception will be made at least for the treatment of advanced cancer patients. A major concern would be whether the product derived from a virus transformed cell might contain virus or contaminating viral nucleic acid. The most satisfactory way to overcome such an objection is to purify the material to homogeneity. The fibroblasts used for production of fibroblast interferon are human diploid cells and have a limited life span, necessitating eventual reisolation and characterization. Perhaps the most serious problem associated with the large-scale production of fibroblast interferon is the fact that the cells are anchorage-dependent. Many different approaches are being utilized to overcome the problem of supplying adequate surface area for growth. Thus, technology is still under development and the application of new methods, such as the use of microcarriers on a large-scale, remain to be established.

A quite separate issue from how difficult it is to make large quantities of the interferons mentioned is their biological equivalence. There is no doubt that leukocyte-and fibroblast-derived interferons are different: they are antigenically distinguishable (74) and have different host cell ranges for their antiviral activity (75). It would be surprising if their efficacy in the treatment of viral infections and/or cancer did not also differ. A more difficult question to answer is how different, if at all, are leukocyte and lymphoblastoid interferons. Antigenically the major components of the two are indistinguishable (74) but recent evidence (13,21) from the partial amino acid sequence of lymphoblastoid and the total amino acid sequence of leukocyte interferon (derived from a cloned gene) show that there are differences. Heterogeneity of interferons may derive from different degrees of glycosylation of the same interferon protein, or from differences in the amino acid sequence of the protein itself. Evidence for this heterogeneity comes from two sources, namely, high resolution separation techniques (SDS-PAGE and HPLC applied to highly purified material), and interferon genes cloned into bacteria.

VI. GENETIC ENGINEERING AND INTERFERON PRODUCTION

The recent ability to clone specific genes from eucaryotes into procaryotes, such as *Escherichia coli*, is particularily interesting in the case of interferon. Mass cultivation of such bacteria containing the appropriate gene could lead to the manufacture of interferon on a scale not possible by the more conventional eucaryotic cell culture methods described in this report. Several laboratories have already reported the successful cloning and expression of the interferon gene in bacteria (18-20). Such strains produce a naked protein whereas normally interferon is a glycoprotein. This interferoid has been shown to have equivalent antiviral activity *in vitro*, to interferon itself. However, no data have yet been published on the behavior of an interferoid *in vivo*. If the naked protein is inactive, or less active *in vivo*, it may prove possible to glycosylate it by chemical means and overcome this limitation. Evidence is now accumulating that interferon preparations previously considered to be homogeneous are in fact mixtures of closely related proteins (44). Gene cloning will probably be the most effective way of producing individual components of such mixtures. When this has been achieved, the different interferons can be separately tested for their efficacy as antiviral and anticancer agents. The possibility cannot be discounted that a mixture of related interferons may be required for the best therapeutic effect.

VII. INTERFERON INDUCTION *IN VIVO*

One way of circumventing the problems associated with the large-scale production of interferon would be to bring about the *in-situ* synthesis of interferon by administration of an appropriate inducer to patients. A wide variety of compounds have been evaluated as *in vivo* inducers, and some measure of success obtained. However, a major limitation to this approach so far has been the unacceptable toxicity of the inducing agents (2).

VIII. CHEMICAL SYNTHESIS OF INTERFERONS

Now that the complete amino acid sequences of several interferons are known, total synthesis of the proteins can be attempted. Current methods (Merrifield solid phase synthesis) make this a technically feasible project and ribonuclease, a protein of similar molecular weight, has already been made by this approach (76). It is doubtful that this approach will represent a viable economic alternative to genetic engineering as a means of producing interferoids. If however, a peptide fragment of interferon retains biological activity, the synthetic approach could become interesting and be an effective means of producing analogues for structure-activity studies.

IX. CONCLUSION

In the short term, traditional methods of producing human interferon by large-scale cell culture will play an important role in supplying material for clinical trials to establish the efficacy of interferon as an antiviral and anticancer agent. Furthermore, interferon produced by these methods will serve as an important control in judging the material produced by genetic engineering. A major issue to be resolved for the protein interferoids produced by recombinant DNA methods or total synthesis is their *in vivo* activity. If they have activity comparable to the native glycoprotein interferon, production by genetic engineering undoubtedly becomes the method of choice.

ACKNOWLEDGMENTS

This work was supported by the U.S. National Cancer Institute under Contract #NO-1-CO-75380 with Litton Bionetics, Inc. The authors would like to acknowledge Mr. Robert Ricketts for help in reviewing the manuscript.

REFERENCES

1. Isaacs, A. and Lindenmann, J., *Proc. Royal Soc. B 147*, 258 (1957).
2. Stringfellow, D.A., *in* "Interferon and Interferon Inducers" (D.A. Stringfellow, Ed.), p. 145, Dekker, New York (1980).
3. Stewart, W.E. III, *in* "The Interferon System", Springer-Verlag, Wien (1979).
4. Revel, M., *in* "Interferon 1" (I. Gresser, Ed.), p. 102, Academic Press, London (1979).
5. Cantell, K. *in* "Interferon 1" (I. Gresser, Ed.), p. 1, Academic Press, London (1979).
6. Gresser, I., Coppey, J., Falcoff, E. and Fontaine, D., *Proc. Soc. Exp. Biol. Med. 124,* 84 (1967).
7. Priestman, T.J., *Cancer Treatment Reviews 6,* 223 (1979).
8. Strander, H., *Blut 35,* 277 (1977).
9. Gidlund, M., Orn, A., Wigzell, H., Senik, A. and Gresser, I., *Nature 273,* 759 (1978).
10. Havell, E.A., Yip, Y.K., and Vilcek, J., *J. Gen. Virol. 38,* 51 (1977).

11. Interferon Nomenclature, *Nature 286,* 110 (1980).
12. Zoon, K.C., Smith, M.E., Bridgen, P.J., Zue Nedden, D. and Anfinsen, C.B., *Proc. Nat. Acad. Sci. USA 76,* 5601 (1979).
13. Zoon, K.C., Smith, M.E., Bridgen, P.J. and Anfinsen, C.B., *Science 207,* 527 (1980).
14. Knight, E. Jr., *Proc. Nat. Acad. Sci. USA 73,* 520 (1976).
15. Knight, E. Jr., Hunkapiller, M.W., Koant, B.D., Hardy, R.W.F. and Hood, L.E., *Science 207,* 525 (1980).
16. Okamura, H., Berthold, W., Hood, L., Hunkapiller, M., Inoue, M., Smith-Johannsen, H. and Tan, Y.H., *Biochem. 19,* 3831 (1980).
17. Maxam, A.M. and Gilbert, W., *Proc. Nat. Acad. Sci. USA 74,* 560 (1977).
18. Mantei, N., Schwarzstein, M., Steuli, M., Panem, S., Nagata, S. and Weissmann, C., *Gene 10,* 1 (1980).
19. Derynck, R., Content, J., DeClerq, E., Vokkaert, G., Tavernier, J., Devos, R. and Fiers, W., *Nature 285,* 542 (1980).
20. Taniguchi, T., Ohno, S., Fujii-Kuriyama, Y. and Muramatsu, M., *Gene, 10,* 11 (1980).
21. Taniguchi, T., Mantei, N., Schwarzstein, M., Nagata, S., Muramatsu, M., and Weissmann, C., *Nature 285,* 547 (1980).
22. Houghton, M., *Nature 285,* 536 (1980).
23. Kauppinen, H.L., Myllyla, G. and Cantell, K. *in* "Human Interferon" (W.R. Stinebring and P.J. Chapple, Eds.), p. 1, Plenum Press, New York (1978).
24. Burke, D.C. and Isaacs, A., *Brit. J. Exp. Path. 39,* 452 (1958).
25. Mogensen, K.E. and Cantell, K., *Pharmac. Ther. C 1,* 369 (1977).
26. Strander, H., Mogensen, K.E. and Cantell, K., *J. Clin. Microbiol. 1,* 116 (1975).
27. Klein, G., Dombos, L. and Gothoskar, B., *Int. J. Cancer 10,* 44 (1972).
28. Klein, F., Ricketts, R.T., Jones, W.I., DeArmon, I.A., Temple, M.J., Zoon, K.C. and Bridgen, P.J., *Antimicrob. Ag. Chemother 15,* 420 (1979).
29. Bodo, G., *Proc. Symposium on Preparation, Standardization and Clinical Use of Interferon,* 49 (1977).
30. Reuveny, S., Bino, T., Rosenberg, H., Traub, A. and Mizrahi, A., *Develop. Biol. Standard 46,* 281 (1980).
31. Johnston, M.D., Christofinis, G., Ball, G.D., Fantes, K.H. and Finter, N.B., *Develop. Biol. Standard 42,* 189 (1979).
32. Zoon, K.C., Bridgen, P.J. and Smith, M.E., *J. Gen. Virol. 44,* 227 (1979).
33. Bridgen, P.J., Anfinsen, C.B., Corley, L., Bose, S., Zoon, K.C. and Ruegg, U.T., *J. Biol. Chem. 252,* 6585 (1977).

34. Zoon, K.C., Buckler, C.E., Bridgen, P.J. and Gurari-Rotman, D., *J. Clin. Immunol. 7,* 44 (1978).
35. Carter, W.A. and Horoszewicz, J.S., *Pharmac. Ther. 8,* 359 (1980).
36. Weiss, R.E. and Schleicher, J.B., *Biotechnol. Bioeng. 10,* 601 (1968).
37. Molin, O. and Heden, C.G., *Progr. Immunobiol. Standard 3,* 106 (1969).
38. House, W., Shearer, M. and Maroudas, N.G., *Exp. Cell Res. 71,* 293 (1972).
39. Van Wezel, A.L., *Nature 216,* 7 (1967).
40. Giard, D.J. and Fleischaker, R.J., *Antimicrob. Ag. Chemother. 18,* 130 (1980).
41. Levine, D.W., Wong, J.S., Wang, D.I.C. and Thilly, W.G., *Somatic Cell Genet. 3,* 149 (1977).
42. Havell, E.A. and Vilcek, J., *Antimicrob. Ag. Chemother. 2,* 476 (1972).
43. Cavalieri, R.L., Havell, E.A., Vilcek, J. and Pestka, S., *Proc. Nat. Acad. Sci. USA 74,* 4415 (1977).
44. Berg, K. and Heron, I., *Scand. J. Immunol. 11,* 489 (1980).
45. Rubinstein, M., Rubinstein, S., Familleti, P.C., Miller, R.S., Waldman, A.A. and Pestka, S., *Proc. Nat. Acad. Sci. USA 76,* 640 (1979).
46. Edy, V.G., Billiau, A. and De Sommer, P., *J. Biol. Chem. 252,* 5934 (1977).
47. Zoon, K.C., Smith, M.E., Bridgen, P.J., Zur Nedden, D., and Anfinsen, C.B., *Proc. Nat. Acad. Sci. USA 76,* 560 (1979).
48. Cantell, K. and Hirvonen, S., *J. Gen. Virol. 39,* 541 (1978).
49. Erickson, J.S. and Paucker, K., *Anal, Biochem. 98,* 214 (1979).
50. Priestman, T.J., *Lancet 113* (1980).
51. Biliau, A., Van Damme, J., Van Leuven, F., Edy, V.G., De Ley, M., Cassiman, J-J., Van Den Bergle, H., De Somer, P., *Antimicrob. Ag. Chemother. 16,* 49 (1979).
52. Rubinstein, M., *Anal. Biochem. 98,* 1 (1979).
53. Regnier, F.E. and Gooding, K.M., *Anal. Biochem. 103,* 1 (1980).
54. Kato, Y., Komiya, K., Sasaki, H. and Hashimoto, T., *J. of Chromatogr. 190,* 297 (1980).
55. Kato, Y., Komiya, K., Sawada, Y., Sasaki, H. and Tsutomu, H., *J. of Chromotogr. 190,* 305 (1980).
56. Kohler, G. and Milstein, C., *Nature 256,* 495 (1975).
57. Secher, D.S. and Burke, D.C., *Nature 285,* 446 (1980).

58. *Human Interferon in the Clinic: Guidelines for Testing,* Workshop Sponsored by the National Cancer Institute, the National Institute of Allergy and Infectious Diseases and The Bureau of Biologics, Food and Drug Administration, Oct. 29, 1979, Bethesda, Maryland, USA.
59. Petricciani, P., *Points to be Considered in the Production and Testing of Interferon Intended for Investigational Use in Humans.* Food and Drug Administration, Personal Communication, July 1, 1980.
60. Buckler, C.E., *Tex. Rep. Biol. Med. 35,* 150 (1977).
61. Oie, H.K., *Tex. Rep. Biol. Med. 35,* 154 (1977).
62. Finter, N.B., *Tex. Rep. Biol. Med. 35,* 161 (1977).
63. Finter, N.B., *J. Gen. Virol. 5,* 419 (1969).
64. Armstrong, J.A., *Appl. Microbiol. 21,* 723 (1971).
65. Kozikowski, E.H. and Habon, N., *J. Gen. Virol. 4,* 441 (1969).
66. Sedmak, J.J., Grossberg, S.E. and Jameson, P., *Proc. Soc. Exp. Biol. Med. 149,* 433 (1975).
67. Green, J.A., Stanton, G.J., Goode, J. and Baron, S., *J. Clin. Microbiol. 4,* 479 (1976).
68. Campbell, J.B., Grunberger, T., Kochman, M.A. and White, S.L., *Can. J. Microbiol. 21,* 1247 (1975).
69. Richmond, J.Y., Polatnick, J. and Knudsen, R.C., *Appl. Environ. Microbial 39,* 823 (1980).
70. Koblet, H., Kohler, U. and Wyler, R., *Appl. Microbiol. 24,* 323 (1972).
71. Finter, N.B., *Virology 24,* 589 (1964).
72. Jameson, P., Dixon, M.A. and Grossberg, S.E., *Proc. Soc. Exp. Biol. Med. 155,* 173 (1977).
73. Stanton, J.G. and May, D.C., *Proc. Soc. Exp. Biol. Med. 143,* 1048 (1973).
74. Paucker, K., *Tex. Rep. Biol. Med. 35,* 23 (1977).
75. Gresser, I., Bandu, M.T., Brouty-Boye, D. and Tovey, M., *Nature 251,* 543 (1974).
76. Gutte, B. and Merrifield, R.B., *J. Amer. Chem. Soc. 91,* 501 (1969).

CHAPTER 9

YEAST ALCOHOL TOLERANCE AND RECOMBINANT DNA FOR IMPROVED ALCOHOL PRODUCTION PROCESS

Nancy W.Y. Ho

Laboratory of Renewable Resources Engineering
Purdue University
West Lafayette, IN.

I. INTRODUCTION

"Yeast" and "alcohol fermentation" are two terms which have been closely associated throughout the history of mankind. Yeast was used for wine-making from the juices of fruits and berries centuries before Van Leeuvenhoek in 1680 discovered the tiny "animalcules" of yeast cells present in fermenting beer. It took another two hundred years until Pasteur to formally establish the fact that a living substance, capable of "la vie sans air", was responsible for the fermentation of sugar to alcohol. However, in the past century, as science in general has grown at an ever increasing pace, many aspects of yeast have also been studied in depth (1-3), including the complete elucidation of the pathway of alcoholic fermentation from glucose to ethanol (4). Along with the enormous knowledge accumulated on various aspects of yeast, the process of alcohol fermentation by yeast was also improved substantially from the ancient empirical process to well equipped factory operation. In the modern society of the twentieth century, production of ethanol by yeast has two major applications: one is for the production of various alcoholic beverages and the other is to produce alcohol for industrial applications. Industrial alcohol is made either by yeast alcohol fermentation of sugar or related substances or by catalytic hydration of ethylene, a product from the petroleum industry. However, in recent years in technologically more advanced countries, industrial alcohol

ISBN 0-12-040304-8

has been produced mostly by the synthetic method. For example, of the roughly 300 million gallons of industrial alcohol produced in the United States in 1974, only 10 million gallons were made by fermentation (5). Since yeast alcohol fermentation was confined to the production of alcoholic beverages in recent years, most research, particularly that being supported by industries, was oriented towards the selection of species of yeasts capable of producing wine or other alcoholic beverages with fine qualities, and relatively little effort was directed towards increasing the efficiency of alcohol production either in terms of the yields or the time involved in making higher concentrations of alcohol. However, with petroleum resources dwindling throughout the world, the importance of the production of alcohol by fermentation has been re-emphasized. Not only does alcohol for ordinary industrial use have to come from the fermentation process, but alcohol produced by fermentation also has to replace or supplement gasoline for transportation to reduce the world's dependence on petroleum for energy. This sudden change in priority will undoubtedly stimulate new initiatives towards improving alcohol production by fermentation.

Yeast has served mankind in many ways. Not only has it been used for making wines and rising breads centuries ago even before history was recorded, but the study of alcoholic fermentation by yeasts, which began in the second half of the nineteenth century, was of fundamental importance in the development of a new science--biochemistry, in general, and enzymology, in particular. "Enzymes", which literally translates as "in yeast", are named after yeast as a tribute to its role in the early development of biochemistry. The names of Pasteur, Buchner, Harden, and Young are historical landmarks of the pioneering work from the 1850s to the early 1900s. In the following decades, the realization of the very close similarities between anaerobic glycolysis in yeast and muscle, despite the difference in final by-products (lactate in muscle), opened the way to the basic and fruitful concept of unity in the living world at the molecular level. Now, once again, yeast has come to the rescue of mankind to lighten the burden of our energy crisis.

II. A NEW ERA FOR ALCOHOL PRODUCTION BY FERMENTATION

In the article "Stimulation of Innovation in the Fermentation Industries" which appeared in Volume 3 of this series, Professor Perlman (6) indicated that there are three major factors which may stimulate innovation:

(1) the occurrence of "crises" and their "resolution",
(2) the availability of new technology, and
(3) interference in the "normal pattern" by outside influences.

These are the precise factors which have made alcohol production by fermentation enter a new era. As indicated above in the introduction, in recent years alcohol for industrial use was almost totally produced by the synthetic method and alcohol fermentation was used mostly for the production of wine and other alcoholic beverages. For wine making, flavor of the end product rather than the efficiency of alcohol production is the major concern. Thus very little effort has been oriented over the years towards improvement of the efficiency in alcohol production. However, the Arab oil embargo and sudden irrational increases in oil prices in the last few years have shaken the world and made it recognize the existence of a world wide shortage of oil and a resulting energy "crisis." The "resolution" for the crisis is to use other replenishable sources of energy to replace oil. Alcohol produced by fermentation of biomass is one of the alternative renewable energy sources available to replace petroleum products for transportation and industry. The occurrence of the world wide energy crisis "interferes" with the "normal pattern" of alcohol production and makes the production of alcohol by fermentation much more in demand than before. This new development also shifts the priority of alcohol fermentation from "flavoring" for wine making to "efficiency" for mass production of alcohol for transportation and industrial need. For example, although yeast normally can produce beverages containing 12% to 18% alcohol through fermentation of glucose or other fermentable carbohydrates, it takes several months to reach the final high concentration of alcohol. For wine production, the long duration does not cause serious problems because the quantity of wine produced by the wine industry is sufficient to supply the demand of the consumers. However, at a time of a world wide shortage of gasoline, and alcohol produced by fermentation being in high demand for industry and for transportation, the several months duration for the production of high concentrations of alcohol is intolerable and it has to be speeded up in order to meet the challenge. Thus, efforts should now be directed towards improving yeast alcohol tolerance as well as other factors which may affect the overall efficiency of alcohol production by fermentation.

Even though the problems have been clearly identified, they may not be solved for a long time to come because of technical difficulties. For example, it has long been recognized that DNA is the chemical basis of the gene and analysis of the nucleotide sequence of DNA is the key to exploring numerous cellular regulation processes. However, it was not until 1977, that the first meaningful DNA sequence was analyzed and reported (7) because, prior to that time, no method was available to degrade the long chain DNA molecules to unique fragments, and no method was available to determine the base order in a DNA molecule. By the same token, the limited "alcohol tolerance" of the yeasts was observed long ago but the problem was not solved in the past partly because no better methods were available to improve yeast genetically other than by mutation or adaptation. The use of these techniques apparently was not enough to make yeast capable of tolerating higher concentrations of alcohol and producing alcohol more efficiently. Fortunately, the recent advances in molecular biology, particularly the development of techniques for molecular cloning of recombinant DNA (8-10) make it possible to explore new approaches to solving the old problems for alcohol production by fermentation. This article is devoted to the exploration of the possible causes which affect yeast alcohol tolerance and efficiency in alcohol production by fermentation. It also covers some of the recent advances in molecular biology which provide the techniques for the development of recombinant DNA technology and for unlocking the secrecy coded in the nucleotide sequences of DNA. And, finally, it will also discuss some of the possibilities, although mostly speculative, that alcohol fermentation can be improved by using the newly developed recombinant DNA techniques. This article is only intended to identify some of the problems faced by alcohol fermentation in order to meet the world demand for mass production of alcohol by fermentation of biomass or industrial wastes as a substitute for oil and to stimulate innovation for research towards improvement of alcohol production by fermentation. By no means should it be considered a complete review on the subject of yeast alcohol tolerand and possible improvement of alcohol production by recombinant DNA techniques.

III. YEAST ALCOHOL TOLERANCE

Although Cruess *et al.* (11) reported that 19% ethanol could be obtained from yeast fermentation, this high concentration of ethanol was obtained by adding concentrated grape juice during fermentation and by controlling the temperature. In practice, much lower yields were obtained and the yield of ethanol production by yeast fermentation varied markedly with the strains of yeast. For example, Castan (12) tested 17 strains of yeasts on a must containing 32% sugar; the maximum ethanol produced varied from 11% to 17%. One major factor which causes low alcohol production by yeast seems to be the inhibitory effect of alcohol. It was observed long ago that alcoholic fermentation normally does not occur in the presence of 18% ethanol (13-16). The inhibitory effect of ethanol on fermentation also increases with an increase in alcohol concentration and/or temperature. According to Schanderl (17), at zero concentration of ethanol 15% of the yeast cells survive heating for one minute at 50°C, whereas at 6% ethanol only 0.28% survive, and at 15% ethanol only 0.05% of the cells survive. However, the way in which ethanol causes the inhibition of fermentation by yeast is not certain. Recently Navarro and Durand (18) studied the influence of the temperature on the accumulation of alcohol by *S. carlsbergensis* and they reported that the decisive factor in retarding growth of yeast during fermentation is the build-up of ethanol inside the cell rather than the concentration of ethanol in the medium. They found that growth stopped very soon after a critical intracellular ethanol concentration had been reached which, under their conditions, amounted to 100 gm/l (2.2 M) at 10°C, 170 gm/l (3.7 M) at 20°C, and 300 gm/l (6.6 M) at 30°C. At the same time, the external ethanol concentrations were 35 gm/l, 25 gm/l, and 13 gm/l respectively. These results implied that the growth inhibitory effect of alcohol was related to its retention within the yeast cells. The higher the temperature, the higher the maximal intracellular alcohol concentration, and the greater the inhibitory effect of the ethanol. In a related study, Steinkraus and co-workers have reported the effect of various factors, including ethanol production and accumulation, on a process called "rapid fermentation" (19,20). They found that, in the presence of 13% dissolved oxygen, 9.5% (wt/vol) ethanol could be produced by *S. cerevisiae* in 3 hr at 30°C when the suspension contained 8×10^8 cells/ml. However, under these conditions, the cells died; the internal ethanol concentration was found to be 2×10^{11} molecules per cell. Keeping the cell concentration at 8×10^8 cells/ml, but lowering the fermentation temperature to 15°C resulted in 9.5% (wt/vol) ethanol production in 6 hr. Under these conditions, the cells remained viable; the internal concentration was 4×10^{11} molecules per cell. Again, the implication may be that the equilibration between the cytoplasm and the surrounding medium with respect to ethanol is slow and the accumulation of a high concentration of ethanol within the yeast affects the yeast's viability as well as its capability to produced ethanol.

Holzberg *et al.* carried out a kinetic study of the alcoholic fermentation of grape juice (21). They found that when fermentation was carried out with 20 g/100 ml initial sugar concentration without aeration but with three 3 in. turbine impellers turning at 350 rpm in a glass jar containing 6 liter buffered pasteurized grape juice, the cell growth stopped at about 19 hr when the sugar concentration was still 15% and when the alcohol concentration was only about 2.4 gm/100 ml as shown in Figure 1. In batchwise studies, the growth rate was unaffected by low concentrations of alcohol. Holzberg *et al.* had also studied the growth rate at various higher alcohol concentrations by operating the fermentor under continuous flow conditions (21). They found that growth should have continued to an alcohol concentration of 6.85 gm/100 ml (Figure 2) rather than 2.4 gm/100 ml (Figure 1). Consequently, instead of an abrupt shift from the exponential growth to a stationary growth (Figure 1) the gradual accumulation of alcohol should have resulted in a transition phase of some 27 hr if alcohol was the only limiting factor. This seems to indicate that one or more factors, in addition to alcohol, may also contribute in limiting alcohol production and cell growth during fermentation. Mounting evidence has accumulated over the years indicating aeration during fermentation affects alcohol production in several ways and the various effects of oxygen on fermentation are discussed below.

Although Pasteur established "fermentation" as "la vie sans air", in reality alcoholic fermentation can be thought of as consisting of two different, but overlapping phases: the aerobic growth phase (respiration) and the anaerobic stationary phase (fermentation). In the initial exponential growth phase of alcoholic fermentation, the effect of oxygen is to stimulate yeast growth since the metabolic pathways associated with respiration are also those which provide precursors of the cell constituents. As early as 1895, Wortmann (22) showed that much higher yields of yeast were obtained in aerated cultures. Kunkee and Amerine (23) have noted that the presence of some air during the beginning of fermentation in nutritionally deficient medium is essential to obtain yeast cells capable of completing the fermentation if the medium contains high concentrations of sugar. Although Pasteur discovered the inhibition of fermentation by respiration, the so-called "Pasteur effect" is only operative when the medium contains a very low concentration of sugar (less than 3%). To the contrary, Wikén and Richard (24) showed that a very small amount of oxygen stimulated fermentation by young growing yeast cultures. Ough and Amerine (25) also reported a delay of two days in the growth of yeasts in grape juice stripped of oxygen. Ricketts and Hough (26) found that in a single-vessel system of continuous

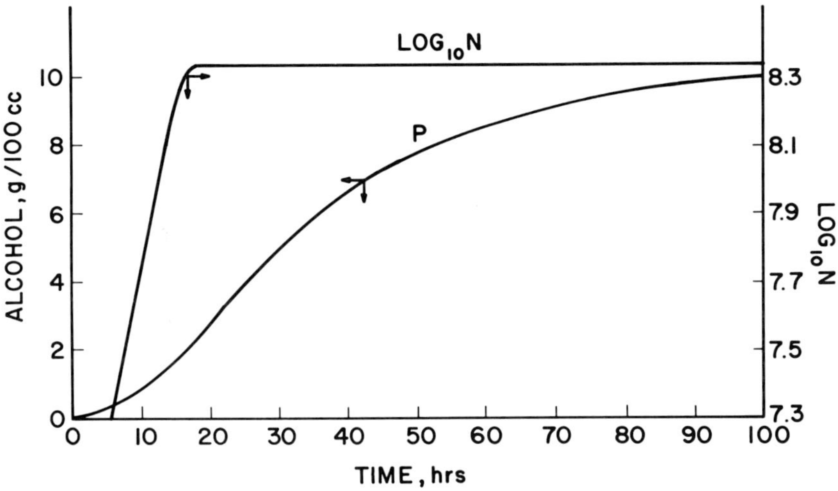

FIGURE 1. Patterns of Growth and Alcohol Formation of Wine Yeasts During the Fermentation of Grape Juice (from Holzberg et al. [21]).

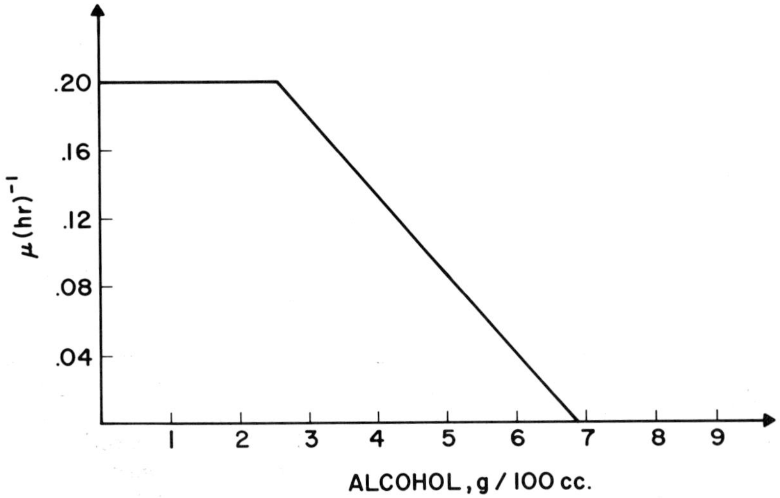

FIGURE 2. Effect of Alcohol Concentration on the Specific Growth Rate of Wine Yeasts (from Holzberg et al. [21]).

fermentation, the rate of beer production was low under virtually anaerobic conditions. They found that increasing the aeration rate to 30 ml air/h caused a marked rise in yeast concentration and rate of beer production. Thereafter, little change was observed with air flows up to 500 ml/h, above which the yeast concentration and rate of beer production again increased. At aeration rates exceeding 500 ml/h, the following facts were observed: (1) yeast concentration rose stepwise and the rate of beer production remained unaltered; (2) at yet higher rates, the rate of beer production rose stepwise, while yeast concentration remained constant. These observations indicated that aeration had independent effects on rate of fermentation and on cell growth. Ricketts and Hough concluded that it is desirable to permit a low rate of aeration in the system in order that there be a faster rate of beer production. Enari and Makinen (27) also observed that restriction of air lowered the rate of beer production. Couland (28) studied the effect of aeration during fermentation of hopped wort in a continuous fermentation system. He found that oxygen continuously supplied to the ingoing wort stimulated the growth rate, so that yeast concentration and consumption of sugar and production of ethanol per unit weight of yeast increased markedly without decreasing the proportion of sugar fermented to ethanol. Improved brewing performance therefore resulted, because ethanol formation was accelerated by the increase in cell population and the high rate of yeast metabolism. The "negative Pasteur effect", in the sense of an inhibition of alcoholic fermentation in the absence of gaseous oxygen and a stimulation of fermentation in the presence of oxygen, has been observed by numerous researchers (29-38).

During the so-called "rapid fermentation" process, a high rate of ethanol production could be achieved by using a high yeast concentration. For example, Steinkraus and Morse (39) reported that at a temperature of 25°C, honey solution or grape juice was fermented to 9.5% (wt/vol) ethanol within a 4 h period. However, a problem that severely limits the possibility of applying rapid fermentation, particularly to continuous fermentation, has been the high death rate of yeast cells. Nagodawithana *et al.* (20), therefore, studied the effect of dissolved oxygen on the viability of *S. cerevisiae* in "rapid fermentation". They found that, as the dissolved oxygen in the medium was increased, there was a progressive increase in the percentage of cells surviving. The ethanol production rate and the total final ethanol concentration were not affected by a dissolved oxygen content of up to 20%. The above results were confirmed by many recent studics (40-42). Furthermore, Moreno and Goma (43) studied the alcohol fermentation under strick anaerobiosis in a plug flow fermentor and concluded that strict anaerobiosis has a negative effect for fermentation for both the specific growth rate and specific ethanol production rates.

It should be kept in mind that yeasts are not true facultative anaerobes because they can only grow a few generations under anaerobic conditions. Molecular oxygen, therefore, is required, not only for the generation of energy through respiration, but also for the synthesis of certain essential cell constituents which do not form in the absence of oxygen. The sterols are among those constituents which require molecular oxygen for their biosynthesis. In the following the effect of sterols on yeast alcoholic fermentation will be discussed.

It has been demonstrated by Andreasen and Stier (44) and by Brechot (45,46), that certain sterols (ergosterol, landosterol, and cholesterol) as well as oleanolic acid are yeast anaerobic growth factors. In the yeast cells, sterols amount to 6% by weight of the protoplasmic membrane dry matter (47). The sterol content of the cells also differs with the strain of yeast (48) and the conditions of culture (49). Demel and Krugff (50) showed that sterols are regulatory; they can liquefy and condense the membrane. It has been shown that molecular oxygen is necessary for yeasts to synthesize sterols (51) and Gordon and Stewart (52) established the fact that there is a definite increase in the sterol content of yeast exposed to air for a long period after being cultured anaerobically. Kirsop (53) demonstrated the relationship between sterol biosynthesis and the oxygen requirements of yeasts. He found that these requirements are satisfied by supplementing the medium with ergosterol in Tween 80. Similarly Lafon-Lafourcade *et al.* (54) showed that certain substances, when added to grape must, aerated and inoculated with aerobically cultivated yeast, contributed to the maintenance of the viability and fermentative activity of the resting cells but not the cellular multiplication. They proposed that these substances should be considered as the "survival factors" of the yeasts. Sterols are among those survival factors. The presence of these "survival factors" was believed to be the reason for some peculiarities of yeast fermentation, especially in grape must of high sugar concentration (55,56). Lipids of the grape cuticle can act as growth promoting factors for the anaerobic growth of yeast. This may also partially explain why the fermentation of red wines in which the grapes are not separated from the skins, proceeds at a faster rate than that of white wines (51). The effect of ergosterol in grape must on the fermentation with aerobically and anaerobically cultivated yeasts was compared and shown in Figures 3 and 4.

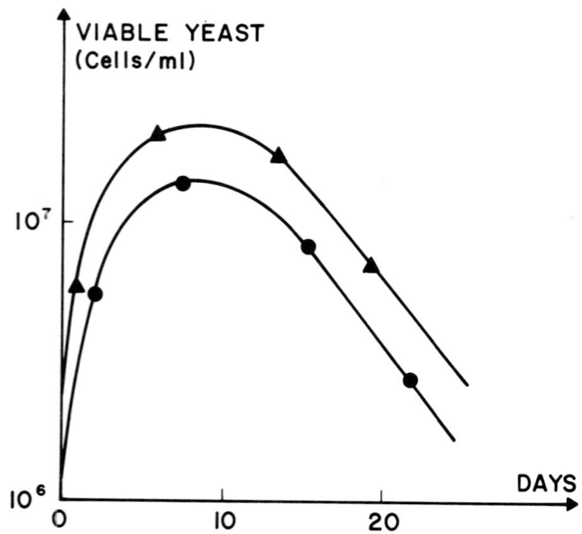

FIGURE 3. Effect of Ergosterol in Grape Must, Initially Deprived of Oxygen by Heating and Inoculated with Anaerobically Cultivated Yeasts (from Lafon-Lafourcade et al. [55]. Symbols: (●) Control; (▲) Plus Ergosterol

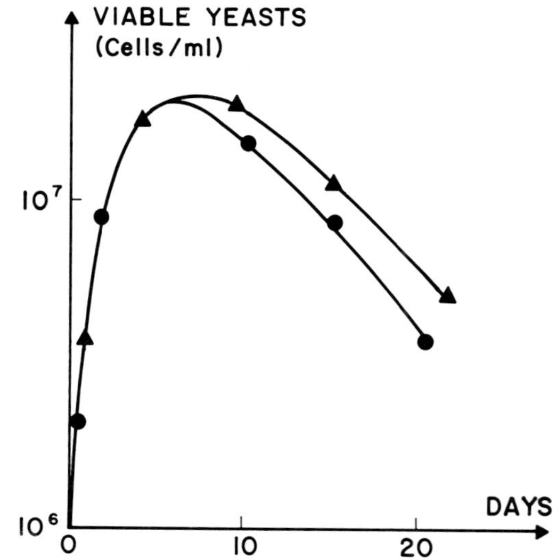

FIGURE 4. "Survival Factor: Action of Ergosterol in Grape Must, Initially Aerated and Inoculated with Aerobically Cultivated Yeasts (from Lafon-Lafourcade et al. [55]). Symbols: (●) Control; (▲) Plus Ergosterol

The fact that oxygen stimulates fermentation is also supported through the study of various respiratory-deficient (RD) petite mutants of the yeasts. A number of chemical agents have been found to induce the respiration-deficient mutants in yeasts (57-59). Subsequently, it was found that protein denaturing agents can also induce RD mutants (60-62). These mutants generally have non-functional mitochondria and display structural abnormalities such as forming smaller "petite" colonies. The early work on the RD mutants of yeasts has been reviewed by Nagai *et al.* (63) and the molecular mechanism for the formation of the RD mutants has also been explored (64-66). Recently Bacila *et al.* (67) have shown that RD mutants of yeasts are more efficient in producing alcohol during fermentation. Thus, oxygen serves more than one function in yeast metabolism. It is apparently required for both aerobic and anaerobic growth.

The work cited above implies that the accumulation of alcohol within the yeast cells as shown by Navarro and Durand (18) may not be the only inhibiting factor which limits the yeast fermentation process. There is no doubt that the presence of alcohol in the medium has an inhibiting effect on the yeast growth and a high concentration of ethanol is fatal to the yeast culture. However, it has been shown that as high as 30% ethanol can be produced within the yeast cells; also, with special brewing techniques, Sáke yeast normally produces Sake containing 20% alcohol. Kinetic studies (Figure 1) indicated that, under normal fermentation conditions, yeast ceases (or at least slows down considerably) to produce alcohol even before reaching 10% ethanol concentration. Thus the possibility should not be ruled out that under anaerobic conditions, certain essential constituents which are required for cell growth (replication) cannot be synthesized. This causes the accumulation of other cell constituents (perhaps ATP, for one) which may trigger a series of feedback mechanisms acting in sequence and lead to the slow down and eventual cessation of the consumption of glucose or other fermentable substrates. Although the presence of alcohol may have enhanced the slowing process, alcohol probably is not "the factor" which shuts off the fermentation process.

IV. RECENT ADVANCES IN MOLECULAR BIOLOGY

Since Avery and his co-workers reported their astonishing results in 1943 showing that DNA was the primary genetic material which controls the expression of various traits of an organism, there have been numerous momentous developments in molecular genetics. Among the more recent developments are the discovery of the restriction endonucleases, the development of molecular cloning of recombinant DNA techniques and the development of rapid DNA sequencing techniques. The significance of these developments will be discussed below.

A. Restriction Endonucleases

As mentioned early in the text, it has long been known that the double stranded DNA molecules are the chemical basis of the gene. The information contained within the genome of every organism provides a precise program upon which its biological processes depend. Analysis of the organization as well as the nucleotide sequence of the gene is essential for the total comprehension of the biological processes at the molecular level. However, these types of analyses were considered to be virtually impossible even less than six years ago. One main factor which hampered the detailed analysis of the molecular structure of the gene is its extremely complex nature; for example, even a simple bacterial chromosome, such as that of *E. coli* consists of a single long chain DNA molecule containing several million nucleotides. The analysis of such a complex molecule requires that it be dissected into discrete segments, amenable to biochemical analysis. The discovery of restriction endonucleases which has made the dissection of DNA into specific segments feasible is a major breakthrough in molecular biology.

The study of restriction enzymes can be traced back to the early 1950s when Luria and his collaborators reported the phenomenon of host-controlled variation (68-70). They showed that the ability of bacteriophages to grow on particular strains was dependent upon the specific "modifications" induced by the host in which they had been propagated previously. It was concluded that these bacteria must contain some "specificity systems" able to restrict the host range of phages. Extensive genetic and biochemical studies of these systems in *E. coli* strains led to the identification of the two components involved (71,72). The first was an endonuclease (restriction endonuclease) which could distinguish between host DNA and foreign DNA. This was made possible by strain specific modification, accomplished by the second component of the system--a modification enzyme. Usually this is mediated by methylation of specific DNA sequences, within either the host DNA or phage DNA grown on that host, which then prevents their cleavage by the restriction endonuclease. Such restriction-modification (R-M) systems are now found to be widespread in bacteria and are thought to play a role in eliminating foreign DNA that gain entrance to the cell via viruses or as naked DNA.

It has been found that at least two different types of restriction endonucleases are present in bacteria. Type I restriction endonucleases require Mg^{2+}, ATP, and s-adenosylmethionine for their endonucleolytic activity (73-75). Although type I restriction endonucleases bind to specific sites on the DNA, they show no cleavage specificity (74,75). On the contrary, type II restriction endonucleases require only Mg^{2+} as a cofactor, recognize a specific sequence, and cleave at a specific site. The first characterized restriction endonuclease is now known to belong to the type I restriction endonucleases.

The type II restriction endonucleases are the enzymes which can digest DNA into specific segments and are important for the analysis of the structure of the gene at the nucleotide level. The first type II "site specific" restriction endonuclease, Hind II, isolated from *Hemophilus influensae* Rd, was reported in 1970 (76,77). Since then, there has been an almost explosive rate of discovery of new site-specific restriction endonucleases. Up to the present, more than 140 such site-specific type II restriction endonucleases have been isolated. Many highly purified restriction endonucleases (type II) are commercially available from companies such as New England Biolabs (Beverly, MA), Bethesda Research Laboratories (Rockville, MD), Boehringer Mannheim Biochemicals (Indianapolis, IN), etc. The type II restriction endonucleases (from hereon referred to as restriction endonucleases) are named according to the proposal of Smith and Nathans (78). For example, EcoRI is the restriction endonuclease isolated from *E. coli* RY13, Pst I from *Providencia stuartii* 164, and DpnI from *Diplococcus pneumoniae*.

Restriction endonucleases normally recognize 4 to 6 nucleotides of a duplex DNA and cleave at these recognition sites in one of three different ways as illustrated below.

1. generating protruding 5'-ends.

```
5'------------GAATTC------------3'
3'------------CTTAAG------------5'
                 |
                 | ECORI
                 ↓
5'------G          AATTC----------3'
               +
3'------CTTAA          G----------5'
```

2. generating protruding 3'-ends.

```
5'------------CTGCAG------------3'
3'------------GACGTC------------5'
                 |
                 | PstI
                 ↓
5'-------CTGCA        G----------3'
3'-------G     +  ACGTC----------5'
```

3. generating even (blunt) ends.

```
5'-------------GATC-------------3'
3'-------------CTAG-------------5'
                 |
                 | DpnI
                 ↓
5'----------GA      TC------------3'
3'----------CT  +   AG------------5'
```

Restriction endonucleases which produce all three types of ends are found useful for gene mapping, DNA sequencing and molecular cloning. Most restriction endonucleases do not degrade duplex DNA containing methylated bases within their recognition sequences. However, there are restriction endonucleases which can degrade DNA containing both methylated or unmethylated bases in their recognition sequences (79), and a restriction endonuclease specific for methylated DNA has also been isolated (80).

Although innovative ideas are essential to any field of research, their exploitation is usually limited by the tools and techniques available. Consequently, when a new tool is discovered, a burst of knowledge usually follows. Such has been the case for the restriction endonucleases which, in less than 10 years since their discovery, have led to at least three major advances in molecular biology. The first concerns the physical mapping of DNA genomes and their correlation to genetic maps. This has resulted in the discovery that some eukaryotic messenger RNAs are not a direct copy of the DNA but subsequently spliced after transcription (81). The second has been the development of extremely rapid techniques for the determination of nucleotide sequences of DNA (82,83), which was considered nearly impossible to accomplish in the past. The third lies in the establishment of a new field, genetic engineering. DNA sequencing and molecular cloning (genetic engineering) will be discussed slightly in detail below. In any event, the discovery of restriction endonucleases has truly made it possible for scientists to explore biology at the molecular level. The various aspects of both types of restriction endonucleases have been extensively reviewed elsewhere (84-87).

B. DNA Sequencing

Since 1943, DNA has been established to be the primary genetic material of the living organism; within the following three decades (up to the early 1970s), a great deal of information about DNA has already accumulated. For example, it has been found that DNA consists of two polydeoxyribonucleotide chains twisted about each other in a double-stranded helix. Each polymeric chain contains a vast number of monomeric units--the nucleotides. There are four different nucleotides in DNA and their sequence along a given chain is very irregular. Each of the four nucleotides contains a deoxyribonucleoside, a phosphate group and a purine and a pyrimidine base. There are two pyrimidine bases (thymine *[T]* and cytocine *[C]*) and two purine bases (adenine *[A]* and guanine *[G]*). In the polynucleotide chain, the sugar-phosphate junction always involves the same chemical groups. Hence, this part of the molecule, called the backbone, is very regular. In contrast, the order of the purine and pyrimidine bases along the chain is highly irregular and varies from one molecule to another. The two chains are joined together by hydrogen bonds between pairs of bases. A is always paired with T and G always with C. The existence of the base pairs means that the sequences of nucleotides along the two chains are not identical but complementary; therefore, if the sequence of one chain is known, that of its partner is automatically known. Since the sugar-phosphate backbone is the same in all DNA molecules, the genetic information of a DNA molecule must reside in the linear sequence of the four bases. Theoretically, a very, very large number of different sequnces can exist, since the number of possible permutations is 4^n, where "n" is the number of nucleotides along a DNA chain. The number of nucleotides in a DNA molecule usually is extremely large; for example, even the smallest bacterial phage, ϕx174, contains more than 5000 nucleotides in its DNA. Although it has long been known that the analysis of the nucleotide sequences of DNA molecules will provide substantial information regarding the control mechanisms of the cell, prior to 1977, the determination of DNA sequences was confined to the analysis of a few pyrimidine tracts and cohesive ends of bacteriophages of less than 30 nucleotides long (88,89).

The main hindrances to sequence analysis of DNA in the past have been the lack of reference points in these exceedingly long chain polymers and the lack of methods to dissect and fractionate them into orderly, manageable fragments. These difficulties have been largely solved by the recent developments in the nucleic acids research. The most significant development naturally has to be the discovery and application of highly specific restriction endonucleases which have been described above. These enzymes permitted the hydrolysis of the DNA molecule to well-defined fragments which could, in principle, be analyzed by a number of means. A second powerful tool has been the use of radioactive labeling for sequencing studies, rather than the dependence on optical and strictly chemical methods for analysis. Consequently, a very small amount of a DNA fragment, such as a few picomols, is sufficient for the analysis of its primary structure. Another recent, major development has been the use of electrophoresis in agarose and acrylamide gels for separating DNA fragments of all sizes, particularly the use of denaturing acrylamide slab gels which permit the separation of fragments containing the same sequences but differing in chain length by only one nucleotide. By taking advantage of these new developments, recently several outstanding methods for rapid DNA sequencing have been developed. Three of the most significant techniques for DNA sequencing are described below.

1. *The Plus and Minus Method.* The plus and minus method was developed by Sanger and Coulson (83). This method is designed for the determination of nucleotide sequences in single-stranded DNA by primed synthesis with DNA polymerase under conditions containing different limiting nucleoside triphosphates. This method was used for the determination of the total nucleotide sequence of ϕx174 DNA which contains more than 5000 nucleotides (7). Although the plus and minus method of Sanger and Coulson retains its historical significance for being the first of its kind capable of rapid sequencing of large DNA molecules, the fact that it requires carrying out 8 different DNA polymerase reactions in order to determine the sequence of a DNA fragment makes it less useful than the chemical method of Maxam and Gilbert for DNA sequencing (82).

2. *The Chain Terminator Method.* Sanger and his colleagues have subsequently modified their plus and minus method by incorporating chain terminators such as 2',3'-dideoxynucleoside or arabinosyl nucleoside triphosphates, analogs of the normal deoxynucleoside triphosphates, to terminate the action of DNA polymerase (90). This method is considerably simpler than the plus and minus method. Also it is advantageous over the Maxam

and Gilbert method in that it does not require mapping restriction fragments prior to sequencing. Thus, it has the potential to become a major technique for DNA sequencing in the future; however, up to now, most DNA sequences have still been analyzed by the Maxam and Gilbert method.

3. *The Chemical Method (Maxam and Gilbert Method)*. The "chemical method" for sequence analysis of DNA was developed by Maxam and Gilbert (82). In this method, a terminally labeled DNA fragment is treated by various chemical reagents under specific conditions so that the terminally labeled DNA is preferentially cleaved at specific bases. Only four such chemical reactions are needed in order to completely deduce the nucleotide sequence of a DNA fragment up to more than 200 nucleotides from a fixed end. This method is simple and accurate, and it can analyze both double-stranded and single-stranded DNA molecules. Thus, despite the fact that it requires mapping of restriction fragments prior to sequence analysis of large DNA fragments, currently it is still the method of choice for DNA sequencing. The detailed procedure for the Maxam and Gilbert technique has been reported recently (91).

All the above described methods for rapid sequence analysis of DNA are based on the same principle; that is to use enzymatic or chemical means to produce a series of ^{32}P labeled oligonucleotides with one common end but varying in length at the other end by a single nucleotide. The labeled oligonucleotides are then resolved by high resolution electrophoresis on denaturing polyacrylamide gels. Each resolved band on the gel specifies the chain length and the nature of the terminal base of a specific fragment, thus the nucleotide sequence can be deduced directly from the ladder-like pattern of the autoradiogram of the sequencing gels. These methods are limited only by the resolving power of the polyacrylamide gel; the development of very thin urea-containing polyacrylamide gels (92) has permitted the resolution of products, resulting from a single sequencing reaction, up to a chain length of 250 to 300 nucleotides per loading. With the availability of labeled nucleotides of high specific activity, it requires only a few picomol of a DNA fragment in order to deduce more than 200 nucleotides of its sequence. An example of an autoradiogram of a sequencing gel, to demonstrate that the DNA sequence can be read directly from the pattern of bands, is shown in Figure 5.

As mentioned above, prior to the development of these rapid DNA sequencing techniques, only a few small fragments less than 30 nucleotides long were sequenced; however, within a few years after the development of the rapid sequencing techniques, particularly that of Maxam and Gilbert, an

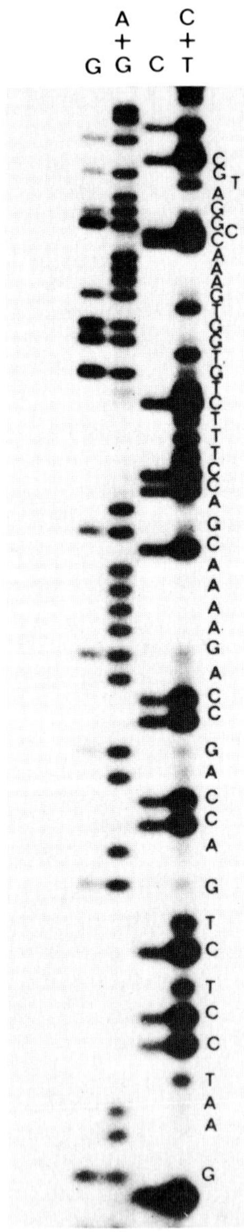

FIGURE 5. A Section of an Autoradiograph of a Sequence Gel. The Data were Produced by the Maxam and Gilbert Procedure (82). The Sequence of the DNA Fragment Can be Directly Derived from the Autoradiograph as Indicated Along the Side of the Ladder-Like Pattern of Bands From Bottom Up.

enormous amount of sequence data has been accumulated, including the complete sequences of a number of viruses such as ϕx174, SV_{40} and fd whose DNA contains several thousand nucleotides. The accomplishments in DNA sequencing in recent yeats is astonishing.

The importance of the analysis of nucleotide sequences of DNA molecules for the advancement of the basic knowledge of molecular biology can be demonstrated by the following examples. From the ϕx174 sequence came the first observation that two or more genes may sometimes be encoded within the same stretch of DNA. A similar phenomenon was also observed from the SV_{40} sequence; thus, the so-called "one gene, one enzyme" theory may not be totally correct. DNA sequencing also played an important part in confirming that many eukaryotic m-RNAs are not a direct product of transcription, but are spliced immediately after transcription. As the sequence information accumulates, many more such surprising phenomena may be discovered. Most significantly, the development of the rapid DNA sequencing techniques together with the development of molecular cloning of recombinant DNA would permit the study of the gene organization of eukaryotes. Knowledge of the control mechanism of the DNA of eukaryotes is the key to understanding many of the eukaryotic molecular mechanisms such as differentiation, transformation, cell proliferation, etc.

DNA sequencing is also significant for molecular cloning of recombinant DNA. Analysis of the nucleotide sequence of the cloned gene would provide the necessary information for structural modification of the cloned gene by further genetic manipulation in order to achieve a higher level of functional expression of the cloned gene. It has been reported, in some cases, that the level of eukaryotic enzyme activities produced in the bacterial host cells from the cloned eukaryotic genes are initially quite low, but spontaneous mutants of the *E. coli* transformants may result in the production of significantly higher levels of eukaryotic enzymes (93-95). The changes are known to be plasmid mediated and are not host specific. Analysis of the nucleotide sequence of the cloned gene would provide information regarding the precise structural change leading to the higher level of functional expression of foreign (eukaryotic) genes in bacteria. Cohen *et al.* (96) have also reported the isolation of spontaneous mutants of yeast transformants harboring plasmids inserted with a bacterial gene capable of producing higher concentrations of bacterial enzyme. The analysis of the nucleotide sequences of the cloned DNA from those transformants would provide information for structural modification of cloned foreign (bacterial) genes in order to overcome the barrier to more efficient expression of a foreign gene in an unrelated host cell.

C. Molecular Cloning of Recombinant DNA

Molecular cloning of recombinant DNA involves cleaving DNA and splicing it into a carrier molecule so that genetic information from one organism may be transferred to an unrelated organism in which the DNA replicates and expresses itself. Molecular cloning of recombinant DNA, which is also often referred to as genetic engineering, has been the most publicized development among the recent advances in molecular biology; it has been subjected to great controversy both among the scientists and the public because the development of such technology can be both beneficial and hazardous. The potential benefits associated with the use of these new techniques are countless. In brief, they may be divided into two main categories.

a. For the isolation and amplification of DNA for basic studies. Molecular cloning of recombinant DNA offers a simple and easy method for isolating and amplifying DNA fragments which are normally difficult to obtain by the conventional methods. Thus, it provides an indispensible tool for the study of the structure and organization of genes in the chromosomes of all living organisms. Since the initiation of recombinant DNA research in 1973 (97), an enormous amount of new information regarding gene organization, particularly in the eukaryotes, has been accumulated and should be attributed solely to the advent of recombinant DNA methodology. One example is the recent application of molecular cloning procedures to the study of the globin gene family which has led to significant progress in the understanding of the structural, chromosomal arrangement, and evolutionary aspects of these genes (98-101). These studies have also provided the necessary molecular evidence to show that genes are often not colinear with their products. Instead genes contain intervening sequences or introns which interrupt the continuity of the genetic information. As a result, a eukaryotic gene coding for a protein may be much larger than is required for the simple coding of the amino acid sequence. The β-hemoglobin gene in mouse, for example, contains two intervening sequences of 116 and 646 base pairs (102) and the coding sequence is only 432 base pairs. Also, for some time it has been known that genomes are not static but that large segments of DNA may move around on the chromosome. Although the genetics of this phenomenon have been studied for at least 40 years, only after the development of molecular cloning techniques it is possible to study the mechanism of these "jumping genes" at the molecular level (103). This is another of many examples demonstrating that molecular cloning can benefit basic research immensely.

b. For the production of desired products useful in medicine, agriculture, commerce, and industry. Genetic manipulation with recombinant DNA techniques allows us to consider the construction of special organisms (mostly bacteria and fungi which can be grown easily and inexpensively) that will synthesize specific products beneficial to mankind. The most publicized examples of this are the production of insulin and interferon by *E. coli* through genetic engineering. It will not be surprising in the future that microorganisms such as bacteria can be "made" for the conversion of sunlight directly into food substances and it is also almost certain that genetic engineering will play an important part in solving our energy problems by converting renewable resources from agriculture into useful fuel.

The potential hazards associated with the development of recombinant DNA techniques, although the nature and likelihood being mostly speculative, are not possible to assess at this time. However, the general feeling is that the possibility of constructing novel organisms containing toxin producing capabilities or antibiotic resistant genes not found in nature or construting microorganisms containing DNA from cancer causing viruses should not be ignored. The accidental release of these new (novel) organisms from the laboratory into the biosphere may increase the incidence of fatal diseeases or cancer. The concern of the potential biohazards associated with recombinant DNA research among scientists has led to a series of important discussions about the subject which resulted in the formation of a National Institutes of Health Recombinant DNA Molecule Program Advisory Committee to establish formal guidelines for U.S. investigators. Similar committees are at work in other nations to devise comparable guidelines so that this research can be conducted worldwide in a manner that would insure maximum benefits while minimizing, if not precluding, manifestation of biohazards.

The historical developments leading to the ability to clone foreign genes in the bacterium *E. coli* K-12 have been reviewed by Cohen (104), the potential biohazards associated with the development of recombinant DNA research have been thoroughly discussed by Curtiss (105), and various other aspects regarding the advances in this new technology have also been reported (9-10). Furthermore, the complete methodology for molecular cloning of recombinant DNA has also been compiled by Wu (106). Thus this section is only partly devoted to a brief discussion on the background leading to the development of recombinant DNA research, the general protocol for molecular cloning followed by most investigators, and the possible strategies that most likely lead to the successful cloning of a desired gene. The rest of this section will be devoted to reporting the recent developments in cloning yeast DNA into other organisms as well as cloning foreign genes into yeast cells.

The advances that led to the successful development of molecular cloning of recombinant DNA technology were many, but most significant were the following developments: (a) the discovery of restriction endonucleases which can degrade DNA molecules to unique fragments with specific ends as described earlier, (b) the discovery of the enzyme, DNA ligase, which can join two separate DNA fragments together (107,108), (c) the discovery of another enzyme, terminal deoxynucleotidyl transferase (109,110), capable of adding a series of identical deoxynucleotides to the single- or double-stranded DNA molecules (this enzyme makes it possible to join two DNA fragments by the poly (dA·dT) or poly (dC·dG) "connector" method (111, 112), (d) the development of favorable conditions that permit the introduction of relatively large pieces of intact DNA into subsequently viable cells of *E. coli* (113), (e) the development of gel electrophoresis techniques which allow the fractionation and isolation of a mixture of DNA fragments of various sizes into individual separate fragments, and (f) the discovery that plasmid DNA can serve as the cloning vehicle to carry the foreign DNA fragment into the recipient cell in which the hybrid DNA molecule propagates autonomously. Most segments of DNA do not have an inherent capacity for self-replication, particularly after being transferred to a new unrelated host, and one requirement for successful cloning is that the cloned DNA fragments are able to replicate in the recipient cells. Thus, the discovery that DNA fragments joined to the plasmid DNA are capable of self-replicating in the recipient cell has certainly solved the problem.

There are five basic requirements for successful cloning of recombinant DNA molecules: (a) a method for the selection of the DNA fragment to be cloned, (b) the selection of a suitable gene carrier or cloning vehicle--the replicon (usually a plasmid or a virus)--capable of replicating itself autonomously in the recipient cell, (c) the selection of a method for joining the gene (or genes) to be cloned to the cloning vehicle to form the hybrid (or chimeric) DNA molecule--the recombinant DNA, (d) a procedure for introducing the chimeric DNA molecules into the recipient cells (generally known as the process of transformation), and (e) a proper host or recipient organism which can be transformed with high frequency (efficiency) by the chimeric molecule and can also provide certain means for the selection of the proper clones acquiring the desired gene (or genes). The protocol for molecular cloning used by most investigators is similar to the one shown in Figure 6.

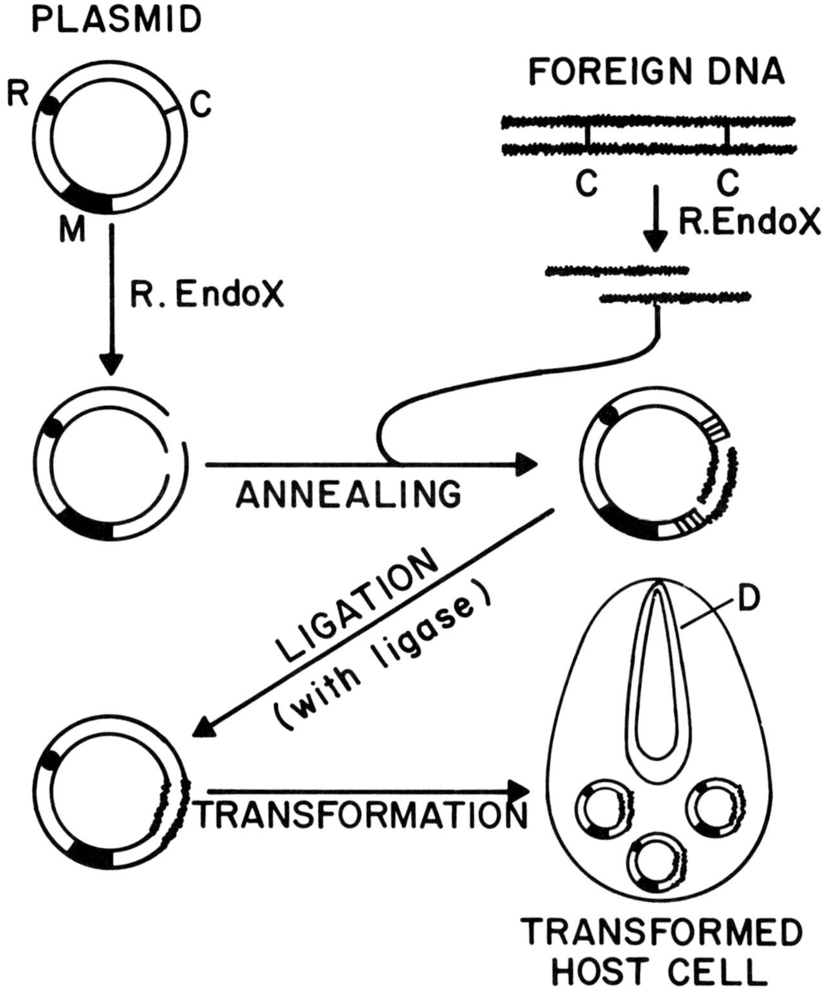

FIGURE 6. DNA Cloning Procedure. R, DNA Replicator; M, Selection Marker; C, Specific Sites for the Hypothetical Restriction Endonuclease X (R. Endo X) in Plasmid and Foreign DNA; D, Host Chromosome.

There are several or even many possible approaches which can be used to fulfill some of the essential requirements listed above for molecular cloning. Thus careful planning of the strategies for cloning is the key leading to success or, at the very least, a poorly planned project will take much longer to accomplish the ultimate goal. The possible options for the fulfillment of each requirement have been discussed by Sinsheimer (8).

Early work in the field of genetic engineering was primarily confined to the cloning of procaryotic and eukaryotic genes into *E. coli*. As early as 1976, Struhl *et al.* (114) reported molecular cloning and functional expression of a yeast gene coding for imidazole glycerol phosphate dehydratase in *E. coli* by complementation of an *E. coli* histidine auxotroph. Subsequently, Kramer *et al.* (115) reported the cloning of yeast ribosomal RNA genes in *E. coli*. In both cases, λ DNA was used as the carrier for the yeast genes. In 1977, Ratzkin and Carbon (116) constructed a collection of hybrid circular DNA molecules *in vitro* using the poly (dA·dT) connector method. Each hybrid molecule contained one molecule of plasmid ColEl and a fragment of sheared yeast DNA. Such hybrid DNA molecules were used to transform *E. coli* cells, and a sufficient number of transformants were obtained so that the hybrid plasmid population was assured of representing the entire yeast genome. Various hybrid ColEl-yeast DNA plasmids capable of complementing *E. coli* auxotrophic mutations were selected from this population. This work demonstrated that complementation of bacterial mutations by yeast DNA segments does not appear to be a rare phenomenon. Since then quite a few yeast genes have been cloned in *E. coli* (117-121). However, it has been reported that, at least in some cases, the levels of complementing enzyme activities produced in the bacterial host cells from the cloned yeast genes are quite low, resulting in slow growth of the transformants on selective media. Thus barriers against efficient utilization of the yeast genes in bacteria still exist. Walz *et al.* (122) showed that the inefficient expression in *E. coli* of a cloned gene could be corrected by spontaneous insertion (*in vivo*, into the yeast gene) of a transposable genetic element which carries an *E. coli* promoter for RNA transcription.

In 1978, Hinnen *et al.* (123) demonstrated for the first time that transformation of yeast cells could be accomplished with bacterial plasmid, although the frequency of transformation was low (10^{-7} transformants per viable cell). The low frequency of transformation was apparently due to the fact that the transforming plasmid was not present in yeast clones in free or non-integrated form but was integrated into yeast chromosomes at more than one site. Subsequently, higher efficiency of transformation was accomplished when the 2-μm DNA or a segment of yeast DNA containing a yeast chromosomal replicator (124-129) was incorporated in the chimeric molecules for transformation. The hybrid DNA molecules containing the 2-μm yeast plasmid or a yeast chromosomal replicator were found capable of replicating autonomously in yeast transformants with the frequency of transformation as high as 10^{-4} transformants per viable cell. With the development of a high frequency transformation system for yeast, molecular cloning and functional expression of yeast or bacterial genes in yeast transformants have been reported (130,96). Barriers against efficient utilization in yeast of bacterial genes have also been observed, and spontaneous mutation, which led to higher levels of functional expression of bacterial genes in yeast, has also been reported (131); however, the nature of such mutation has not been explored.

A system of biological containment for yeast transformation with recombinant DNA molecules has also been developed--the establishment of the "shy strains" (131). The theoretical basis of this biological containment is based on the hypothesis that the principal route of "escape" for a segment of DNA cloned in a yeast host is the possibility of a mating between the *in vitro* recombinant and another hardier yeast strain. Thus the primary basis of the biological containment established for the "shy" strains is sterility: i.e., the inability for these strains to mate with other yeasts.

V. THE IMPROVEMENT OF ALCOHOL PRODUCTION BY RECOMBINANT DNA TECHNIQUES

The recent advances in recombinant DNA and other techniques have opened up an entirely new spectrum of possibilities for the improvement of alcohol production by fermentation. Most important of all, the availability of efficient systems for the transformation of *E. coli* as well as yeast cells makes it possible to isolate yeast DNA containing a specific gene (or genes) of interest for structural analyses. Knowledge of the gene organization is the blueprint for the construction of tailor-made microorganisms for our service. For example, yeast is limited to carrying out predominantly fermentation or respiration at any one time. The total understanding of the molecular mechanism responsible for switching the processes "on" and "off" will provide the information necessary to design experiments to alter or even totally abolish the control mechanisms which regulate these processes. As a result, fermentation and respiration may proceed hand in hand in yeast at the same time; there will be no more "glucose" or "Pasteur" effect and yeast may continue to produce alcohol and to grow (replicate) simultaneously forever and ever. This may appear to be imaginary, but certainly it is not an impossible dream to accomplish.

The advent of recombinant DNA technology also makes it possible to introduce extra genetic information into yeast which has never been present before and would make yeast capable of directly fermenting substrates such as cellulose, pentose, and others which are considered non-fermentable by yeast at the present time.

The development of recombinant DNA technology also permits the investigation of the possibility of using organisms other than yeast for fermentation of carbohydrates to alcohol. Although yeast has been considered to be the most effective microorganism for fermentation of glucose to ethanol up to the present, there are organisms, such as *E. coli*, which are missing but one or two enzymes required for alcoholic fermentation. With the availability of genetic engineering technology, the necessary genes responsible for the production of the missing enzymes can be cloned into the organism and thus a new organism more suitable for alcoholic fermentation of carbohydrates may then be found.

VI. CONCLUSION

This past decade has been full of dramatic developments. On the one hand, we are facing a world wide, unprecedented problem - the energy crisis which presses for solution. An important component of the solution to the energy problem is to derive energy from replenishable resources. Alcohol produced by fermentation of carbohydrates with microorganisms is one such source of energy that the world relies on. This sudden increase in demand for alcohol makes the traditional process for the production of alcohol by fermentation of sugar with yeast insufficient to meet the challenge. Substantial improvements are desired in terms of the alcohol concentration, the duration required for the production of a high concentration of alcohol as well as the limited types of carbohydrates that yeast can normally ferment. On the other hand, we are blessed with sprouting new discoveries in the field of molecular biology, one after the other, which have resulted in the development of the new technology for molecular cloning of recombinant DNA (genetic engineering). This new technology allows us to peek into the molecular mechanisms controlling the functions of the genes inside the cell, to alter the genetic make-up of an organism precisely according to our desire, and to make the tailor-made organism producing the desired product cheaply and effectively. As Professor Perlman indicated in his article (6), the occurence of "crisis" and the availability of "new technology" are some of the usual factors stimulating innovation. There is no doubt that alcohol production by fermentation under these circumstances will be improved; the only question is to what extent and how long it takes (to make substantial improvements). It is up to us to make it soon and revolutionary.

REFERENCES

1. Rose, A.H. and Harrison, J.S., "The Yeasts" Vol. 1 (1969), Vol. 2 (1971), and Vol. 3 Academic press.
2. Phaff, J.H., Miller, M.W. and Mrak, E.M., "The Life of Yeasts", Harvard University press (1978).
3. Horecker, B.L. in "Biochemistry and Genetics of Yeasts", (M. Bacila, B.L. Horecker, and A.D.M. Stoppani, Eds), Academic Press. (1978).
4. Newberg, C. and Reinfurth, E., *Biochem. Z. 92,* 234 (1910).
5. Chapter 10 in "The Life of Yeasts", (H.J. Phaff, M.W. Miller, and E.M. Mrak Eds), Harvard University Press (1978).
6. Perlman, D., *Ann. Reports Ferm. Processes 3,* 1. (1979).
7. Sanger, F., Air, G.M., Barrell, B.G., Brown, N.L., Coulsan, A.R., Fiddes, J.C., Hutchison, C.A., Slocombe, P.M., and Smith, M., *Nature 256,* 687 (1977).
8. Sinsheimer, R.L., *Annual Rev. of Biochem. 46,* 415 (1977).
9. Cohen, S.N., Cabellow, F., Casadaban, M., Chang, A.C.Y. and Timmis, K., in "Molecular Cloning of Recombinant DNA", (W.A. Scott and R. Werner Eds), P. 35, Academic Press (1977).
10. Morrow, J.F., in "Methods in Enzymology, Vol. *68*" (Ray Wu, Ed), Academic Press (1979).
11. Cruess, W.V., Brown, E.M. and Flossfeder, F.C., *J. Ind. Engng Chem. 8,* 1124 (1916).
12. Castan, R., *Landw. Jb. Schweiz. 41,* 311 (1927).
13. Delle, P.N., Odessa. Otch. "Vinodiel'cheskoi Stantsii Russiakh Vinogr. Vindo." 1908g i 1909g, 118-160 (1911).
14. Amerine, M.A. and Kukee, R.E., *Vitis 5,* 187 (1965).
15. Kunkee, R.E. and Amerine, M.A., *Appl. Microbiol. 16,* 1067 (1968).
16. Amerine, M.A. and Kunkee, R.E., *Ann. Rev. Microbiol. 22, 323* (1968).

17. Schanderl, H. in "Mikrobiologie des Mostes and Weines," 2nd Ed, Verlag. E. Ulmer, Stuttgart (1959).
18. Navarro, M.J. and Durand, G., *Ann. Microbiol. 129 B,* 215 (1978).
19. Nagodawithana, T.W., Castellano, C., and Steinkraus, K.H. *Appl. Microbiol. 28,* 283 (1974).
20. Nogodawithana, T.W. and Steinkraus, K.H., *Appl. Environ. Microbiol. 31,* 158 (1976).
21. Holzberg, I., Finn, R.K., and Steinkraus, K.H., *Biotechnology and Bioengineering 9,* 413 (1967).
22. Wortmann, J., *Mitt Weinb Kellerw 7,* 65 (1895).
23. Kunkee, R.E. and Amerine, M.A., Chapter 2, in "The Yeasts", Vol. 3 (A.H. Rose and J.S. Harrison Eds). Academic Press (1970).
24. Wikén, T. and Richard, O., *Antonie van Leeuwenhock 21,* 337 (1955).
25. Ough, C.S. and Amerine, M.A., *Bull. Calif. Agric. Exp. Stn. 827,* 1 (1966).
26. Ricketts, R.W. and Hough, J.S., *J. Inst. Brew. 67,* 29 (1961).
27. Enari, T.M. and Makinen, V., *Brauwissen-Schaft 14,* 253 (1961).
28. Couland, T.W., *J. Inst. Brew. 73,* 542 (1967).
29. Wikén, T. and Richard, O. *Enperimentia 9,* 417 (1953).
30. Wikén, T. and Richard, O., Schweiz, *Z. Path Bakt. 17,* 475 (1954 a).
31. Wikén, T. and Richard, O., *Antonie van Leeuwenhoek 20,* 385 (1954 b).
32. Wikén, T. and Richard, O., *Schweiz. Z. Path. Bakt 18,* 970 (1955 a).
33. Wikén, T. and Richard, O., *Antonie van Leeuwenhoek 21,* 337 (1955 b).
34. Wikén, T., and Pfennig, N., *Antonie van Leeuwenhoek 23,* 113 (1957).
35. Wikén, T., and Pfennig, N., *Antonie van Leeuwenhoek 25,* 193 (1959).
36. Wikén, T.O., *Sci, Rep. 1st Super. Santia 1,* 309 (1961).
37. Idziak, E.S. and Wikén, T.O., *Path. Microbiol. 25,* 653 (1962).
38. Wikén, T.O., Van Nieuelt, M.C.A., and Knetemann, J.C.A., Doodewaard, J. and Scheffers, W.A., in "Yeast Proceedings of the 2nd Symposium on Yeasts, 1966", Vydavatelsvo slovenskej Academic vied Bratislava (1969).
39. Steinkraus, K.H. and Morse, R.A., *J. Appicutt, Res. 5,* 17 (1966).
40. Crysewski, G.R. and Wilke, C.R., *Biotech. Bioeng., 19,* 1125 (1977).
41. Ghose, T.K. and Tyagi, R.P., *Biotech. Bioeng. 21,* 1287 (1979).

42. De Rosario, E.J., Lee, K.J., and Rogers, P.L., *Biotech. Bioeng. 21,* 1477 (1979).
43. Moreno, M. and Goma, G., *Biotechnol. Letters, 1,* 483 (1979).
44. Andreasen, A.A. and Stier, T.J.B., *J. Cell Comp. Physiol. 41,* 23 (1953).
45. Brechot, P., Vignes et Vins, N° Special Mans p 26 (1974).
46. Brechot, P., Chauvet, J., Bupuy, P., Croson and Rabatu, A., *C. R. Acad. Sci. Ser. D. 272,* 890 (1971).
47. Longley, R.P., Rose, A.H., and Knights, B.S., *Biochem. J. 108,* 401 (1967).
48. Dulaney, L.M., Stapley, E.D., and Simpf, K., *Appl. Microbiol. Chem. 226,* 497 (1954).
49. Maugenet, J. and Dupug, P., *Ann. Technol. Agric. 13,* 329 (1964).
50. Demel, R.A. and de Kruyff, B., *Biochem. Biophys. Acta. 457,* 109 (1976).
51. Radler, F., *Ann. Technol. Agric. 27,* 203 (1978).
52. Gordon, P.A., and Stewart, P.R., *J. Gen. Microbiol 72,* 231 (1972).
53. Kirsop, B.H., *J. Inst. Brew. 80,* 252 (1974).
54. Lafon-Lafourade, S., Larue, F., Brechot, P., and Ribereau-Gayon, P., *C.R. Acad. Sci. Ser D. 284,* 1939 (1977).
55. Lafon-Lafourcade, S., Larue, F., and Ribereau-Gayon, P., *Appl. Enviro. Microbiol. Vol. 38 ,* 1069 (1979).
56. Larue, F., Lafon-Lafourcade, and Ribereau-Gayon, *Appl. Environ. Microbiol. 29,* 808 (1980).
57. Moaf, A.G., Peters, G.N., and Srb, A.M., *J. Bacteriol. 77,* 673 (1959).
58. Slonimski, P.P., Perrodin, G., and Croft, J.H., *Biochem. Biophys. Res. Commun. 30,* 232-239 (1968).
59. Sugimura, R., Okabe, K., and Kodama, M., *J. Bacteriol. 97,* 964 (1969).
60. Costa, S.O.P. and Bacila, M., *An. Acad. Bras. Cienc. 43,* 711 (1971).
61. Juliani, M.H., Costa, S.O.P. and Bacila, M., *Biochem. Biophys. Res. Comm. 53,* 531 (1973).
62. Xavier, A.W., Costa, S.O.P., and Bacila., M., *An. Acad. Bras. Ciens. 47,* 557 (1975).
63. Nagai. S.N. Yanagishima, and Nagai, H., *Bacteriol. Rev. 25,* 404 (1961).
64. Bernardi, G., Carnevali, F., Nicolaieff, A., Piperno, G., and Tecce, G., *J. Mol. Biol. 37,* 493 (1968).
65. Bernardi, G., Prunell, A., and Kopecka, H., in "Molecular Biology of Nucleocytoplasmic Relationships", p. 85, Elsevier Amsterdam (1975).

66. Bernardi, G., Culard, F., Fonty, G., Coursot, R. and Prunell, A., in "Biochemistry and Genetics of Yeats" (M. Bacila, B.L. Horecker, and A.O.M. Stoppani, Eds), p 85, Academic Press (1978).
67. Bacila,M., Xavier, A.W. and Horii, J. in "Biochemistry and Genetics of Yeasts" (M. Bacila, B.L. Horecker and A.O.M. Stoppani, Eds), p 241, Academic Press (1978).
68. Luria, S.E. and Hurman, M.L., *J. Bacteriol. 64,* 557 (1952).
69. Bertani, G. and Weigle, J.J., *J. Bacteriol. 65,* 113 (1953).
70. Luria, S.E., *Cold Spring Harbor Symp. Quant Biol., 18,* 237 (1953).
71. Linn, S. and Arber, W., *Annu. Rev. Biochem. 28,* 467 (1969).
72. Meselson, M. and Yuan, R., *Nature, 217* (1968).
73. Roulland-Dussoix, D. and Boyer, H.W., *Biochim. Biophys. Acta. 195,* 219 (1969).
74. Horiuchi, K. and Zinder, N.D., *Proc. Natl. Acad. Sci. 69,* 3220 (1972).
75. Murray, N.E., Batten, P.L. and Murray, K., *J. Mol. Biol. 81,* 395 (1973).
76. Kelly, T.J. Jr. and Smith, H.O., *J. Mol. Biol. 51,* 393 (1970).
77. Smith, H.O. and Wilcox, K.W., *J. Mol. Biol. 51,* 379 (1970).
78. Smith, H.O. and Nathans, D.J., *J. Mol. Biol. 81,* 419 (1973).
79. Waallwijk, C. and Flavell, R.A., *Nucl. Acids Res. 5,* 3231 (1978).
80. Vovis, G.F. and Lacks, S., *J. Mol. Biol. 115,* 525 (1977).
81. Darnell, J.E. Jr., *Science 202,* 1257 (1978).
82. Maxam, A.M. and Gilbert, W., *Proc. Natl. Acad. Sci. 74,* 560 (1977).
83. Sanger, F. and Coulson, A.R., *J. Mol. Biol. 94,* 441 (1975).
84. Meselsan, M., Yuan, R. and Heywood, J., *Annu. Rev. Biochem. 41,* 447 (1972).
85. Arber, W., *Prog. Nucleic Acids Res. Mol. Biol. 14,* 1 (1974).
86. Nathans, D. and Smith, H.O., *Annu. Rev. Biochem. 44,* 273 (1975).
87. Zabeau, M. and Roberts, R.J., in "Molecular Genetics, Part III" (Taylor, J.H., Ed), Academic Press (1979).
88. Ling, V., *J. Mol. Biol. 64,* 87 (1972).
89. Barrell, B.G. and Clark, B.F.C., "Hand Book of Nucleic Acid Sequences", Joynson-Bruvvers (1974).
90. Sanger, F., Nicklen, S. and Coulson, A.R., *Proc. Natl. Acad. Sci. 74,* 5463 (1977).

91. Maxam, A.M. and Gilbert, W. in "Methods in Enzymology" Vol. 65 (L. Grossman and K. Moldave, Eds), Academic Press (1980).
92. Sanger, F. and Coulsan, A.R., FEBS Lett. 87. 107 (1978).
93. Carbon, J., Ratzkin, B., Clarke, B. and Richardson, D., Brook haven Sym. Biol. 29, 277 (1977).
94. Clarke, L. and Carbon, J., *J. Mol. Biol., 120,* 517 (1978).
95. Walz, A., Ratzkin, B. and Carbon, J., *Proc. Matl, Acad. Sci. 75,* 6172 (1978).
96. Cohen, J.D., Eccleshall, T.R., Needleman, R.B., Federoff, H., Buchferer, B.A. and Marmur, *Proc. Natl. Acad. Sci. 77,* 1078 (1980).
97. Cohen, S.N., Chang, A.C.Y., Boyer, H.W. and Helling, R.B., *Porc. Natl. Acad. Sci. 70,* 3240 (1973).
98. Maniatis, T. in Cell Biology, A. Comprehensive Treatis (L. Goldstein and D.M. Prescott, Eds), Academics Press (1980).
99. Efstratiadis, A. *et al, Cell 21,* 653 (1980).
100. Weatherall, D.J. and Clegg, J.B., *Cell 16,* 467 (1979).
101. Bunn, H.F., Forget, B.G. and Ranney, H.M., "Human Hemoglobins", Saunders, (1977).
102. Konkel, D.A., Tilghman, S.M. and Leder, P., *Cell 18,* 865 (1978).
103. Simon, M., Zieg, J., Silverman, M., Mandel, G. and Doolittle, R., *Science 209,* 1370 (1980).
104. Cohen, S.N., *Scientific American 233,* 25 (1975).
105. Curtiss, R. III, *Ann. Rev. Microbiol. 30,* 507 (1976).
106. Wu, R., "Recombinant DNA, Methods in Enzymology, Vol. 68", Academic Press (1979).
107. Harvey, C.L. and Wright, R., *Biochemistry 11,* 2667 (1972).
108. Segaramella, V. and Khorana, H.G., *J. Mol. Biol. 72,* 493 (1972).
109. Lobban, P.E. and Kaiser, D., *J. Mol. Biol. 78,* 453 (1973).
110. Roychoudhury, R., Jay, E. and Wu, R., *Nucleic Acids Res. 3,* 863 (1976); Brutlag, D., Fry, K., Nelsen, T. and Hung, P., *Cell 10,* 509 (1977).
111. Jackson, D.A., Symons, R.H. and Berg, P., *Proc. Natl. Acad. Sci. 69,* 2904 (1972).
112. Clarke, L. and Carbon, J., *Proc. Natl. Acad. Sci. 72,* 4361 (1975).
113. Mandel, M. and Higa, A., *J. Mol. Biol. 53,* 159 (1970); Taketo, A., *Z. Naturforsch 30 C,* 520 (1975).
114. Struhl, K., Cameron, J.R. and Davis, R.W., *Proc. Natl. Acad. Sci. 73,* 1471 (1976).
115. Kramer, R.A., Camerson, J.R. and Davis, R.A., *Cell 8,* 227 (1976).

116. Ratzkin, B. and Carbon, J., *Proc. Natl. Acad. Sci. 74,* 487 (1977).
117. Dickson, R.C. and Markin, J.S., *Cell 15,* 123 (1978).
118. Clarke, L. and Carbon, J., *J. Mol. Biol. 120,* 517 (1978).
119. Schell, M.A. and Wilson, D.B., *Gene. 5,* 291 (1979).
120. Citron, B.A., Feiss, M. and Donelson, J.E., *Gene 6,* 251 (1979).
121. Holland, M.J., Holland, J.P., and Jackson, M.A. in "Methods in Enzymology," vol. 68 (R. Wu, Ed), Academic Press (1979).
122. Walz, A., Ratzkin, B., and Carbon, J., *Proc. Natl Acad. Sci, 75,* 6172 (1978).
123. Hinnen, A., Hicks, J.B. and Fink, G.R., *Proc. Natl. Acad. Sci. 75,* 1929 (1978).
124. Begg, J.D., *Nature 275,* 104 (1978).
125. Stinchcomb, D.T., Struhl, K. and Davis, R.W., *Nature 282,* 39 (1979).
126. Gerbaud, C., Fournier, P., Blanc, H., Aigle, M., Heslot, H. and Guerineau, M., *Gene 5,* 233 (1979).
127. Struhl, K., Stinechcomb, D.T., Scherer, S. and David, R.W., *Proc. Natl. Acad. Sci. 76,* 1035 (1979).
128. Broach, J.R., Strathern, J.N. and Hicks, J.B., *Gene 8,* 121 (1979).
129. Tschumper, G. and Carbon, J., *Gene 10,* 157 (1980).
130. Nasmyth, K.A. and Reed, S.I., *Proc. Natl. Acad. Sci. 77,* 2119 (1980).
131. Botstein, D., Folco, C., Stewart, S.E., Brennan, M., Seherer, S., Stinechcomb, D.T., Struhl, K. and Davis, R.W., *Gene 8,* 17 (1979).

CHAPTER 10

AN EVALUATION OF N_2 FIXATION AND H_2 PRODUCTION IN FERMENTATION CULTURE

Michael W.W. Adams
Robert G. Upchurch
and
Leonard E. Mortenson

Department of Biological Sciences
Purdue University
West Lafayette, Indiana

I. INTRODUCTION

The title of this review appears to associate two unrelated processes which by their very nature depict on the one hand, the ability to use the earth's atmosphere as a nutrient source, while on the other, a seemingly futile reaction undertaken by a variety of organisms. Indeed, H_2 is produced by several bacteria incapable of N_2 fixation and these are considered in this article. However all N_2-fixing bacteria also evolve H_2 - an inescapable consequence of the N_2-fixing process. This is taken to the extreme in some microorganisms which, in the absence of N_2, utilize the N_2-fixing apparatus in the exclusive production of H_2. As will be discussed, the allocation of electrons from reductant for these two reactions is dependent on the organism and the culture conditions.

Only two enzymes are known which are capable of reducing protons to H_2. The simplest, both structurally and mechanistically, is the enzyme, hydrogenase, which catalyzes a reversible reaction (eqn. 1). Representatives of almost all types of bacteria have been found to possess hydrogenase activity. Its presence enables organisms to either dispose of excess reductant using protons as the terminal electron acceptor as in H_2 evolution, or alternatively, it catalyzes the consumption of H_2 which may then be used as a source of energy and/or

ISBN 0-12-040304-8

$$2H^{+} + 2e^{-} \rightleftharpoons H_2 \tag{1}$$

reductant. The production of H_2 is also catalyzed by the enzyme complex responsible for the reduction of N_2 to ammonia, nitrogenase. In contrast to hydrogenase, nitrogenase-catalyzed H_2 evolution requires energy (ATP) and it is an "irreversible" reaction (eqn. 2). Although the enzyme is found in anaerobic,

$$2H^{+} + 2e^{-} + 4ATP \longrightarrow H_2 + 4ADP + 4Pi \tag{2}$$

aerobic and photosynthetic bacteria, nitrogenase appears to be much more limited in its distribution in the microbial world. However, almost all N_2-fixing organisms possess hydrogenase activity.

Of the microorganisms that have a H_2 metabolism some, such as the methanogenic bacteria and most species of aerobic hydrogen bacteria, only consume H_2; they possess hydrogenase activity but they do not fix N_2. The majority evolve H_2 and these may be subdivided into three groups. First there are those which produce H_2 solely via the nitrogenase system although they may also contain hydrogenase. This group includes the cyanobacteria and the aerobic N_2-fixing bacteria e.g. *Rhizobium* and *Azotobacter*. Secondly, there are some N_2-fixing bacteria which, depending on the growth conditions, evolve H_2 via both hydrogenase and nitrogenase e.g. the photosynthetic bacteria and some obligate anaerobes such as species of *Clostridium*. Into the third category fall the eucaryotic algae and some facultative anaerobes such as *Escherichia coli*. These do not fix N_2 but exhibit hydrogenase-catalyzed H_2 evolution.

In all bacterial species that fix N_2, the nitrogenase system is usually repressed when readily utilizable N-sources are present in the growth medium. Also, the nitrogenases so far examined from a wide variety of bacteria are all very similar in activity and molecular composition. In contrast, some organisms possess an inducible hydrogenase whereas in others the enzyme is apparently constitutive. Furthermore, hydrogenases from a variety of genera are very diverse in their physical and catalytic properties and *in vivo* hydrogenase functions exclusively in H_2 production in some bacteria whereas in others it catalyzes only H_2 consumption. The ability of a particular organism to fix N_2 and/or evolve H_2 is very dependent on the culture conditions and the type of hydrogenase they possess.

The aim of this article is to describe the conditions required to observe N_2 fixation and H_2 evolution in anaerobic, aerobic and photosynthetic organisms, with an emphasis on the metabolic relevance of these reactions and the interrelationships between them. To do justice to this task, it is felt necessary to first describe some of the properties of the two enzymes involved, namely nitrogenase and hydrogenase.

II. PROPERTIES OF NITROGENASE

A. *Physical Properties.*

Biological N_2 fixation is the reduction of N_2 to NH_3 catalyzed by the enzyme nitrogenase. Chemical reduction of N_2 can also be catalyzed by transition metal complexes (1,2) but both chemical and biological reductions require energy and a source of reductant. Nitrogenase is found only in procaryotes and unlike hydrogenases (Section III) nitrogenases from a variety of bacteria are remarkably similar in structural, mechanistic and regulatory properties (3,4). The principles of nitrogenase purification have been recently reviewed (5). The proteins that comprise the nitrogenase complex, the MoFe protein (component I, dinitrogenase) and the Fe protein (component II, dinitrogenase reductase), are both extremely sensitive to O_2 and are usually isolated separately although a complex of the two proteins has been isolated from *Azotobacter vinelandii* (6,7). Neither protein alone has nitrogenase activity when supplied with reductant and ATP. Because of the complexity of the N_2-fixing reaction and the dissociating nature of the nitrogenase complex, no official (EC) name has been given.

The Fe protein is a dimer of molecular weight around 60000 and electron paramagratic resonance (EPR) spectroscopy and ligand displacement techniques show the protein to contain one [4Fe-4S] cluster which is thought to be liganded between the two monomers (3). EPR and Mössbauer spectroscopy indicate that the 4Fe-cluster is of the classical ferredoxin-type and transfers one electron. The half-life of the Fe protein in air is about 30 seconds thus its purification is performed under stringent anaerobic conditions. The amino-acid sequence of the protein was recently completed (273 residues; 3, 8-10 and shows it to have a flexible conformation reflecting the known changes that take place during catalysis. The Fe protein, but not the MoFe protein, binds MgATP (K_D = 60 μM) and MgADP but only two of the probable four adenine nucleotide binding sites appear to be involved in nitrogenase catalysis (3). On binding MgATP, the Fe protein undergoes a conformational change as evidenced by the fact that its midpoint potential becomes 120mV more negative, its sensitivity to O_2 increases and its iron atoms become sensitive to chelation (3). MgADP is a competitive inhibitor of the functioning of MgATP in the catalytic reaction.

The MoFe protein is a tetramer of molecular weight around 220,000 that consists of two dissimilar subunits (50,000 and 60,000). The MoFe protein contains 2 Mo atoms, *ca.* 30 Fe atoms and an equal number of acid-labile sulfur atoms. EPR spectroscopy, extrusion techniques and Mössbauer spectroscopy have indicated the presence of four [4Fe-4S] clusters, two uncharacterized centers of composition $MoFe_8S_6$, and possibly a [2Fe-2S] cluster (3). Anaerobic treatment of the MoFe protein with N-methyl formamide releases a component termed the iron-molybdenum cofactor (FeMoco) which contains Fe, Mo and S^{2-} in the ratio 8:1:2 but no amino-acids (3,11). FeMoco restores nitrogenase activity to crude extracts of a mutant of *A. vinelandii* (UW45) which contains inactive Mo-free MoFe protein (11). This cofactor appears to be the major part of the active site of nitrogenase. Data obtained using the new technique of X-ray absorption spectroscopy called extended X-ray absorption fine structure (EXAFS) shows that the nitrogenase MoFe protein has no Mo-O or Mo-Mo bonds and Mo is primarily liganded to sulfur and possibly to Fe through sulfur. The valence of Mo in nitrogenase is either 3^+ or 4^+ (3).

B. Mechanism

The requirements for nitrogenase to catalyze (a) N_2 reduction, (b) H_2 evolution, or (c) reduction of non-physicological substrates e.g. C_2H_2, CN^- or N_3^-, are: active MoFe and Fe proteins, a source of reductant (reduced ferredoxin or flavodoxin *in vivo* or sodium dithionite *in vitro*), the energy source (MgATP), and a reducible substrate. Assays *in vitro* usually include an ATP-generating system (creatine phosphate plus creatine phosphokinase) to prevent inhibition by the product of ATP utilization, ADP (see next section). Nitrogenase activity is conveniently assayed by measuring C_2H_2 reduction, a simple rapid assay that uses gas chromatography to detect C_2H_4. However caution must be exercised when relating C_2H_2 reduction to NH_3 production, especially in whole cell systems (see 12).

A simple equation for N_2 reduction by nitrogenase is given below and the mechanism of electron flow is depicted in Figure 1. The evolution of H_2 seems to be an intrinsic

$$N_2 + 8H^+ + 8e^- + 16ATP \longrightarrow 2NH_3 + H_2 + 16ADP + 16Pi \quad (3)$$

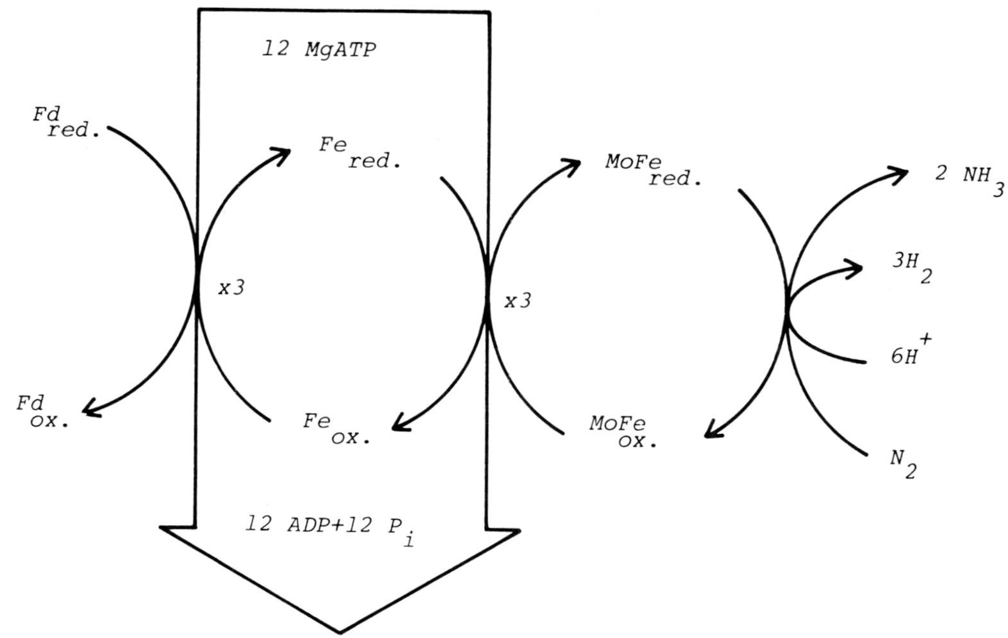

Figure 1. Scheme of electron flow during catalysis of N_2 reduction by nitrogenase. Fd, ferredoxin; Fe, Fe protein; Mo, MoFe protein; ox. and red., oxidized and reduced proteins.

property of the enzymatic reduction of N_2 and both activities are ATP dependent (3). Several factors *in vitro* and presumably *in vivo* control the ratio of N_2 reduced to H_2 evolved. These include the Fe protein: MoFe protein ratio, the reductant concentration, the ADP/ATP ratio and the absolute ATP concentration (13). In the absence of added reducible substrates, nitrogenase catalyzes only ATP-dependent H_2 evolution (H^+ reduction). Carbon monoxide inhibits the reduction of all nitrogenase substrates except H^+ and H_2 inhibits N_2 reduction. The former reaction is important in differentiating the *in vivo* functions of nitrogenase and hydrogenase (see Sections VII and IX).

The mechanism of nitrogenase catalysis involves the transfer of electrons from the electron donor e.g. reduced ferredoxin, to the Fe protein and then to the MoFe protein (Figure 1). Although the Fe protein must bind MgATP in order to be in the conformation which can accept and transfer electrons, the exact sequence of the binding of ATP, the hydrolysis of ATP, the binding of MoFe protein and transfer of electrons is not fully established. Two ATP molecules are hydrolyzed per electron passed to the MoFe protein when an ATP-generating system is present. Recent evidence (14) indicates that the Fe and MoFe proteins dissociate after each one election transfer i.e. the binding of the Fe protein to the MoFe protein is transient. Several studies indicate that the ratio of ATP hydrolyzed to electron pair transferred through the system is a minimum of 4, and this varies with reaction conditions. Components of the reaction have been examined by EPR spectroscopy and the appearance and disappearance of the EPR signals in response to the electronic state of the FeS and MoFeS centers have been used in the establishment of the electron flow (3).

C. Regulation of Activity

The N_2-fixing system, because of its great energy demands on the cell, is under fairly tight control both in activity and biosynthesis (for reviews, see 2-4, 15). Work with both crude extracts (16) and purified proteins (4,17) have shown that ADP, the product of ATP utilization, inhibits nitrogenase activity and studies both *in vitro* and *in vivo* have shown that nitrogenase is regulated by the ADP/ATP ratio. The curve relating activity to increasing ADP/ATP is asymptotic rather than linear. For example, Seto and Mortenson (4) found that at ADP/ATP ratios of 0.2 and 1.0, the C_2H_2 reduction activity of *C. pasteurianum* nitrogenase was inhibited by 40% and 90%, respectively. N_2-fixing cells have a lower energy charge (ATP + 0.5 ADP/ATP + ADP + AMP, 0.65) and a higher ratio of ADP/ATP (0.70) than ammonia grown cells (0.83 and 0.25, respectively;

for discussion, see 17). Based on the intracellular ADP/ATP ratio, the nitrogenase of N_2-fixing cells probably operates at about one-third of its theoretical maximum activity (17).

Aerobic N_2-fixing bacteria appear to have a mechanism to protect nitrogenase from high O_2 concentrations. The nitrogenase activity of *Azotobacter* in chemostat cultures decreased when incubated with a $pO_2 > 0.2$ but normal activity was restored upon returning to atmospheric pO_2 (18). High O_2 concentrations appear to 'switch off' nitrogenase by converting it to an O_2-resistant conformation. This seems to result from its complexing with an iron-sulfur protein (Fe-S protein II) which interupts electron flow from reductant (19).

Carbamyl phosphate inhibits the *in vitro* nitrogenase activity of *Azotobacter* and *Clostridium* up to 60% and the *in vivo* nitrogenase activity of *C. pasteurianum* in chemostat cultures was inhibited about 70% (4,20). Unlike ATP, carbamyl phosphate binds to the MoFe protein but maximum binding only occurs in the active complex of Fe and MoFe proteins or with the MoFe protein in its reduced form. Carbamyl phosphate appears to be a regulatory agent since its intracellular concentration (a value close to its K_m) is sufficient to affect nitrogenase activity: in N_2-fixing cells, it is about 50 μM compared with 100 μM in ammonia grown cells.

It has been known for many years that H_2 is an inhibitor of N_2 fixation, but it was only recently shown that H_2 is a competitive inhibitor of N_2, but not of other reducible substrates (4). During N_2 reduction H_2 inhibition results from a reaction between H_2 and a postulated diimide intermediate of nitrogen fixation such that enzyme$\cdot N_2H_2$ + H_2 yields $2H_2$ + N_2 + oxidized enzyme and ammonia is not produced (4). A paradox exists since H_2 is a normal product of N_2 reduction by nitrogenase and some N_2-fixing bacteria such as *C. pasteurianum* evolve large amounts of H_2 during fermentation. That N_2-fixing organisms have a way to prevent this energy-requiring futile cycle seems likely since clostridial nitrogenase is relatively insensitive to H_2 (K_i = 0.5 atm.). Nitrogenase activity of *Azotobacter* is much more sensitive to H_2 but its cells do not evolve H_2 during their normal metabolism (see Section VII).

In vitro experiments (14) have shown that electron allocation to mixed substrates of nitrogenase (H^+, N_2, C_2H_2) is controlled by the electron flux (rate of electron flow) through the enzyme to substrates. This is an example of affinity regulation since the ratio of N_2 reduced/H_2 evolved is controlled by the rate of electron flow through the enzyme, which in turn can be regulated by the Fe protein: MoFe protein ratio as well as by the reductant concentration in the reaction system. When reductant is limited as described above, reduction of substrates by two electrons is favored over reductions requiring more than two (N_2). The significance of these very interesting findings have not yet been explored *in vivo*.

Until recently, the addition of fixed N to N_2-fixing cells was thought to inhibit only the further synthesis of nitrogenase, and possibly to activate nitrogenase-degrading proteases. Nitrogenases of the photosynthetic bacteria *Rhodopseudomonas capsulata* and *Rhodospirillum rubrum* differ from those of the non-photosynthetic heterotrophic bacteria in that their activity also is subject to an additional mechanism of regulation by fixed N (21,22). Nitrogenase obtained from N-starved cells (N_2-ase A) is fully active *in vivo* and *in vitro*, whereas nitrogenase from N_2 or glutamate-grown cells (N_2-ase R, regulatory) is always inactive *in vitro* because of a modification of the Fe protein. The latter can be activated however, by a Mn^{2+}-ATP dependent protein activating factor. Changing the N supply of these bacteria greatly influences the cellular concentration of nitrogenase A and R and is accompanied by changes in whole cell N_2-fixing activity. Little is known about the conversion mechanism but kinetic and inhibitor studies indicate that *de novo* protein synthesis is not required and glutamine synthetase is involved in the conversion.

D. *Regulation of Synthesis*

Nitrogenase has been estimated to compose 2-5% of total cell protein (23). The intracellular ratio of Fe to MoFe protein seems to be 1:1 or 2:1 (3) however Fe protein in excess of this has been estimated from recent kinetic data (24). Obviously, the synthesis of an enzyme system produced in such quantities and requiring so much ATP for activity is under rigorous control.

It has been known for about thirty years that ammonia and other fixed N sources repress the synthesis of nitrogenase. N_2-fixing cells growing on ammonia under an atmosphere of N_2 use ammonia preferentially. When such cultures consume the ammonia, a diauxic lag phase of variable time ensues depending on the nutrient supply, during which nitrogenase genes are transcribed and nitrogenase protein synthesized (4). Clostridial cultures exhausted of ammonia and supplied with argon (rather than N_2) still synthesize nitrogenase so N_2 itself is not an inducer (4). The Fe and MoFe proteins are coordinately synthesized in *Azotobacter* but the MoFe protein is transcribed before the Fe protein in *Clostridium* and *Rhizobium* (4,25).

The regulation of nitrogenase synthesis by ammonia has been the subject of several reviews (4,23,26). Certain mutants of *K. pneumoniae* which lack the protein of glutamine synthetase are unable to de-repress ('switch on') nitrogenase (23). However, other mutants in other genes involved in ammonia assimilation (but not the structural gene for glutamate synthetase) also may result in the derepressed phenotype (23). Apparently ammonia itself is not the actual effector of repression but rather a fixed N product such as glutamate or glutamine since inhibitors of glutamine synthetase activity (27) allow *Klebsiella* to synthesize high levels of nitrogenase even in the presence of ammonia. Nitrogenase is regulated by the unadenylated form of glutamate synthetase but the regulation seems independent of the biosynthetic activity of glutamine synthetase (23). Further discussion of nitrogenase regulation by fixed-N appears in following sections.

There are also other factors which regulate nitrogenase biosynthesis. O_2 represses nif gene expression although the aerobe, *A. vinelandii,* is much less sensitive than the facultative anaerobe, *K. pneumoniae* (28). Temperature is also a factor in the control of nitrogenase synthesis in *Klebsiella* (29). Carbamyl phosphate both inhibits the activity of and partially represses the biosynthesis of the nitrogenase of *Clostridium* (4).

Several experiments have suggested a regulatory role for molybdenum in nitrogenase biosynthesis. Without added Mo *K. pneumoniae* grows well on ammonia but does not grow on N_2 (4). *C. pasteurianum* and *K. pneumoniae* do not synthesize either the MoFe or Fe protein in the absence of Mo whereas *A. vinelandii* synthesizes normal levels of the Fe protein but not the MoFe protein (4). Like nitrogenase biosynthesis Mo accumulation in *Clostridium* is regulated by ammonia. Protein synthesis is required for Mo accumulation and the transport of Mo is energy-dependent (30). Recent results indicate that nitrate regulates the synthesis of nitrogenase in *Rhizobium* bacteroids and *Klebisiella pneumoniae* (31) via a process mediated through nitrate metabolism and through a control system operating on the oxidation-reduction pathways.

E. Genetics

Our understanding of the genetics of N_2 fixation has largely come through studies of the N_2-fixing system (nif) of *Klebisiella pneumoniae* (for review, see 23). Analysis of the nif genes has been studied by phage P1 transduction, phage Mu mutagenesis, transposon mutagenesis and complementation analysis (23). All nif^- mutations cluster near the his operon and at least 17 nif loci make up this cluster with no non-nif genes interspersed. Complementation analysis with polar mutants and studies with lac fusions indicate that the genes are organized into seven operons. The nif region is approximately 24 kilobases long (32) and a physical map of cloned nif DNA has been produced by restriction enzyme analysis (33). Nif DNA can be easily obtained since the entire nif region has been ligated to amplifiable plasmids (23). Two-dimensional gel electrophoresis has been used to determine gene-product relationships in *K. pneumoniae* and most nif genes have been identified (34). Some code for proteins functioning in electron transport to nitrogenase and one of these gene products, a flavoprotein, has been purified (35).

The nif genes of *K. pneumoniae* have been transferred by conjugation to non-N_2 fixing bacteria such as *E. coli*, *Salmonella typhimurium* and *Agrobacterium tumefaciens* and to nif$^-$ mutants of *Azotobacter* and *Rhizobium* species (23). With the exception of *Azotobacter* where the nif genes are expressed aerobically, the other hybrid bacteria fix N_2 only under anaerobic and N-limited conditions.

Cloned N_2 fixation genes of *K. pneumoniae* when isolated hybridize to the DNA of a wide variety of N_2-fixing bacteria but only with the genes for the Fe and MoFe proteins (23). Even so, it is remarkable that such relatedness exists in the structural genes considering the diverse N_2-fixing bacteria examined. If these bacteria are recipients of nif DNA passed around by gene transfer, it might be that the capacity to catalyze N_2 reduction places such stringent restraints on protein secondary and tertiary structure that only a small variation in amino-acid sequence can be tolerated.

III. PROPERTIES OF HYDROGENASE

Hydrogenase (EC class 1.12) was first identified as the enzyme capable of activating molecular H_2 by Stephenson and Stickland in the 1930's (37) who observed that crude cell extracts of *Escherichia coli* could use H_2 to reduce a variety of substrates. In subsequent years, investigations into the H_2 metabolism of microorganisms suggested that only a few bacterial species possessed hydrogenase activity and the nature of the enzyme and its prosthetic group became a great source of controversy (38,39). It soon became apparent though that hydrogenase was present in a wide range of both bacteria and algae and its presence fulfilled a variety of metabolic roles (40,41). Research prior to the 1970's had shown that hydrogenase in the cell-free state was usually very sensitive to inactivation by O_2 and often associated with electron transport proteins and cell membranes - properties which severely hampered purification. Thus it is only in the last five years or so that hydrogenase has been isolated and purified, mainly as a result of an increase in the types of organisms examined and an improvement in protein purification techniques, particularly the use of strict anaerobic conditions. However, a comparison of the properties of the enzymes so far characterized reveals only one common feature, all are non-heme iron sulfur proteins. They differ considerably in molecular composition, electron carrier specificity, sensitivity to O_2 and cellular location. In many ways, this diversity reflects the various physiological roles of hydrogenase in the different taxonomic groups, which is discussed in subsequent sections. Hydrogenase has recently been the subject of a detailed review (42).

Some properties of some of purified hydrogenases are given in Table 1. These in many ways are properties designed for the special needs of the different organisms. It should be emphasized that hydrogenase has also been purified and characterized from the anaerobes *Megasphera elsdenii* (50) and *Desulfovibrio vulgaris* strains Hildenborough (51) and Miyaziki (52), from the photosynthetic bacteria *Thiocapsa* (53) and *Chromatium* (54,55), from the facultative anaerobe *Proteus mirabilis* (56) and from the areobic hydrogen bacteria *Paracoccus denitrificans* (57), *Pseudomonas pseudoflava* (58) and *Alcaligenes eutrophus* Z1 (59).

The hydrogenases of the anaerobic bacteria, *C. pasteurianum* and *M. elsdenii*, are both soluble and ferredoxin is the natural electron carrier (which is replaced by flavodoxin in iron-deficient media). Both enzymes are very sensitive to irreversible inactivation by O_2, for example the clostridial enzyme has a half-life in air of only a few minutes (60). The purification of this enzyme is thus carried out under strictly anaerobic conditions, and all buffers contain sodium dithionite to remove any traces of O_2 (61). Both hydrogenases are fairly high molecular weight monomers each containing twelve iron and twelve acid-labile sulfur atoms per molecule. In the case of *C. pasteurianum*, experiments in which the iron-sulfur centers of the hydrogenase are displaced or extruded by thiol reagents have demonstrated the presence of three [4Fe-4S] centers. In contrast to the latter two enzymes, the hydrogenases of the sulfate-reducing bacteria all couple specifically to a low potential cytochrome c_3, the physiological electron carrier. The enzymes purified from three species of *Desulfovibrio* are all periplasmic (exist between the membrane and the cell wall) or membrane-associated) but they vary in their sensitivity to O_2. Hydrogenase from *D. vulgaris* (strain Hildenborough) for example, is relatively stable and can be purified aerobically. Like the

TABLE 1. Properties of Bacterial Hydrogenases

	Anaerobic		*Photosynthetic*	*Facultative Anaerobic*	*Aerobic Hydrogen*	*Aerobic N_2-Fixing*
	Clostridium pasteurianum	Desulfovibrio gigas	Rhodospirillum rubrum	Escherichia coli	Alcaligenes eutrophus H16	Rhizobium japonicum
Location	*-cytoplasm*	*-periplasmic*	*-membranes*	*-membranes*	*-cytoplasm*	*-membranes*
Physiological role	*H_2 evolution*	*H_2 uptake*	*H_2 uptake*	*H_2 evolution*	*H_2 uptake*	*H_2 uptake*
Electron carrier	*ferredoxin*	*cytochrome c_3*	*-unknown*	*-unknown*	*NAD*	*-unknown*
O_2 sensitivity	*-very sensitive*	*-sensitive*	*-relatively stable*	*-sensitive*	*-stable*	*-very sensitive*
Mole. Wt. *Subunits*	*60,000* *1x60,000*	*89,000* *1x62,000* *1x26,000*	*66,000* *1x66,000*	*113,000* *2x56,000*	*205,000* *1x68,000* *1x60,000* *2x29,000*	*65,000* *1x65,000*
Fe/mole	*12*	*12*	*4*	*12*	*12*	
S^{2-}/mole	*12*	*12*	*4*	*12*	*12*	
Specific activity[a]	*4000 (V_m)*	*91 (1mM)*	*33 (V_m)*	*136 (V_m)*	*48 (2.4mM)*	*0.07 (1mM)*
Fe-S center	*3 x [4Fe-4S]*	*3 x [4Fe-4S]*			*2 x [4Fe-4S]* *2 x [2Fe-2S]*	
Reference	*43, 44*	*45*	*46*	*47*	*48*	*49*

[a] H_2 evolution activity with dithionite-reduced methyl viologen at the indicated concentration as the electron donor, expressed as μmoles H_2 evolved/min/mg protein. V_m indicates maximum velocity.

other strict anaerobes, two of these enzymes also contain twelve iron and twelve acid-labile sulfur atoms per molecule but the enzyme of *D. vulgaris* (strain Miyazaki) was reported to contain only eight atoms of each. Again, extrusion experiments have shown the presence of *[*4Fe-4S*]* centers in these hydrogenases.

The hydrogenases purified from the three photosynthetic bacteria *Rhodospirillum, Chromatium* and *Thiocapsa* are distinct from the enzymes described above. They are all intimately associated with the cell membrane, are very resistent to inactivation by O_2 and all three contain only four iron and four acid-labile sulfur atoms per mole. Also these hydrogenases tend to have specific activities in H_2 evolution much lower than the enzymes from the strict anaerobes. The natural electron carriers for the hydrogenases of the photosynthetic bacteria are not known, and this is also true of the enzymes isolated from the facultative anaerobes *E. coli* and *P. mirabilis*. Indeed none of these enzymes show much interaction in the purified state with ferredoxins, flavodoxins, low potential cytochromes or nicotinamide or flavin nucleotides. The hydrogenases of *E. coli* and *P. mirabilis* are both membrane-bound but in contrast to the *E. coli* enzyme (Table 1), *P. mirabilis* hydrogenase is relatively stable to O_2, and contains 24 iron and acid-labile sulfur atoms per mole. The hydrogenase of the aerobic N_2-fixing bacterium *Rhizobium japonicum,* though not fully characterized, differs from the others so far purified in that it has extremely low H_2 evolution activity and has a very high affinity for H_2, which relates to its physiological role (Section VII).

Hydrogenase has also been purified and characterized from the membrane fraction of two hydrogen bacteria, *P. pseudoflava* and *A. eutrophus* H16 (62). These two enzymes are stable to O_2 and are purified aerobically. The natural electron carriers for these ensymes are not known. In contrast to all the hydrogenases previously discussed they contain six iron and six acid-labile sulfur atoms per mole, although the types of iron-sulfur centers are undetermined. *A. eutrophus* H16 also possesses a soluble hydrogenase (Table 1) which is quite distinct from the membrane-bound enzyme, and indeed from all other hydrogenases so far purified. It is specific for NAD as the natural electron acceptor, contains two FMN (flavin mononucleotide) prosthetic groups per mole and although it contains twelve iron and twelve acid-labile sulfur atoms per mole, these comprise two *[*4Fe-4S*]* and two *[*2Fe-2S*]* centers. The enzyme is also unique with respect to inactivation by O_2, and also by carbon monoxide. CO is a potent inhibitor of the activity of all the hydrogenases tested so far, except the soluble enzyme from *A. eutrophus* H16. The O_2 sensitivity of the various hydrogenases listed in Table 1 refers to irreversible inactivation during purification or storage. However all hydrogenases are sensitive to inhibition by O_2 during the catalytic reaction and this is a reversible process. The exception is *A. eutrophus* hydrogenase which can catalyze the evolution or consumption of H_2 in the presence of O_2.

All the hydrogenases so far isolated in high states of purity fall into the molecular weight range 50,000 to 110,000, or around 200,000, and a common feature is that all are either of molecular weight 60,000 or contain a subunit of that size. All are acidic proteins containing a high content of aromatic and hydrophobic residues. Purified hydrogenase preparations are yellow-brown in color and spectroscopically show a broad shoulder in the 400nm region, typical of proteins with a [4Fe-4S] chromatophore. Absorption coefficients are usually around 4,000 $M^{-1}cm^{-1}$ per Fe atom in the oxidized enzyme. Most of the purified hydrogenases have been examined by electron paramagnetic resonance (EPR) spectroscopy and all show strong EPR absorption in the oxidized state which is typical of bacterial high potential (> + 300mV) iron-sulfur proteins (Hipip's). Based on present knowledge, it seems that the H_2 catalytic center is a novel [4Fe-4S] cluster which exhibits the described EPR signal but has a low midpoint potential (around -400mV) and is capable of binding a hydride ion (see below). The other [4Fe-4S] centers in the multi-cluster hydrogenases are of the classical ferredoxin type (in most cases) and appear to function in relaying electrons to (from) the electron carrier for the enzyme from (to) the catalytic center. Again, a notable exception is the soluble *A. eutrophus* hydrogenase which contains 2Fe- and 4Fe- clusters all of the ferredoxin type.

All the isolated hydrogenases catalyze the activation of H_2 and the reaction is reversible to a greater or lesser extent. The evolution or consumption of H_2 is usually measured *in vitro* using artificial electron carriers. For example, hydrogenases readily use H_2 to reduce methylene blue (MeB, E_m + 11mV):

$$MeB_{ox} + H_2 \longrightarrow MeB_{red} + 2H^+ \qquad (4)$$

or catalyze H_2 evolution with dithionite-reduced methyl viologen (MV, E_m - 440mV):

$$2MV_{red} + 2H^+ \longrightarrow 2MV_{ox} + H_2 \qquad (5)$$

The midpoint potentials of these electron carriers allows H_2 evolution or H_2 consumption to be thermodynamically favorable.

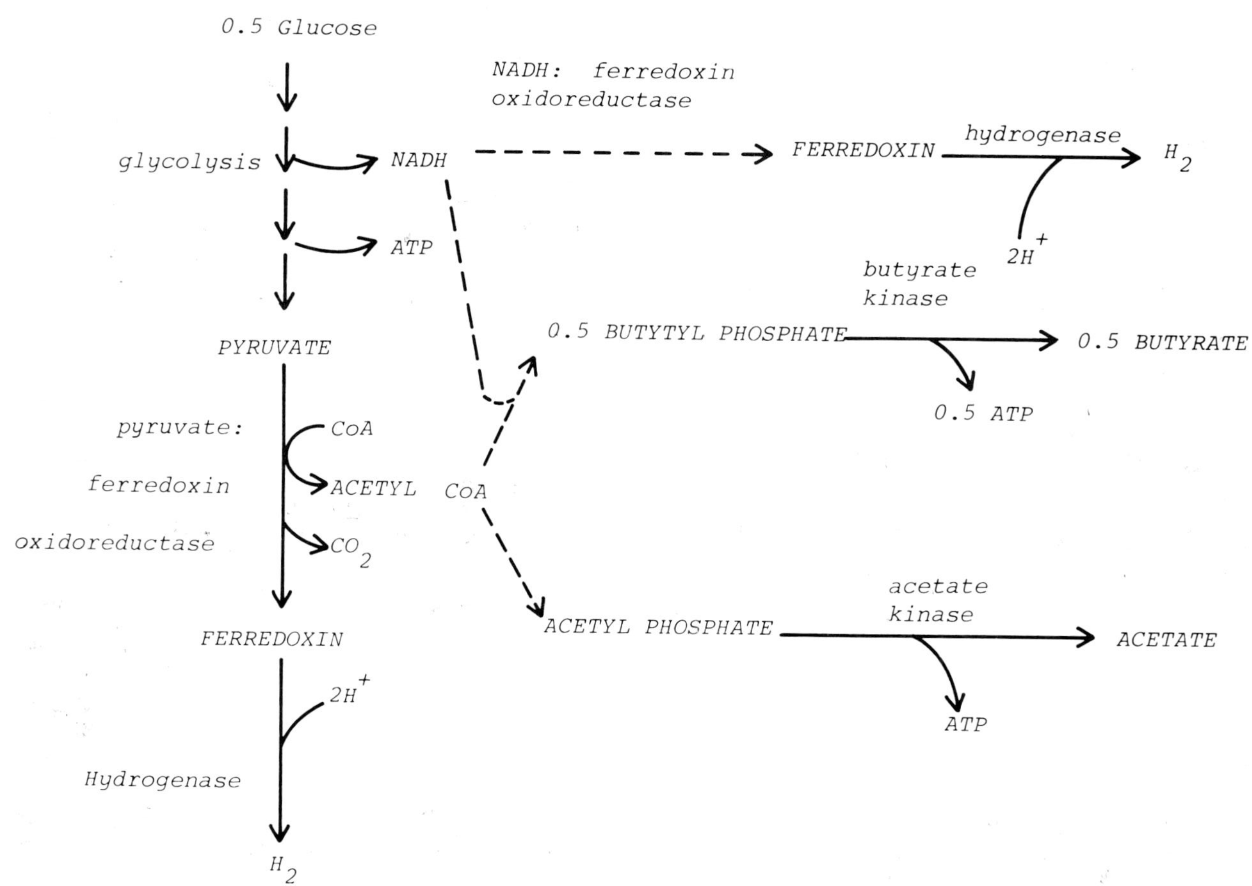

Figure 2. H_2 *production by* Clostridium pasteurianum. *Modified from (69).*

The appearance or disappearance of H_2 may be followed manometrically, spectrophotometrically, amperometrically or by gas chromatography.

Normal H_2 gas is a mixture of two isomers in thermodynamic equilibrium. They differ in their nuclear spins (ortho- or para-) and hydrogenase catalyzes their rapid equilibration if the mixture is moved to a different temperature. This is a direct assay of hydrogenase activity without added electron carriers but the method is slow and insensitive. Hydrogenase catalyzes the exchange reaction between H_2 and D_2O (or T_2 and H_2O etc.) according to eqn. 6. This reaction is also of

$$H_2 + D_2O \rightleftharpoons HD + HDO \tag{6}$$

importance since it showed that H_2 activation occurred primarily by a heterolytic cleavage of the H_2 molecule (eqn. 7, where E is hydrogenase). In D_2O the back reaction of eqn. 7

$$H_2 + E \rightleftharpoons EH^- + H^+ \tag{7}$$

with D^+ replacing H^+ produces HD rather than D_2.

IV. ANAEROBIC N_2 FIXATION

Bacteria capable of fixing N_2 under anaerobic conditions include both the strict anaerobes and the facultative anaerobes (see Table 2). Facultative organisms grow well on fixed N sources under aerobic conditions, slowly on N_2 under microaerophilic conditions (<1% O_2) and maximally under strict anaerobic conditions. The anerobic N_2-fixing bacteria have not evolved an O_2 protective mechanism for intracellular nitrogenase since growth requires the strict absence of O_2 (no selective pressure). *Clostridium pasteurianum* for example has been found to contain only very low levels of such 'aerobic' protective enzymes as catalase and superoxide dismutase. Anaerobic metabolism has recently been reviewed (63). Representatives of the class of anaerobic N_2-fixing organisms are *C. pasteurianum* and *Klebsiella pneumoniae*. Highly purified nitrogenase components have been obtained from these anaerobes and from *Bacillus polymyxa* (63-65)

C. pasteurianum ferments sugars to the end-products butyrate, acetate, CO_2 and H_2 (Figure 2). Reduced ferredoxin is the reductant for the nitrogenase system as well as for hydrogenase. The supply of ATP for N_2 fixation is closely linked

TABLE 2. Representative N_2-Fixing Bacteria

Type of Metabolism	
I. Obligate anaerobic	Clostridium pasteurianum Desulfovibrio vulgaris
II. Facultative anaerobic	Klebsiella pneumoniae Erwina herhicola Citrobacter freundii Enterobacter cloacae Bacillus polymyxa
III. Aerobic	Azotobacter vinelandii Azotobacter chroococcum Xanthobacter flavum Mycobacterium flavum Xanthobacter autotrophicus
IV. Microaerophilic	Rhizobium Azospirillum Actinomycetes
V. Photosynthetic	Rhodospirillum Rhodopseudomonas Nostoc muscorum Plectonema boryanum Anabaena cylindrica

to H_2 production by hydrogenase (Section V*(A)*). Like the clostridia, *K. pneumoniae* also evolves large amounts of H_2 and CO_2 during carbohydrate fermentation. This organism is closely related to *Escherichia coli*; both are enteric bacteria and metabolize pyruvate produced from glycolysis to organic acids, CO_2 and H_2 (Section V*(B)*). However, *Klebsiella* also produces 2,3-butanediol from pyruvate as a characteristic end-product of fermentative growth. This is produced via α-acetolactate synthetase and decarboxylase, and 2,3-butanediol dehydrogenase. The pathway of electron flow from oxidized substrate to nitrogenase is not known. Ferredoxin (or flavodoxin) is the electron donor to nitrogenase and recently a flavoprotein was isolated that also participates in this reaction (35). The molar growth yield (Y_{ATP}) of *K. pneumoniae* is about 29 (66). The net gain in ATP for its butanediol fermentation is 2 to 2.5 moles per mole of glucose. More recent work with *Klebsiella* mutants altered in ammonia assimilation such that they fix N_2 but excrete the product, ammonia, showed that *in vivo* 7.5 to 8.0 moles of glucose were required per mole of N_2 fixed. The $Y_{sucrose}$ value for *C. pasteurianum* supplied with ammonia as the N source is about 79: this is 1.7 times higher than that of N_2-grown cells (67).

Growing *Clostridium* and *Klebsiella* under N_2-fixing conditions is relatively simple and has been described by several workers (64,67). *C. pasteurianum* is easily maintained in chemostat culture which allows for pH control (the optimum is 5.9) together with its requirement for anaerobiosis. Large batch fermentors must be equipped with a pH controller and must be sparged with N_2. A 160-200ℓ fermentor is innoculated with cells from an 8ℓ culture innoculated from the chemostat. The medium has been described (67); biotin is also a growth requirement. Late log phase cells are harvested by centrifugation, washed and then quick frozen in liquid N_2 and then stored at -20°. Batch culture procedures for *Klebsiella* are similar. However *Klebsiella* may be maintained in an aerobic shake culture. Prior to fermentative growth, cells are transferred to an anaerobic medium containing ammonia which is used as an inoculum for the fermentor medium which contains a limited ammonia supply (usually supplemented with PABA and biotin). The cells are then allowed to exhaust the ammonia and continue growth by N_2 fixation. This derepression method results in a higher yield of cells. Much less information exists concerning the anaerobic culturing of other N_2-fixing organisms. A medium has been described for the N_2-fixing sulfate-reducing *Desulfovibrio vulgaris* in which the atmosphere comprises N_2 supplemented with H_2 (68, see also Section V).

V. ANAEROBIC H_2 PRODUCTION

In natural environments the majority of microorganisms evolve H_2 under anaerobic conditions and H_2 plays a central role in the anaerobic food chain. For example, most of the H_2 evolved by fermentative and photosynthetic bacteria is immediately consumed by the methane-producing bacteria to reduce CO_2 which is released as methane. In many ecosystems the association between H_2 evolving and H_2 consuming organisms is so close that the extra-cellular concentration of H_2 is often undetectable. The laboratory culture of many anaerobic bacteria is characterized by the production of large quantities of both H_2 and CO_2, and here we describe the process for some obligate and facultative anaerobes. The photosynthetic bacteria, which could be included here, are discussed in Section IX (B).

A. Obligate Anaerobic Bacteria

The production of H_2 by microorganisms during anaerobic culture represents their special ability to use protons as one of the terminal electron acceptors in anaerobic energy metabolism. In all cases H_2 evolution is closely linked to ATP generation (see 69). Many clostridia for example perform a fermentation of soluble carbohydrates, starch or pectin, with the formation of acetic and butyric acids, H_2 and CO_2. These saccharolytic clostridia grow poorly (if at all) in complex media devoid of fermentable carbohydrate. These organisms lack a cytochrome system and a mechanism for oxidative phosphorylation, and obtain energy (ATP) by substrate level phosphorylation during fermentation. In *Clostridium pasteurianum* one mole of glucose is metabolized via the glycolytic pathway to yield two moles each of pyruvate, NADH (from glyceraldehyde-phosphate dehydrogenase) and ATP. Pyruvate is further metabolized to acetyl CoA, CO_2 and H_2 via pyruvate: ferredoxin oxidoreductase, ferredoxin and hydrogenase. As shown in Figure 2, both acetyl CoA and NADH occypy branch point positions in the overall fermentation process. Acetyl CoA is converted to acetyl phosphate via phosphotransacetylase and then to acetate via acetate kinase yielding one mole of ATP per acetyl CoA. Alternatively, two moles of acetyl CoA are condensed to acetoacetyl CoA, reduced to butyryl CoA (which requires NADH) and converted to butyrate by phosphotoransbutrylase and butyrate kinase. This yields only 0.5 ATP per acetyl CoA. Butyrate production quantitatively could consume all the NADH from glycolysis but since some of the acetyl CoA is used to produce acetate, excess NADH is available. This NADH is oxidized to H_2 via NADH: ferredoxin oxidoreductase, ferredoxin and hydrogenase. Sole acetate production would yield 4 moles of ATP per glucose whereas if butyrate is the sole carbon product, only 3 moles of ATP are produced. The key regulatory enzyme in this system is NADH: ferredoxin oxidoreductase which is activated by acetyl CoA but inhibited by CoA. Thus the acetyl CoA/CoA ratio regulates H_2 production from NADH and indirectly controls ATP production, such that fermentation produces both acetate and butyrate with 3.3 moles of ATP per mole of glucose. H_2 production is thus a means of maintaining the intracellular redox balance without the necessity of terminal electron carriers other than protons.

Some strict anaerobes such as those in the rumen generate only acetate during glucose fermentation. For example in *Ruminococcus albus* (see 70) all reductant from the glycolytic pathway and pyruvate oxidation is transferred ultimately to H_2 via hydrogenase, and all the acetyl CoA is used to produce acetate. However, monocultures of *R. albus* although evolving H_2 also produce ethanol as well as acetate during glucose fermentation. As in *C. pasteurianum,* only 3.3 moles of ATP are produced per mole of glucose since the high extracellular concentration of H_2 prevents all of the reductant forming H_2. Some of the electrons are thus 'trapped' as ethanol by the reduction of acetyl CoA. In the rumen however, very little H_2 is generated and the same is observed in chemostat cocultures of *R. albus* and the H_2-utilizing bacterium *Vibrio succinogenes*. In the latter case *R. albus* generates more acetate than in monoculture, ethanol is not produced and one mole of glucose yields 4 moles of ATP. *V. succinogenes* uses H_2 to reduce fumarate to succinate and ATP is produced by cytochrome-dependent electron transport phosphorylation. This interspecies H_2 transfer dramatically reduces the H_2 concentration and effectively raises the potential of the H_2/H^+ redox couple. This enables the H_2-producing anaerobes to dispose of all electrons from NADH as H_2 rather than as carbon compounds, thereby maximizing the amount of ATP formed during fermentation. Other examples of this type of association include cocultures of *Selenomonas ruminantium* and *Methanobacterium ruminantium,* and *Clostridium thermocellum* and *M. thermoautotrophicum,* except here the H_2 generated by fermentation is used to reduce CO_2 to CH_4, with the concommitant production of ATP driven by electron transport.

The fermentation of other substrates can also lead to ultimate H_2 generation. A number of clostridia can grow well in complex media containing peptone or yeast extract, in the absence of fermentable carbohydrates. Some of these organisms are responsible for the decomposition of nitrogenous compounds naturally i.e. putrefaction, and include the pathogenic clostridia e.g. *C. botulinum* and *C. tetani*. However, their metabolism involves the fermentation of amino-acids with acetate, CO_2 and ammonia, rather than H_2, as end-products. One exception is the dissimilation of glutamate which produces acetate and pyruvate. The latter is further degraded to acetate, butyrate, CO_2 and H_2 as in the saccharolytic fermentors.

The H_2-producing non-saccharolytic mode of fermentation exhibited by *C. kluyveri* deserves special mention as this organism grows only at the expense of ethanol and acetate with the production of butyrate, caproate and H_2. The metabolism of ethanol produces the exact amount of reductant required if all acetyl CoA were used in butyrate (or caproate) synthesis:

$$CH_3CH_2OH + CoA + 2NAD \longrightarrow CH_3CO.CoA + 2NADH + 2H^+ \quad (8)$$

$$CH_3CO{\cdot}CoA + 2NADH + 2H^+ + CH_3COOH \longrightarrow CH_3CH_2CH_2COOH + 2NAD + H_2O + CoA \quad (9)$$

Since ATP can only be generated by the acetate kinase reaction, obviously not all the acetyl CoA and NADH produced from ethanol can be used in organic acid synthesis. Like *C. posteurianum* the excess NADH is oxidized via ferredoxin and hydrogenase to evolve H_2. For every mole of acetyl CoA used in ATP production one mole of H_2 is evolved.

The genus *Desulfovibrio* characteristically possess the ability to use sulfate as the terminal electron acceptor of a complex electron transport chain which uses either organic compounds or H_2 as the initial electron donor. Energy (ATP) is generated by electron transport phosphorylation with ultimate sulfate reduction to H_2S. However these organisms can also obtain energy by sybstrate level phosphorylation analogous to the clostridia. ATP is produced via acetyl CoA and acetyl phosphate and the excess reductant is disposed of as H_2. In contrast to the clostridia reduced cytochrome c_3 (not reduced ferredoxin) donates the electrons to hydrogenase (Figure 3). For example, under the usual conditions employed for the culturing of these bacteria (lactate plus sulfate) some H_2 is produced initially during the early stages of growth (71). The first step in the reduction of sulfate is its ATP-dependent conversion to sulfite and the initial H_2 burst is thought to result because of the need to generate sufficient ATP to activate sulfate. Once sulfite is produced, it replaces protons as the electron acceptor in a more energy efficient pathway.

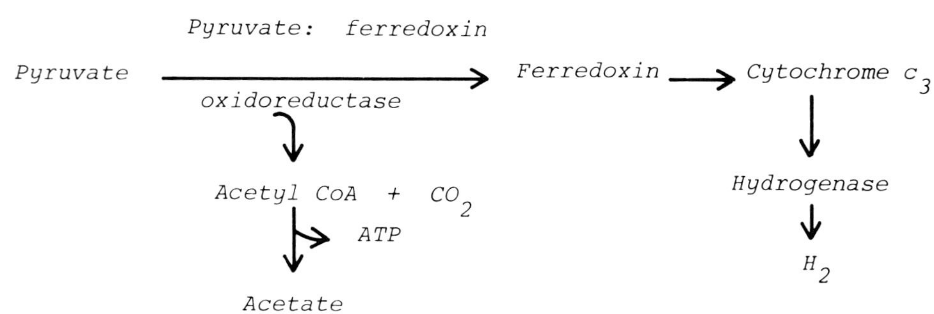

Figure 3. H_2 Production by Desulfovibrio.

B. Facultative Anaerobic Bacteria

In the anaerobic clostridia formate is not a major intermediate in the breakdown of pyruvate. Although such organisms contain formate dehydrogenase, the enzyme catalyzes formate synthesis rather than its degradation (72). In contrast the facultative anaerobic coli-aerogenes bacteria e.g. *Escherichia coli*, which also produce large quantities of H_2 during fermentation, metabolize pyruvate in two separate stages. Pyruvate is first converted to formate by pyruvate: formate lyase with the production of acetyl CoA, and energy is conserved by ATP formation via acetyl phosphate (Figure 4). The problem is the removal of formate. In the absence of other suitable electron acceptors, formate is metabolized to H_2 and CO_2 by a membrane-bound multienzyme system, formate hydrogenlyase. This consists of formate dehydrogenase and hydrogenase linked by unknown electron carriers. The activity of the formate hydrogenlyase system and thus H_2 production is repressed if nitrate is present in the growth medium. Under anaerobic conditions formate oxidation can be coupled to nitrate reduction to nitrite, via nitrate reductase, or with fumarate reduction to succinate, via fumarate reductase. These oxidants, nitrate and fumarate, act as the terminal electron acceptors. In both cases electron transport generates ATP and these reactions are 'preferred' to the disposal of electrons as H_2 since formate hydrogenlyase catalyzes a non-energy yielding reaction (H_2/H^+, E_m -420mV; formate/CO_2, E_m -420mV). Under aerobic conditions O_2, via the aerobic respiratory chain, is the ultimate electron acceptor. As shown in Figure 4, *E. coli* is also able to grow anaerobically in a non-fermentative mode; cells will grow with fumarate (or malate) as the carbon source if H_2 (or formate) is present as the electron donor. Evidence at the present time suggests that *E. coli* produces two distinct hydrogenases, one comprises part of the formate hydrogenlyase complex while the other functions in H_2 consumption (see 42).

The production of H_2 in the facultative anaerobes is thus an efficient way of removing formate thus facilitating energy metabolism. The production of both H_2 and CO_2 during fermentation is a characteristic of the genera *Escherichia*, *Proteus* and most species of *Salmonella* although these organisms may also produce lactate, acetate, ethanol or succinate depending on the strain and the culture conditions i.e. mixed acid fermentation. However, the ability to evolve H_2 is absent in the pathogens of the *Shigella* group and *Salmonella typhi*, because they lack the formate hydrogenlyase complex. In the clostridia H_2 evolution is a major and obligatory means of regenerating oxidized electron carriers thereby allowing continued energy generation. The facultative anaerobes however

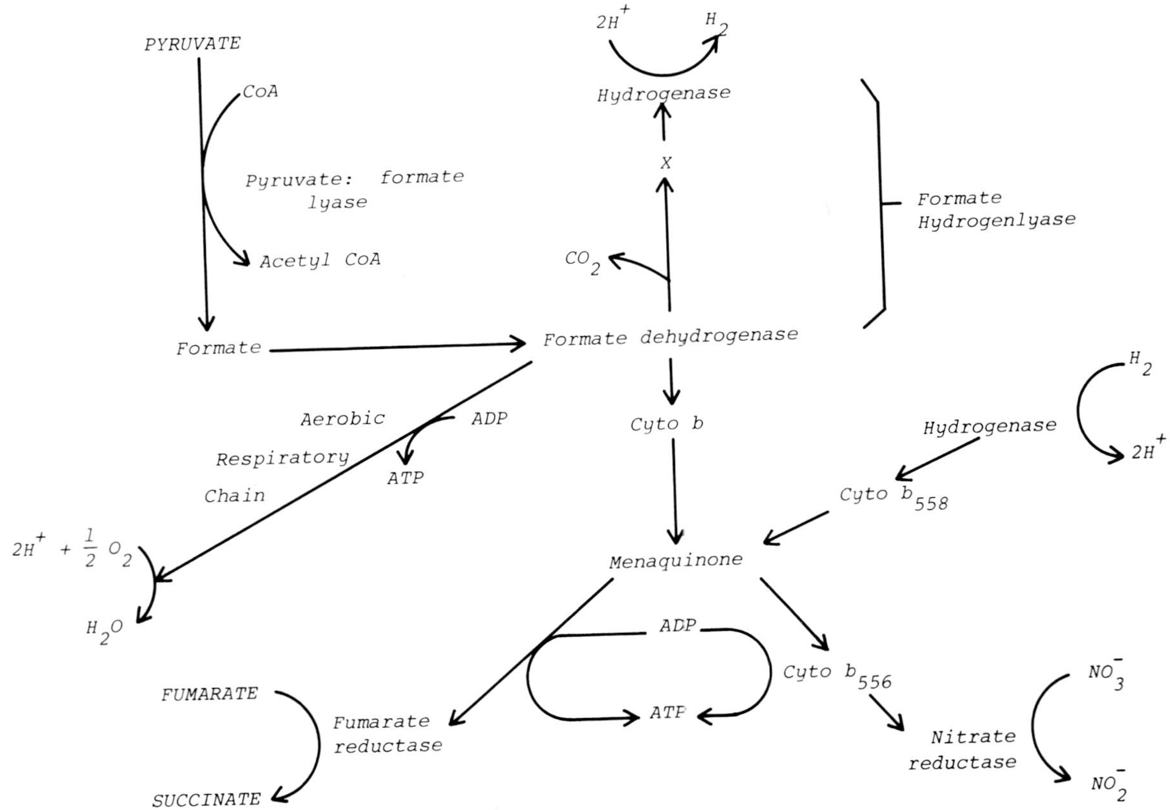

Figure 4. *Pathways of H_2 metabolism in* E. coli. *Modified from (73,74). X, unknown electron carrier(s).*

have several options and H_2 production may be regarded as a 'last resort', occuring in the absence of other electron acceptors which themselves afford an additional energy supply by virtue of electron transport phosphorylation.

VI. AEROBIC N_2 FIXATION

Aerobic N_2-fixing bacteria are diverse in their metabolism and natural living state (Table 1). Like *Clostridium* and *Klebsiella* these bacteria are usually found in soil and/or aquatic environments. None of these organisms form heat resistant spores as do *C. pasteurianum* and *B. polymyxa*, but *Azotobacter* is capable of forming dessication-resistant microcysts. *Rhizobium, Actinomycetes* and *Azospirillum* are areobic bacteria in the sense that they have a respiratory chain and perform electron-transport coupled phosphorylation, but nitrogenase is synthesized only under reduced O_2 concentrations. Most of our knowledge of aerobic N_2 fixation comes from studies with *Azotobacter* and this is detailed below. Highly purified nitrogenase components have been isolated from *Azotobacter vinelandii* and *A. chroococcum* (75,5). Purified MoFe protein and partially purified nitrogenase have been isolated from *Rhizobium japonicum* (76) and from a pure culture of an Actinomycete (77), respectively.

Azotobacter is cultured in a fermentor in a similar manner to *C. pasteurianum,* except pH control is not needed and the gas phase is air. *Azotobacter* metabolizes carbohydrate to pyruvate and a complete tricarboxylic acid cycle furnishes NADH to the respiratory chain for ATP production. Calculations indicate that whole cells require about 4 - 5 moles of ATP for the reduction of one mole of N_2 to ammonia (73). It was previously postulated that the driving force for electron flow through the respiratory chain was a high NADH + NADPH: NAD + NADP ratio (79). However recent evidence indicates that electron flow is tightly coupled to a highly energized membrane maintained by a large pH gradient (80). The immediate effect of ammonia addition on nitrogenase activity may be explained by the finding that ammonia (as the NH_4^+ ion) interupts the electrochemical gradient thus interrupting the flow of electrons through the electron transport system to nitrogenase (81). Apparently electrons flow from the membrane gradient to the Fe protein of nitrogenase via ferredoxin and azotoflavin (79). The observation that c-AMP accelerates derepression by ammonia in *Azotobacter* remains to be clarified (82).

The best understood symbiotic associations are those of *Rhizobium* with legume roots and the cyanobacterium *Anabaena* with the fern *Azolla* (see Section VIII). Not until 1975 (83-86) was there a direct demonstration that nitrogenase was synthesized in the rhizobial endophyte, and not because of some association with plant nodule cells. The prime condition for the expression of nitrogenase in cultures of free-living *Rhizobium* is low O_2 tension (< 1%); this environmental factor is controlled by the heme protein leghemoglobin in the nodule. Another important finding was that rhizobia export their fixed N_2 as ammonia to the surrounding nodule (87,88). The plant supplies the largely non-growing, immobilized rhizobial bacteroid with carbohydrate (for energy) and fixed N (for cell maintenance). Bacteroids have a leghemoglobin-coupled terminal oxidase for ATP generation (89) and generate reductant from the organic acid fraction of supplied photosynthetate (succinate, malate) via a complete TCA cycle (90). Recently a ferredoxin was isolated from bacteroids which apparently serves as a reductant for rhizobial Fe protein (91).

It would seem that one need only to culture free-living rhizobia under reduced microaerophilic conditions in a medium with succinate as carbon source and amino-acids as N sources to derepress nitrogenase synthesis. This is true when examining dilute cultures of certain rhizobia but not true in large batch cultures of high cell density. Researchers have apparently not learned how to maintain the proper O_2 concentration in such cultures. In addition NH_4^+ accumulates in the medium with this type of culture and some strains of *R. japonicum* are more sensitive than others to ammonia repression. Autotrophic growth of rhizobia on $H_2/CO_2/O_2$ (see Section VII) also does not allow nitrogenase biosynthesis.

The problem of O_2 protection of nitrogenase has been solved in several ways. The genus *Rhizobium* forms symbiotic associations with legumes and leghemoglobin controls the flow of O_2 to the bacteroids such that N_2 fixation is not affected by free O_2 and yet oxidative phosphorylation is supported (92). *Azotobacter* species protect their nitrogenase by two methods: respiratory protection via two special branched chain pathways and nitrogenase conformational protection as described previously (Section II*(A)*). The way in which O_2 concentration is controlled in the non-legume actinomycete-nodule symbiosis remains to be determined since these nodules do not contain leghemoglobin. Control of nitrogenase synthesis via ammonia assimilation and glutamine synthetase in *Rhizobium* has recently been reviewed (93,94).

VII. AEROBIC H_2 PRODUCTION

The metabolism of H_2 under arobic conditions may be divided into the aerobic N_2-fixing bacteria and the areobic hydrogen bacteria. The cyanobacteria which could be included here are discussed in Section IX(A). The aerobic hydrogen bacteria are characterized by their ability to grow autotrophically using H_2 as the sole electron donor and CO_2 as the sole carbon source (see 95). H_2 may be used both as a reductant and as a source of energy (ATP). These bacteria comprise a rather heterogeneous group taxonomically and include both gram-negative genera e.g. *Alcaligenes, Pseudomonas, Paracoccus, Aquaspirillum* and *Xanthobacter,* and gram-positive genera such as *Nocardia, Mycobacterium* and *Bacillus.* However of these only the coryneform bacteria e.g. *Xanthobacter,* have been shown to fix N_2 (96) and may thus be included with the aerobic N_2-fixing organisms. The aerobic hydrogen bacteria are facultative autotrophs. They will use organic substrates but with the availability of O_2 as a terminal electron acceptor the energy inefficient reduction of protons is not required and H_2 is not produced.

All H_2 production under aerobic conditions is mediated by the nitrogenase enzyme (except in some photosynthetic organisms, see next section). Thus H_2 evolution is restricted to N_2-fixing organisms and obviously is only observed under culture conditions where nitrogenase activity is not repressed and/or inhibited by the presence of fixed N and/or high concentrations of O_2. This varies between species to some degree as already discussed (see previous section). The reduction of N_2 to NH_3 is a very energy intensive process and there exists general agreement that 4-5 moles of ATP are required per pair of electrons transferred, independent of the substrate reduced (see Section II*(A)*). During N_2 fixation nitrogenase always evolves at least as much H_2 as N_2 reduced, according to eqn. 3 (Section II*(A)*). Thus at least 25% of the reductant and energy (ATP) may be wasted in H_2 production. This seemingly inefficient process is an intrinsic property of the nitrogenase catalytic reaction and is the price that must be paid by the N_2-fixing organism (for mechanistic consideration, see 97). If we include the energy that must also be expended by an aerobic organism to generate low potential reductant (such as reduced ferredoxin or flavodoxin), it is apparent that the reduction of N_2 is very energy consuming, especially if a significant proportion of reductant is released as H_2. Do cultures of aerobic N_2-fixing organisms release substantial quantities of H_2? To answer this question we will consider *Azotobacter* and *Rhizobium,* which are the most studied of the aerobic N_2-fixing bacteria.

Aerobic N_2-fixing cultures of *Azotobacter chroococcum* grown on a minimal medium with mannitol as the carbon source do not evolve significant amounts of H_2, although they are actively fixing N_2 (98). Some clue to the reason behind this apparent anomaly is given by the fact that if CO is also present, H_2 evolution is observed. Further insights were obtained from chemostat cultures of *A. chroococcum* (99). Cells limited for O_2, or for N_2, produced large quantities of H_2 which corresponded to up to 50% of the reductant used by the nitrogenase system. With *Rhizobium* however, the situation is not so well defined. Nodules or bacteroid suspensions containing rhizobia show a great diversity in H_2 loss by nitrogenase. Alfalfa, clover, soybeans and peas innoculated with pure strains of *Rhizobium* lost between 25 and 34% of nitrogenase electron flow as H_2, whereas various strains of cowpea rhizobia usually lost less than 10% as H_2. Of 32 *R. japonicum* strains tested, H_2 production increased if N_2 was omitted (see 100). Schubert and Evans (101) summed up the situation by showing that while some root nodule associations evolved no H_2 others partitioned up to 60% of the electrons to proton reduction by the nitrogenase system.

This great variability in H_2 production by these aerobic N_2-fixing organisms can be largely attributed to the presence or absence of hydrogenase. Dixon (102,103) postulated that in these bacteria hydrogenase could increase the efficiency of the N_2-fixing process by recycling the H_2 lost by the nitrogenase. The H_2 could be used either as a respiratory substrate with O_2 as the terminal electron acceptor thereby recouping some of the ATP expended in H_2 production, or the electrons from H_2 may be recycled back to the nitrogenase increasing the levels of low potential reductant. In addition, the use of H_2 in the so-called oxyhydrogen reaction (eqn. 10) would reduce

$$2H_2 + O_2 \longrightarrow 2H_2O \tag{10}$$

the O_2 concentration and thus protect the nitrogenase system from high O_2 concentrations. Another postulate was that H_2 consumption by hydrogenase would prevent H_2 inhibition of the nitrogenase (and thus N_2 fixation). Evidence in support of these theories has been obtained with both *Azotobacter* and *Rhizobium*.

A scheme for the recycling of H_2 by aerobic N_2-fixing bacteria is shown in Figure 5. Considering first *Azotobacter* which contains hydrogenase, H_2 is not evolved under the usual conditions of culture (see above) since it is effectively recycled by hydrogenase. It has been estimated that the organism regains about 12% of its overall energy expenditure for N_2

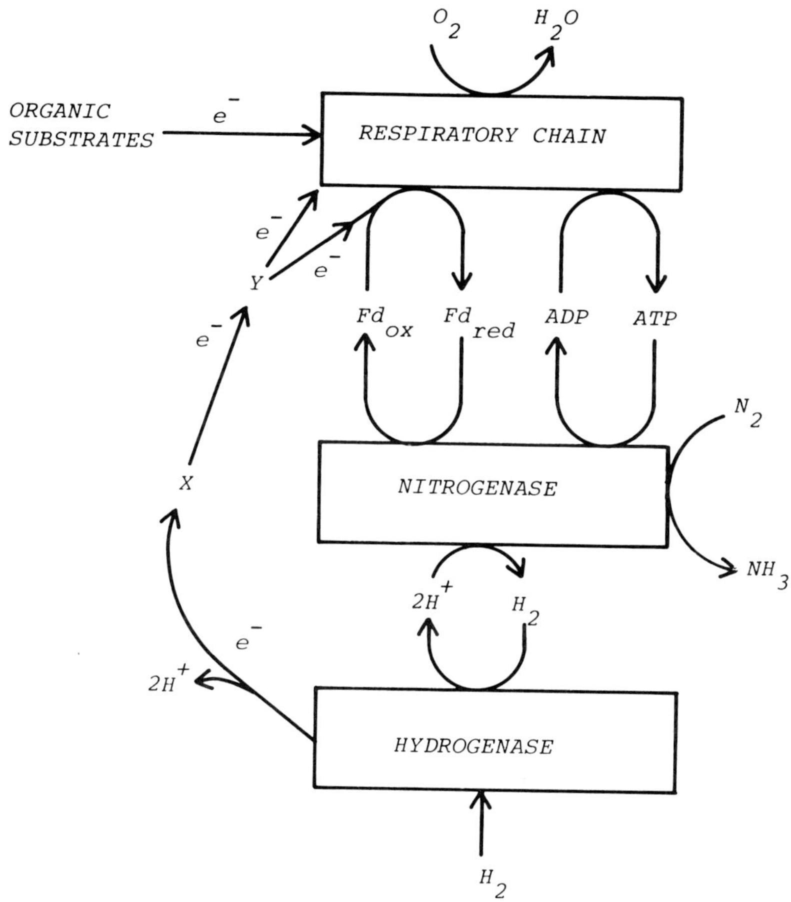

Figure 5. Role of hydrogenase in aerobic N_2-fixing bacteria. Fd, ferredoxin; X, Y, unknown electron carriers.

fixation by the recycling of evolved H_2 (104). If CO is present, this inhibits the hydrogenase and now one can detect the H_2 evoluted by nitrogenase. *A. chroococcum* can use H_2 (if supplied exogenously and so presumably also when recycled from the nitrogenase) as a source of energy (for ATP synthesis) via the oxyhydrogen reaction and as an electron donor for N_2 fixation. The presence of H_2 also increases the optimal pO_2 for N_2 fixation although there is no evidence with *Azotobacter* that hydrogenase protects nitrogenase from H_2 (99). The latter mechanism of nitrogenase protection was reported to occur in *Xanthobacter autotrophicus* (105). Under N_2-limiting conditions, the increased capacity of nitrogenase to evolve H_2 (protons are reduced in the absence of other reducible substrates) supercedes the ability of hydrogenase to recycle H_2 and under O_2-limited conditions, H_2 reutilization by hydrogenase is limited by the terminal electron acceptor O_2. It should perhaps be noted here that we are only just beginning to get some idea of the pathways of electron flow and the electron carriers involved in the oxyhydrogen reaction and in H_2-dependent nitrogenase activity (see 106). Also although hydrogenase has been partially purified from *A. chroococcum* (107), we do not know the physiological acceptor for the enzyme.

The postulates of Dixon outlined above have also been verified to a large extent with *Rhizobium*. Bacteroid suspensions of *R. japonicum* which contain hydrogenase activity can use H_2 as a source of ATP and the nitrogenase system is less sensitive to O_2 in the presence of H_2 (see 100,108). The difference in this case is that only a few strains of *Rhizobium* are known that possess hydrogenase activity and this correlates with whether or not H_2 is released by nodules or bacteroids. Those containing hydrogenase-less wild-type rhizobia or mutants lacking hydrogenase evolve substantial amounts of H_2 during N_2 fixation whereas little or no H_2 is evolved from nodules containing hydrogenase (see 109). The benefit of H_2 recycling to the N_2-fixing ability of the plant is shown by the observation that soybeans innoculated with *R. japonicum* strains containing hydrogenase activity fixed more N_2 and showed greater yields than plants containing rhizobial strains lacking hydrogenase (see 110). This shows the increase in the efficiency that hydrogenase has on the N_2-fixing process and obviously is of great importance from an agricultural point of view.

The conditions required for growth of *Rhizobium* in laboratory cultures, referred to as free-living rhizobia, have been under great scrutiny in the last few years with some very encouraging results so far. Evans and coworkers (see 111) made the important discovery that under conditions where *R. japonicum* expresses hydrogenase activity (low O_2 concentrations, low concentrations of carbon substrates and pre-incubation with H_2), the cells also have the capacity to fix CO_2 via ribulose-bisphosphate (RuBP) carboxylase. Subsequent work (see 112) determined the culture conditions required for autotrophic growth of *R. japonicum* on H_2 and CO_2, showing that rhizobia have a close evolutionary link with the aerobic hydrogen bacteria. Such cells have a generation time of about 20 hours under autotrophic growth conditions and ammonia is the best N source: so far, *Rhizobium* will not grow using only H_2, CO_2 and N_2.

VIII. PHOTOSYNTHETIC N_2 FIXATION

Procaryotic photosynthesis is performed by two fundamentally different mechanisms, exhibited by the cyanobacteria and the photosynthetic bacteria. N_2-fixing organisms are found in both classes. The cyanobacteria were formerly known as the blue-green algae until it was realized that their cellular properties were of the procaryotic type although they contain two photosystems, photolyze water and evolve O_2, characteristic of photosynthetic eucaryotes (green algae and higher plants). In contrast the photosynthetic bacteria contain a single photosystem and do not evolve O_2. Reductant for CO_2 fixation is obtained from organic and inorganic substrates rather than water. The genetics of the photosynthetic bacteria have recently been reviewed (113).

Most of the work on the N_2 metabolism of the cyanobacteria has been carried out on the N_2-fixing heterocystous species *Anabaena cylindrica* (114). Nitrogenase has been partially purified from this organism (115,116). The vegetative cells of the heterocystous cyanobacteria contain both photosystem I and photosystem II, evolve O_2 and fix CO_2 but not N_2. The specially differentiated cells (heterocysts) do not fix CO_2, lack photosystem II so do not evolve O_2, and they fix N_2 (some non-heterocystous cyanobacteria also fix N_2, see 117). The nitrogenase system is located within the heterocysts and the differentiation of vegetative cells into heterocysts is induced by starvation of fixed N. The nitrogenase of these cells can withstand the low O_2 concentrations because of an active oxidative metabolism, an impervious cell wall and a highly developed internal membrane structure. ATP is produced in the heterocysts by cyclic photophosphorylation catalyzed by photosystem I. Ferredoxin carries electrons to the nitrogenase system and is in turn reduced by ferredoxin:NADP oxidoreductase utilizing the high concentration of NADPH generated from the oxidation of fixed carbon compounds after their transport from the vegetative cells. Potentially, ferredoxin may be reduced by photosynthetic electron transport, as shown in Figure 6.

Experiments have linked the control of heterocyst formation with c-AMP and nitrate. Intracellular concentrations of c-AMP increased 3-4 fold when *Cyanobacter variabilis* was starved for fixed N prior to heterocyst formation (119). The structure and biochemistry of heterocysts has recently been reviewed (120). Glutamine synthetase is apparently involved in the control of heterocyst development and nitrogenase synthesis in both filamentous and unicellular cyanobacteria (121). The physiology and regulation of the N_2-fixing symbiosis between *Anabeana* and *Azolla* has been recently reviewed (122).

Of the photosynthetic bacteria the purple non-sulfur *Rhodospirillaceae*, which includes *Rhodopseudomonas capsulata* and *Rhodospirillum rubrum*, are the most studied with regard to N_2 metabolism (see also Section IX). Pure nitrogenase preparations have been obtained from *R. rubrum* (123). In spite of their ability to also grow in the dark and under aerobic conditions, active nitrogenase is only present during anaerobic photosynthetic growth in the absence of fixed N. Organic acids usually serve as the electron donors and ATP for CO_2 fixation arises from cyclic photophosphorylation. Nitrogenase is protected from O_2 since these cells do not evolve O_2 photosynthetically. *R. rubrum* and *Rps. capsulata* are easily grown under N_2-fixing conditions on a minimal medium in glass bottles bathed with 'cool' white florescent light. The gas mixture is usually N_2 with 3-5% CO_2. H_2 may also be included. The growth medium has been described (124) and consists of basal salts supplemented with an organic acid e.g. malate. Cyanobacteria e.g. *A. cylindrica,* are grown in a similar manner except that the gas phase is usually air with CO_2 added. The medium (25) is supplemented with $NaHCO_3$ and some *Anabaena* species have a sodium requirement for N_2-fixing growth (126).

IX. PHOTOSYNTHETIC H_2 PRODUCTION

In this section we will discuss H_2 production by the cyanobacteria, the photosynthetic bacteria and the eucaryotic algae. An essential point to be made here is that the cyanobacteria evolve H_2 via nitrogenase, the photosynthetic bacteria exhibit both hydrogenase- and nitrogenase-catalyzed H_2 production, whereas all H_2 evolved by the algae is hydrogenase-mediated. Nitrogenase-catalysed H_2 evolution is obviously only observed

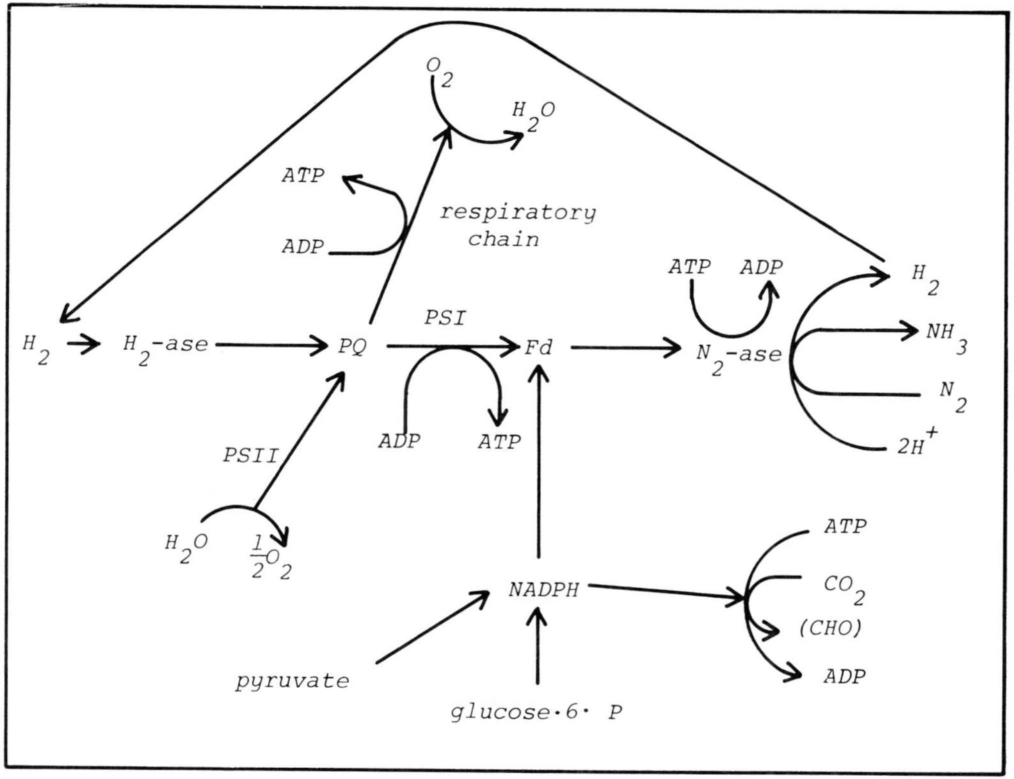

Figure 6. N_2 *fixation and* H_2 *production in* Anabaena. *Data from (118). PQ, plastoquinone; Fd, ferredoxin;* H_2*-ase, hydrogenase;* N_2*-ase, nitrogenase.*

in cell cultures capable of N_2 fixation and thus the culture conditions described above are relevant for the observation of H_2 production, with modifications as described below.

A. *Cyanobacteria*

A considerable amount of research has been carried out on the H_2 metabolism of the cyanobacteria (see 114,127,128). In spite of this, there exists considerable controversy on the capacity of these organisms to evolve H_2. Benemann and Weare (129) first observed that cultures of the cyanobacterium *Anabaena cylindrica* evolve H_2 and O_2 simultaneously in a light-dependent reaction. The only culture requirement was an atmosphere devoid of high concentrations of N_2 or O_2. Further work (130) showed that the rates of H_2 evolution could be increased up to three times that originally observed to about 300 μmoles H_2/hour/mg chlorophyll, if cultures were maintained under argon-CO_2 in the absence of fixed N, factors known to increase nitrogenase activity and heterocyst frequency (the site of nitrogenase activity). In continuous cultures, both H_2 and O_2 (in a ratio of between 1.0 and 4.0) were reported to be evolved for up to 19 days, and under low light intensities similar cultures evolved H_2 continuously for 30 days (131). However, there have been many reports that cultures of cyanobacteria (including *A. cylindrica*) do not produce substantial quantities of H_2 (114,127,132-135). For example, Bothe *et al.* (114) found that the rates of H_2 evolution from cultures of *A. cylindrica* were less than 1 μmole H_2/hour/mg chlorophyll, regardless of the age of the culture or the presence or absence of CO_2, O_2 or N_2. Further, Jones and Bishop (132) found that cultures of *Anabaena* evolved H_2 and O_2 in a ratio of only about 0.025.

Some possible reasons for the observed discrepancies in H_2 production activities are discussed below but what is not in doubt is that H_2 evolution is catalyzed by the nitrogenase system. H_2 is only produced under N_2-fixing conditions and if fixed N, e.g. ammonia or nitrate, is added to the medium both H_2 production and C_2H_2 reduction (a measure of nitrogenase) decrease in parallel. The reaction is not inhibited by the photosynthetic electron transport inhibitor DCMU and so photosystem II is not involved and the reductant is not supplied by water. The production of H_2 is light-dependent and this is thought to arise because of the requirement of nitrogenase for ATP, supplied mainly by cyclic photophosphorylation (photosystem I only). If N_2 or C_2H_2 is added, H_2 evolution decreases by about 90% since these substrates compete with protons for reductant in the nitrogenase reaction. The residual H_2

evolution activity is the result of the intrinsic capacity of nitrogenase to evolve H_2 independent of the substrate reduced. H_2 evolution is also sensitive to uncouplers of phosphorylation showing the absolute dependence on ATP. There have been reports of significant rates of H_2 production in the dark but only if low concentrations of O_2 were present suggesting that respiration also can supply sufficient ATP to support the nitrogenase reaction.

Almost all species of cyanobacteria so far tested, whether having the capacity to fix N_2 or not, have been found to contain hydrogenase activity. The cellular concentration of hydrogenase increases if cultures are grown under H_2 and this has been shown for both N_2-fixing (136) and non-N_2-fixing (128) species. In the cyanobacteria the *in vivo* role of hydrogenase is exclusively H_2 consumption (with one unusual exception, see below). H_2 may be used either as a source of energy (ATP) via the oxyhydrogen reaction, or alternatively it may be used as a source of reductant. For example, *Anabaena* catalyses both H_2-dependent O_2 uptake with the concomitant phosphorylation of ADP, and H_2-dependent C_2H_2 reduction i.e. nitrogenase activity. Utilization of H_2 is a light-dependent reaction (photosystem I only). Thus as shown in Figure 6, hydrogenase has been shown to play a similar role in the N_2-fixing cyanobacteria as it does in the aerobic N_2-fixing bacteria, namely it recycles H_2 lost by the nitrogenase complex and thus increases the efficiency of the N_2-fixing process. As in the non-photosynthetic organisms, nitrogenase activity is much less sensitive to O_2 if H_2 is also present. This may be of particular significance in the cyanobacteria which, unlike the aerobic N_2-fixers, also produce O_2 (via photosystem II). However, this capacity of hydrogenase to recycle H_2 can mask the expected measurement of H_2 production by N_2-fixing cell cultures and may be a factor in the anomalous activities described above. The cellular concentration of hydrogenase depends on the growth conditions and differences are also observed between species and among strains of cyanobacteria. Together with the sensitivity of the nitrogenase system to the culture conditions, these factors explain (to some extent) the wide discrepancies in H_2 production activities of the cyanobacteria.

As already mentioned, hydrogenase does catalyze H_2 evolution *in vivo*, albeit in a very specialized reaction. Many cyanobacteria are capable of facultative anoxygenic photosynthesis in which sulfide (Na_2S) is the electron donor and energy (ATP) is supplied by photosystem I, analogous to the photosynthetic bacteria (137). Of these organisms at least two, *Oscillatoria limnetica* and *Aphanothece halophyta*, have been shown to contain hydrogenase activity and this enables them to use H_2 to photoreduce CO_2 (135). It should be pointed out here that these non-heterocystous non-N_2-fixing cyanobacteria, which also includes such species as *Anacystis*, can use H_2 as a reductant in CO_2 assimilation or for nitrite reduction, whereas *Anabaena* species do not use H_2 to support CO_2 fixation and they cannot reduce nitrite (see 138,139). One of these non-N_2-fixing species, *O. limnetica*, also exhibits H_2 evolution in a sulfide- and light-dependent reaction (140). Aerobically grown cells, transferred to anaerobic conditions in the presence of sulfide and absence of CO_2, evolve H_2 after a few hours. Neither N_2 nor uncouplers of phosphorylation have any inhibitory effect on H_2 production. Inhibitor studies have shown that electrons from sulfide reduce ferredoxin in a photosystem I -dependent reaction, and reduced ferredoxin couples to hydrogenase to evolve H_2 (the intermediate electron carriers are unknown). The production of H_2 is inhibited by CO_2 since this competes for the reductant which now is diverted at the level of reduced ferredoxin to CO_2 fixation. H_2 evolution in the presence of CO_2 occurs at sulfide concentrations too high for optimal CO_2 photoassimilation (> 4mM sulfide). This is the first demonstration of light-dependent H_2 evolution via hydrogenase in either the cyanobacteria or the photosynthetic bacteria (see below). Although hydrogenase has not been fully characterized in an homogeneous state from a cyanobacterium, preliminary results with *Spirulina maxima* (141) and *A. cylindrica* (142) show the enzyme to be fairly similar to those of the anaerobic bacteria but again the physiological electron carrier is not known.

B. *Photosynthetic Bacteria*

In contrast to the cyanobacteria and algae, bacterial photosynthesis is characterized by its anaerobic nature and by the fact that reductant for CO_2 fixation is obtained from organic and/or inorganic substrates rather than water. The ability of some photosynthetic bacteria to produce large quantities of both H_2 and CO_2 during photoheterotrophic growth on organic compounds has been known for several decades (see 39,40,143). However only in the last few years has it been shown that H_2 production is mediated by the nitrogenase system. H_2 evolution only occurs in the absence of ammonia and is inhibited to a large extent by N_2. The production of H_2 is light-dependent and is inhibited by uncouplers of phosphorylation showing it to be energy-dependent. All these facts indicated that nitrogenase was responsible for the H_2 evolution. Conclusive evidence (144) was obtained when it was found that nitrogenase-deficient mutants of *Rhodopseudomonas capsulata* were unable to photoevolve H_2 even though they possessed hydrogenase activity and could utilize H_2 (see below). Furthermore, revertant strains or strains to which nitrogenase activity had been transferred by gene transfer, had the capacity to both fix and to evolve H_2.

Maximal H_2 production by the photosynthetic bacteria occurs during photoheterotrophic growth with certain amino-acids as sources of nitrogen (see 145-147). For example, cultures of *Rps. capsulata* growing on a minimal medium containing lactate and glutamate under an atmosphere of argon evolve H_2 at a rate of about 5 mmoles/hour/g (dry weight) of cells. Over 70% of the substrate (lactate) is dissimilated to H_2 and CO_2. Of the 19 amino-acids tested by Hillmer and Gest (146), only lysine and cysteine supported rates of H_2 evolution of less than 1 mmole/hour/g of cells in the presence of lactate. Lactate could be replaced by pyruvate, succinate, fructose or malate. H_2 production is influenced by the N/C ratio of the growth medium since an N/C ratio above a critical value caused ammonia to appear in the medium and H_2 evolution to stop.

What is the physiological significance of H_2 production by the photosynthetic bacteria? It occurs under conditions where the cells generate excess ATP (by cyclic photophosphorylation) and reductant (from organic substrates) in relation to the demands of the biosynthetic pathways. These bacteria depend on light as the primary source of energy thus both H_2 evolution and N_2 fixation (which require ATP) are light-dependent. The production of H_2 is thought to be a regulatory device to reduce the levels of both ATP and reductant to those more in line with the cells needs. If N_2 is present, H_2 production by nitrogenase is inhibited and reductant and ATP are used for N_2 fixation and for the reductive biosynthesis of the amino-acids. Similarly if ammonia is added, nitrogenase is both inhibited and repressed and again energy and reductant are conserved and channelled to the anabolic pathways. The production of H_2 is used as a fine tuning mechanism to balance energy levels and biosynthetic requirements. One apparent anomaly though is that H_2 evolution is also substantial when growth is limited by the light intensity and therefore when the supply of ATP is limiting. In this case, it would appear that the reductant and ATP used for H_2 production could be utilized for biosynthesis.

All photosynthetic bacteria so far examined have the ability to use H_2 as a source of reductant for CO_2 fixation and some can grow autotrophically with H_2 as the electron donor and CO_2 as the sole carbon source (148). Research has shown that H_2 consumption is mediated by the hydrogenase enzyme (149-151). This H_2 utilization is independent of fixed N and ATP, and is sensitive to inhibition by CO. *Rhodospirillum rubrum* and *Rps. capsulata* photoproduce H_2 under an atmosphere of 99% H_2 showing that nitrogenase is unable to utilize H_2. Also, in contrast to H_2 production, photoreduction using H_2 is stimulated by ammonia. The consumption of H_2 is inhibited by organic substrates suggesting the cells 'prefer' readily usable carbon compounds (fixed carbon) to the energy 'expensive' C-compounds produced by reduction of CO_2. As in the aerobic N_2-fixing bacteria, hydrogenase has been shown to be able to recycle the H_2 lost by nitrogenase during N_2 fixation and H_2 can be used as a source of electrons for reducing N_2 (152,153). *Rps. capsulata* can grow photoautotrophically on H_2, CO_2 and N_2 thus H_2 can serve as an electron donor for both N_2 fixation and CO_2 fixation (82). It should also be mentioned that this organism can also use H_2 as a source of reductant in the dark and also during autotrophic growth under aerobic conditions (see 154, 155).

Some photosynthetic bacteria such as species of *Rhodospirillaceae* e.g. *R. rubrum,* can grow anaerobically in the dark by the fermentation of such substrates as glucose and pyruvate with the production of both H_2 and CO_2. However, it was only recently realized that dark fermentative growth involves a pyruvate: formate lyase and formate hydrogenlyase metabolic pathway analogous to that in the facultative anaerobes such as *E. coli* (156,157). Energy (ATP) is generated from acetyl CoA as in other anaerobic fermentors. The production of H_2 does not require light or nitrogenase activity (which is absent in dark-grown cultures) and is inhibited by CO, indicative of hydrogenase activity. It is not known if this is the same hydrogenase that is involved in the light-dependent reduction of CO_2 by H_2.

C. Algae

Growing cells of many of the photosynthetic eucaryotic algae also evolve substantial quantities of H_2 but unlike the cyanobacteria and the photosynthetic bacteria, the algae lack a N_2-fixing apparatus and so all H_2 metabolism exhibited by these organisms is mediated by hydrogenase. About half the algal species so far examined contain hydrogenase activity but research has mainly been directed towards the *Chlorophyceae,* in particular *Chlamydomonas,Chlorella* and *Scenedesmus* (see 117, 158-160). Algal hydrogenases are extremely sensitive to O_2 and a dark anaerobic 'adaptation' period is required before maximal hydrogenase activity is observed. Adaptation appears to be a result of the activation of existing hydrogenase (mechanism unknown) in addition to *de novo* protein synthesis (161). The sensitivity of the enzyme to O_2 means that inadequate anaerobiosis could give negative results when the activity of algal hydrogenases is assessed.

Algae produce H_2 both in a light-dependent and light-independent manner. In the dark, a slow production of H_2 is observed concomitant with CO_2 production. This is a result of the dark fermentation of substrates such as glucose, lactate or pyruvate. As in the strict anaerobes, H_2 production is a means of disposing of excess electrons without the need for terminal electron acceptors other than protons. Further, more ferredoxin is also involved since H_2 evolution decreases in the presence of a ferredoxin inhibitor. In addition to H_2 and CO_2, *Chlorella fusca* under dark anaerobic conditions also produces lactate, acetate, formate and ethanol from glucose. Similarly, *Chlamydomonas moewusii* degrades cellular starch to H_2, CO_2, acetate, glycerol and ethanol.

Photo-induced H_2 production by algae has been studied in much detail and it is characterized by two phases. After the adaptation of cells, illumination stimulates a rapid rate of H_2 evolution concomitant with O_2 production. This fast initial phase lasts only a matter of minutes but rates of H_2 evolution as high as 1.5 mmoles H_2/hour/g (dry weight) of cells have been reported under optimal conditions (see below). With continued illumination the rate of H_2 evolution decreases rapidly to a more steady rate but over prolonged periods (several hours), H_2 production eventually ceases due to inhibition by evolved O_2. Low light intensities are essential to observe H_2 production by algal cultures since high intensities lead to the rapid accumulation of O_2 because of the resumption of normal photosynthesis.

Light-induced H_2 production by algae seems to occur by two distinct mechanisms. The initial rapid phase has all the characteristics of photosystem II - mediated electron transport. Both H_2 and O_2 are evolved and H_2 production is sensitive to inhibitors of photosynthetic electron transport which indicate that water is the source of electrons and that both photosystems participate. Some mutants with inactive photosystem II do not evolve H_2. However, in the same algal species, H_2 production has been reported to occur in photosystem II mutants and in the presence of electron transport inhibitors specific for photosystem II. Furthermore, the simultaneous evolution of CO_2 (rather than O_2) is observed and H_2 production is stimulated by uncouplers of phosphorylation and by glucose. The latter seems to be the main mechanism of H_2 production in older or starved cell cultures where the metabolism of carbon substrates generates the reductant which is channelled to photosystem I (only) which then couples to the hydrogenase to evolve H_2. It seems probable that both mechanisms operate to some extent depending on the culture conditions and that the same hydrogenase catalyzes H_2 evolution in both cases. Cell-free preparations of hydrogenase have been obtained from several green algae (162-164) but the enzyme has yet to be purified and characterized.

Quantitation of the capacity of algae to evolve H_2 during culture is complicated by two factors. Firstly, photosynthetically evolved O_2 (and perhaps O_2 in the medium if stringent anaerobic conditions are not maintained) inhibits the hydrogenase and secondly, algae have the ability to utilize H_2, thereby "masking" H_2 production. In contrast to the cyanobacteria and most of the photosynthetic bacteria, algae can generate low potential reductants from H_2 without photosynthetic activity. Thus H_2 can be used to reduce ferredoxin, nitrite, nitrate, sulfite and also both flavin and nicotinamide nucleotides. Furthermore, algae also perform the oxyhydrogen reaction although only at low O_2 concentrations because of the sensitivity of the hydrogenase to O_2. The net (measured) rate of H_2 production from algal cultures may thus be the result of several conflicting reactions all of which are very dependent on the culture conditions.

X. CONCLUSIONS AND PERSPECTIVES

Nitrogenase proteins from a variety of N_2-fixing bacteria have similar structural and mechanistic properties. This may seem remarkable in view of the metabolic diversity of these organisms but most likely the specific needs of structure and function has put an evolutionary constraint on the enzyme. The requirements for nitrogenase to function (reductant, ATP and anaerobiosis) are the same whether the organism is an anaerobe, aerobe or photosynthetic bacterium. Both nitrogenase biosynthesis and activity are under strict cellular control by a variety of mechanisms.

Of great importance in meeting the goals of greater world food productivity is increasing the availability of utilizable nitrogen for food crops. Present ammonia production technology (Haber-Bosch process), economic considerations and energy reserves however, will not permit most countries to depend solely on ammonia fertilizer as the means to improved N fertility in the future. Research into the genetics, biochemistry and chemistry of N_2 fixation may supply answers for increasing the N available to crops. Increased soybean yields have already been achieved by use of rhizobia with superior characteristics (increased N_2 fixation and ammonia export) in symbiosis with the legume host. Doubtless, other strains and mutants with increased efficiencies in N_2 fixation will be found. Biochemical and physico-chemical techniques are now being used to probe the structure of the MoFe protein of nitrogenase in order to design a more efficient N_2 reduction catalyst. An important development was the recent isolation in the authors' laboratory of crystals of the MoFe protein from *C. pasteurianum* and *A. vinelandii* which are suitable for X-ray diffraction analysis. Only with the three dimensional model of the nitrogenase active site will we be able to get an insight into the mechanism of N_2 reduction (and H_2 production). It may then be possible to construct simple and efficient models of the N_2-fixing system, which would have enormous impact on industrial ammonia production.

A wide variety of both bacteria and algae evolve H_2 during normal metabolism and this may be catalysed by the enzyme, hydrogenase, or by nitrogenase as an intrinsic part of the N_2-fixing process. Hydrogenase-catalysed H_2 evolution serves to dispose of excess reductant during oxidative metabolism. In contrast to nitrogenase, the hydrogenases purified so far have very diverse properties which reflect their physiological roles. The differences in molecular weight, iron-sulfur content, specificity for electron carriers and O_2 sensitivity of these enzymes makes it difficult to elucidate a common mechanism of action. H_2 catalysis seems to reside with a novel *[*4Fe-4S*]* center although one hydrogenase does not contain this type of center and also has *[*2Fe-2S*]* clusters. The fact that some hydrogenases e.g. from the photosynthetic bacteria, contain only one *[*4Fe-4S*]* center is encouraging from the point of view of the synthesis of model hydrogenase compounds. Simple synthetic Fe_4S_4 clusters do not have hydrogenase activity but as models of the enzyme we can envisage these clusters with complexed oligopeptides containing functional groups (see 42).

There has been a considerable amount of research into the use of photosynthetic H_2-evolving organisms as solar energy conversion systems. Any system which can split water using only visible light and generate H_2, which may be considered an ideal fuel, is of enormous importance as a renewable energy source. Cell cultures of cyanobacteria and algae have this capacity but there are many drawbacks. As already discussed (Section IX), there is great controversy as to the ability of the cyanobacteria to produce large amounts of H_2. Even so, the difficulties in maintaining the stringent conditions required to observe H_2 production together with controlling the nutrients, temperature and gas exchange of a culture over a wide surface area, are obvious. Also, the efficiency of H_2 evolution is too low to be practical (see 131). Green algae are even more limited in their production of H_2 from water since the (water-splitting) photosystem II -dependent reaction is not the main mechanism of H_2 evolution. Also, prolonged H_2 production from algae is only observed in the presence of

O_2 scavengers because of the sensitivity of the hydrogenase to O_2 and then only over a period of hours rather than days. The problem with whole cell systems is that there are few opportunities to manipulate, substitute and stabilize the H_2-evolving apparatus. For this reason alone, it is felt that cell-free systems using purified components, or eventually, stable analogs of these components, are much more applicable to solar energy conversion systems (see 42). Thus in this respect the study and stabilization of the purified hydrogenase is very relevant.

ACKNOWLEDGMENTS

The work reported here from the authors' laboratory was supported by grants PCM 77-24655 and AIO 4865-19 from the National Science Foundation and the National Institutes of Health.

REFERENCES

1. Chatt, J., in "Molybdenum Chemistry of Biological Significance" (Newton, W.E. and Otsuka, S., eds.), p. 241, Plenum Press, New York, 1980.
2. Zumft, W.G. and Mortenson, L.E., *Biochim. Biophys. Acta 416,* 1 (1975).
3. Mortenson, L.E. and Thorneley, R.N.F., *Ann. Rev. Biochem. 48,* 387 (1979).
4. Mortenson, L.E., *Current Topics in Cell. Regulation 13,* 179 (1978).
5. Eady, R.R., in "Methods in Enzymology" (San Pietro, A., ed.), p. 753, Academic Press, New York, 1980.
6. Hadfield, K.L. and Bulen, W.A., *Biochemistry 8,* 5103 (1969).
7. Bulen, W.A. and LeComte, J.R., in "Methods in Enzymology" (San Pietro, A., ed.), p. 456, Academic Press, New York, 1972.
8. Tanaka, M., Hainu, M., Yasunobu, K.T., and Mortenson, L.E., *J. Biol. Chem. 252,* 7081 (1977).
9. Tanaka, M., Hainu, M., Yasunobu, K.T., and Mortenson, L.E., *J. Biol. Chem. 252,* 7089 (1977).
10. Tanaka, M., Hainu, M., Yasunobu, K.T., and Mortenson, L.E., *J. Biol. Chem. 252,* 7093 (1977).
11. Shah, V.K. and Brill, W.J., *Proc. Natl. Acad. Sci. U.S.A. 74,* 3249 (1977).

12. Burris, R.H., in "Nitrogen Fixation" (Newton, W.E. and Orme-johnson, W.H., eds.), Vol. I, p. 7, University Park Press, Baltimore, 1980.
13. Hageman, R.V. and Burris, R.H., *Biochim. Biophys. Acta 591,* 63 (1980).
14. Hageman, R.V. and Burris, R.H., *Proc. Natl. Acad. Sci. U.S.A. 75,* 2699 (1978).
15. Zumft, W.G., *Structure and Bonding 29,* 1 (1976).
16. Bui, P.T. and Mortenson, L.E., *Proc. Natl. Acad. Sci. U.S.A. 61,* 1021 (1968).
17. Upchurch, R.G. and Mortenson, L.E., *J. Bacteriol. 143,* 274 (1980).
18. Dalton, H. and Postgate, J.R., *J. Gen. Microbiol. 54,* 463 (1969).
19. Scherings, G., Haaker, H., and Veeger, C., *Eur. J. Biochem. 77,* 621 (1977).
20. Seto, B.L. and Mortenson, L.E., *J. Bacteriol. 117,* 805 (1974).
21. Ludden, P.W. and Burris, R.H., *Science 194,* 424 (1976).
22. Yoch, D.C., *Biochem. J. 187,* 273 (1980).
23. Brill, W.J., *Microbiol. Rev. 44,* 449 (1980).
24. Davis, L.C. and Wang, Y., *J. Bacteriol. 141,* 1230 (1980).
25. Bisseling, T., Moen, A.A., Van den Bos, R.C., and Van Kammen, A., *J. Gen. Microbiol. 118,* 377 (1980).
26. Tyler, B., *Ann. Rev. Biochem. 47,* 1127 (1978).
27. Gordon, J.K. and Brill, W.J., *Biochem. Biophys. Res. Commun. 59,* 967 (1974).
28. Robson, R.L., *FEMS Microb. Lett. 5,* 259 (1979).
29. Hannecke, H. and Shanmugam, K.T., *Arch. Microbiol. 123,* 259 (1979).
30. Elliot, B.B. and Mortenson, L.E., *J. Bacteriol. 127,* 770 (1976).
31. Hom, S.S.M., Hennecke, H., and Shanmugam, K.T., *J. Gen. Microbiol. 117,* 169 (1980).
32. Cannon, F.C., Riedel, G.E., and Ausubel, F.M., *Mol. Gen. Genet. 174,* 59 (1979).
33. Riedel, G.E., Ausubel, F.M., and Cannon, F.C., *Proc. Natl. Acad. Sci. U.S.A. 76,* 2866 (1979).
34. Roberts, G.P., MacNeil, T., MacNeil, D., and Brill, W.J., *J. Bacteriol. 136,* 267 (1978).
35. Nieva-Gomez, D., Roberts, G.P., Klevickis, S., and Brill, W.J., *Proc. Natl. Acad. Sci. U.S.A. 77,* 2555 (1980).
36. Morris, J.G., *Advs. Microbiol. 12,* 169 (1978).
37. Stephenson, M. and Stickland, L.H., *Biochem. J. 25,* 205 (1931).
38. Gest, H., in "Phosphorus Metabolism" (McElroy, W.D. and Glass, B., eds.), Vol. II, p. 522, John Hopkins Press, Baltimore, 1952.

39. Gest, H., *Bacteriol. Rev. 18,* 43 (1954).
40. Gray, C.T. and Gest, H., *Science 148,* 186 (1965).
41. Mortenson, L.E. and Chen, J-S., in "Microbial Iron Metabolism" (Neilands, J.B., ed.), p. 231, Academic Press, New York, 1974.
42. Adams, M.W.W., Mortenson, L.E., and Chen, J-S., *Biochim. Biophys. Acta* in the press (1980).
43. Chen, J-S., Mortenson, L.E., and Palmer, G., in "Iron and Copper Proteins" (Yasunobu, K.T., Mower, H.F., and Hayaishi, O., eds.), p. 68, Plenum Press, New York, 1976.
44. Gillum, W.O., Mortenson, L.E., Chen, J-S., and Holm, R.H., *J. Amer. Chem. Soc. 99,* 584 (1977).
45. Hatchikian, E.C., Bruschi, M., and LeGall, J., *Biochem. Biophys. Res. Commun. 82,* 451 (1978).
46. Adams, M.W.W. and Hall, D.O., *Arch. Biochem. Biophys. 195,* 288 (1979).
47. Adams, M.W.W. and Hall, D.O., *Biochem. J. 183,* 11 (1979).
48. Schneider, K., Cammack, R., Schlegel, H.G., and Hall, D.O., *Biochem. Biophys. Acta 578,* 445 (1979).
49. Arp, D.J. and Burris, R.H., *Biochim. Biophys. Acta 570,* 221 (1979).
50. van Dijk, C., Mayhew, S.G., Grande, H.J., and Veeger, C., *Eur. J. Biochem. 102,* 317 (1980).
51. van der Westen, H.M., Mayhew, S.G., and Veeger, C., *FEBS Lett. 86,* 122 (1978).
52. Yagi, T., Kimura, K., Daioji, H., Sakai, F., Tamura, S., and Inokuchi, H., *J. Biochem. (Tokyo) 79,* 661 (1976).
53. Gogotov, I.N., Zorin, N.A., Serebriakova, L.T., and Kondratieva, E.N., *Biochem. Biophys. Acta 523,* 335 (1978).
54. Gitlitz, P.H. and Krasna, A.I., *Biochemistry 14,* 2561 (1975).
55. Kakuno, T., Kaplan, N.O., and Kamen, M.D., *Proc. Natl. Acad. Sci. U.S.A. 74,* 861 (1977).
56. Schoenmaker, G.S., Oltmann, L.F., and Stouthamer, A.H., *Biochim. Biophys. Acta 567,* 511 (1979).
57. Sim, E. and Vignais, P.M., *Biochim. Biophys. Acta 570,* 43 (1979).
58. Weiss, A.R., Schneider, K., and Schlegel, H.G., *Current Microbiol. 3,* 317 (1980).
59. Pinchukova, E.E., Varfolomeev, S.D., and Kondratieva, E.N., *Biokhimiya 44,* 477 (1979).
60. Erbes, D.L. and Burris, R.H., *Biochim. Biophys. Acta 525,* 45 (1978).
61. Chen, J-S. and Mortenson, L.E., *Biochim. Biophys. Acta 371,* 283 (1974).
62. Schink, B. and Schlegel, H.G., *Biochim. Biophys. Acta 567,* 315 (1979).
63. Zumft, W.G. and Mortenson, L.E., *Biochem. J. 35,* 401 (1973).

64. Eady, R.R., Smith, B.E., Cook, K.A., and Postgate, J.R., *Biochem. J. 128,* 655 (1972).
65. Emerich, D.W. and Burris, R.H., *Biochim. Biophys. Acta 536,* 172 (1978).
66. Hill, S., *J. Gen. Microbiol. 95,* 297 (1976).
67. Daesch, G. and Mortenson, L.E., *J. Bacteriol. 96,* 346 (1967).
68. Riederer-Henderson, M. and Wilson, P.W., *J. Gen. Microbiol. 61,* 27 (1970).
69. Thauer, R.K., Jungermann, K., and Decker, K., *Bacteriol. Rev. 41,* 100 (1977).
70. Wolin, M.J. and Miller, T.L., *Arch. Microbiol. 124,* 137 (1980).
71. Usuji, K. and Yagi, T., *Arch. Microbiol. 125,* 35 (1980).
72. Scherer, P.A. and Thauer, R.K., *Eur. J. Biochem. 85,* 125 (1978).
73. Haddock, B.A. and Jones, C.W., *Bacteriol. Rev. 41,* 47 (1977).
74. Bernhard, T. and Gottschalk, G., in "Hydrogenases, Their Catalytic Activity, Structure and Function" (Schlegel, H.G. and Schneider, K., eds.), p. 199, E. Goltze KG, Göttingen, 1978.
75. Shah, V.K. and Brill, W.J., *Biochim. Biophys. Acta 305,* 445 (1973).
76. Israel, D.W., Howard, R.L., Evans, H.J., and Russel, S.A., *J. Biol. Chem. 249,* 500 (1974).
77. Benson, D.R., Arp, D.J., and Burris, R.H., *Science 205,* 688 (1979).
78. Dalton, H. and Postgate, J.R., *J. Gen. Microbiol. 56,* 307 (1969).
79. Benemann, J.R., Yoch, D.C., Valentine, R.C., and Aaron, A.I., *Biochim. Biophys. Acta 226,* 205 (1971).
80. Haaker, H., De Kok, A., and Veeger, C., *Biochim. Biophys. Acta 357,* 344 (1974).
81. Laane, C., Krone, W., Konnings, W., Haaker, H., and Veeger, C., *Eur. J. Biochem. 103,* 39 (1980).
82. Lepo, J.E. and Wyss, O., *Biochem. Biophys. Res. Commun. 60,* 76 (1974).
83. Keister, D.L., *J. Bacteriol. 123,* 1265 (1975).
84. McComb, J.A., Elliot, J., and Dilworth, M.J., *Nature 256,* 409 (1975).
85. Pagan, J.D., Child, J.J., Scrowcroft, W.R., and Gibson, A.H., *Nature 256,* 406 (1975).
86. Tjepkema, J. and Evans, H.J., *Biochem. Biophys. Res. Commun. 65,* 625 (1975).
87. O'Gara, F. and Shanmugam, K.T., *Biochim. Biophys. Acta 437,* 313 (1976).
88. Upchurch, R.G. and Elkan, G.H., *J. Gen. Microbiol. 104,* 219 (1978).

89. Bergersem, F.J. and Turner, G.L., *J. Gen. Microbiol. 118,* 235 (1980).
90. Ratcliffe, H.D., Drozd, J.W., Bull, A.T., and Daniel, R.M., *FEMS Microbiol. Lett. 8,* 111 (1980).
91. Carter, K.R., Rawlings, J., Orme-Johnson, W.H., Becker, R.R., and Evans, H.J., *J. Biol. Chem. 255,* 4213 (1980).
92. Wittenberg, J.B., in "Nitrogen Fixation" (Newton, W.E. and Orme-Johnson, W.H., eds.), Vol. II, p. 53, University Park Press, Baltimore, 1980.
93. Dilworth, M.J., in "Nitrogen Fixation" (Newton, W.E. and Orme-Johnson, W.H., eds.), Vol. II, p. 3, University Park Press, Baltimore, 1980.
94. Boland, M.J., Farnden, K.J.F., and Robertson, J.G., in "Nitrogen Fixation" (Newton, W.E. and Orme-Johnson, W.H., eds.), Vol. II, p. 33, University Park Press, Baltimore, 1980.
95. Schlegel, H.G., *J. Microbial. Serol. 42,* 181 (1976).
96. Gogotov, J.N. and Schlegel, H.G., *Arch. Microbiol. 97,* 359 (1974).
97. Mortenson, L.E., *Biochemie 60,* 219 (1978).
98. Smith, L.A. and Yates, M.G., *Nature 262,* 209 (1976).
99. Walker, C.C. and Yates, M.G., *Biochemie 60,* 225 (1978).
100. Evans, H.J., Emerich, D.W., Ruiz-Argüeso, T., Albrecht, S.L., Maier, R.J., Simpson, F., and Russell, S.A., in "Hydrogenases, Their Catalytic Activity, Structure and Function" (Schlegel, H.G. and Schneider, K., eds.), p. 287, E. Goltze KG, Göttingen, 1978.
101. Schubert, K.R. and Evans, H.J., *Proc. Natl. Acad. Sci. U.S.A. 73,* 1207 (1976).
102. Dixon, R.O.D., *Arch. Microbiol. 85,* 193 (1972).
103. Dixon, R.O.D., *Biochemie, 60,* 233 (1978).
104. Yates, M.G. and Walker, C.C., in "Nitrogen Fixation" (Newton, W.E. and Orme-Johnson, W.H., eds.), Vol. I, p. 95, University Park Press, Baltimore, 1980.
105. Berndt, H. and Wölfe, D., in "Hydrogenases, Their Catalytic Activity, Structure and Function" (Schlegel, H.G. and Schneider, K., eds.), p. 327, E. Goltze KG, Göttingen, 1978.
106. Veeger, C., Laane, C., Scherings, G., and van Zeeland-Wolbers, L., *Biochemie 60,* 237 (1978).
107. van der Werf, A.N. and Yates, M.G., in "Hydrogenases, Their Catalytic Activity, Structure and Function" (Schlegel, H.G. and Schneider, K., eds.), p. 307, E. Goltze KG, Göttingen, 1978.
108. Emerich, D.W., Ruiz-Argüeso, T., Ching, T., and Evans, H.J., Abst. Proc. Steenbock-Kettering Intl. Symp. on Biological Nitrogen Fixation, B-97, Madison, Wisconsin, 1978.

109. Maier, R.J., Postgate, J.R., and Evans, H.J., *Nature 276,* 494 (1978).
110. Albrecht, S.L., Maier, R.J., Hanus, F.J., Russell, S.A., Emerich, D.W., and Evans, H.J., *Science 203,* 1255-1257 (1979).
111. Simpson, F.B., Maier, R.J., and Evans, H.J., *Arch. Microbiol. 123,* 1 (1979).
112. Lepo, J.E., Hanus, F.J., and Evans, H.J., *J. Bacteriol. 141,* 774 (1980).
113. Saunders, V.A., *Microbiol. Rev. 42,* 357 (1978).
114. Bothe, H., Distler, E., and Eisbrenner, G., *Biochemie 60,* 277 (1978).
115. Tsai, L.B. and Mortenson, L.E., *Biochem. Biophys. Res. Commun. 81,* 280 (1978).
116. Hallenbeck, P.C., Kostel, P.J., and Benemann, J.R., *Eur. J. Biochem. 98,* 275 (1979).
117. Bishop, N.I. and Jones, L.W., *Current Topics in Bioenergetics 8,* 3 (1978).
118. Eisbrenner, G. and Bothe, H., *Arch. Microbiol. 123,* 37 (1979).
119. Hood, E.E., Armour, S., Ownby, J.D., Handa, A.K., and Bressan, R.A., *Biocheim. Biophys. Acta 588,* 193 (1979).
120. Haselkorn, R., *Ann. Rev. Plt. Physiol. 29,* 319 (1978).
121. Haselkorn, R., Mazur, B., Orr, J., Rice, D., Wood, N., and Rippka, R., in "Nitrogen Fixation" (Newton, W.E. and Orme-Johnson, W.H., eds.), Vol. II, p. 259, University Park Press, Baltimore, 1980.
122. Peters, G.A., Ray, T.B., Mayne, B.C., and Toia, Jr., R.E., in "Nitrogen Fixation" (Newton, W.E. and Orme-Johnson, W.H., eds.), Vol. II, p. 293, University Park Press, Baltimore, 1980.
123. Ludden, P.W. and Burris, R.H., *Biochem. J. 175,* 251 (1978).
124. Ormerod, J.G., Ormerod, K.S., and Gest, H., *Arch. Biochem. Biophys. 94,* 449 (1961).
125. Haystead, A. and Stewart, W.D.P., *Arch. Microbiol. 82,* 325 (1972).
126. Apte, S.K. and Thomas, J., *Current Microbiol. 3,* 291 (1980).
127. Peterson, R.B. and Burris, R.H., *Arch. Microbiol. 116,* 125 (1978).
128. Eisbrenner, G., Distler, E., Floener, L., and Bothe, H., *Arch. Microbiol. 118,* 177 (1978).
129. Benemann, J.R. and Weare, N.M., *Science 184,* 174 (1974).
130. Weissman, J.C. and Benemann, J.R., *Appl. Environ. Microbiol. 33,* 123 (1977).
131. Jeffries, T.W., Timourian, H., and Ward, R.L., *Appl. Environ. Microbiol. 35,* 704 (1978).

132. Jones, L.W. and Bishop, N.I., *Plt. Physiol. 57,* 659 (1976).
133. Bothe, H., Tennigkeit, J., and Eisbrenner, G., *Planta 133,* 237 (1977).
134. Tel-Or, E., Luijk, L.W., and Packer, L., *Arch. Biochem. Biophys. 185,* 185 (1978).
135. Belkin, S. and Padan, E., *Arch. Microbiol. 116,* 109 (1978).
136. Tel-Or, E., Luijk, L.W., and Packer, L., *FEBS Lett. 78,* 49 (1977).
137. Garlick, S., Oren, A., and Padan, E., *J. Bacteriol. 129,* 623 (1977).
138. Bothe, H. and Eisbrenner, G., in "Hydrogenases, Their Catalytic Activity, Structure and Function" (Schlegel, H.G. and Schneider, K., eds.), p. 353, E. Goltze KG, Göttingen, 1978.
139. Peschek, G.A., *Arch. Microbiol. 119,* 313 (1978).
140. Belkin, S. and Padan, E., in "Hydrogenases, Their Catalytic Activity, Structure and Function" (Schlegel, H.G. and Schneider, K., eds.), p. 381, E. Goltze KG, Göttingen, 1978.
141. Llama, M.J., Serra, J.L., Rao, K.K. and Hall, D.O., *FEBS Lett. 98,* 342 (1979).
142. Hallenbeck, P.C. and Benemann, J.R., *FEBS Lett. 94,* 261 (1978).
143. Ormerod, J.G. and Gest, H., *Bacteriol. Rev. 26,* 51 (1962).
144. Wall, J.D., Weaver, P.F., and Gest, H., *Nature 258,* 630 (1975).
145. Gest, H., *Advs. Microbial. Physiol., 7,* 243 (1972).
146. Hillmer, P. and Gest, H., *J. Bacteriol. 129,* 724 (1977).
147. Hillmer, P. and Gest, H., *J. Bacteriol. 129,* 732 (1977).
148. Klemme, J.H., *Arch. Microbiol. 64,* 29 (1968).
149. Schick, H.J., *Arch. Microbiol. 75,* 89 (1971).
150. Schick, H.J., *Arch. Microbiol. 75,* 102 (1971).
151. Schick, H.J., *Arch. Microbiol. 75,* 110 (1971).
152. Meyer, J., Kelley, B.C., and Vignais, P.M., *Biochemie 60,* 245 (1978).
153. Kelley, B.C., Meyer, C.M., Gandy, C., and Vignais, P.M., *FEBS Lett. 81,* 281 (1977).
154. Madigan, M.T. and Gest, H., *J. Bacteriol. 137,* 524-530 (1979).
155. Jouanneau, Y., Kelley, B.C., Berlier, Y., Lespinat, P.A. and Vignais, P.M., *J. Bacteriol. 143,* 628 (1980).
156. Schon, G. and Voelskow, H., *Arch. Microbiol. 107,* 87 (1976).
157. Gorrell, T.E. and Uffen, R.L., *Photochem. Photobiol. 27,* 351 (1978).

158. Kessler, E., in "Algal Physiology and Biochemistry" (Stewart, W.D.P., ed.), p. 456, Blackwell, Oxford, 1976.
159. Kessler, E., in "Hydrogenase, Their Catalytic Activity, Structure and Function" (Schlegel, H.G. and Schneider, K., eds.), p. 415, E. Goltze KG, Göttingen, 1978.
160. Hallenbeck, P.C. and Benemann, J.R., in "Topics in Photosynthesis. Photosynthesis in Relation to Model Compounds" (Barber, J., ed.), Vol. III, p. 331, Elsevier/North Holland Biomedical Press, 1979.
161. Klein, U. and Betz, A., *Plt. Physiol. 42,* 1 (1978).
162. Ward, M.A., *Phytochemistry 9,* 259 (1970).
163. Ben-Amotz, A. and Gibbs, M., *Biochem. Biophys. Res. Commun. 64,* 355 (1975).
164. King, D., Erbes, D.L., Ben-Amotz, A., and Gibbs, M., in "Research in Photobiology" (Castellani, A., ed.), p. 329, Plenum Press, New York, (1977).

Index